Decline of Mediterranean Fruit Crops and Forests Associated with Fungal Trunk Pathogens

Decline of Mediterranean Fruit Crops and Forests Associated with Fungal Trunk Pathogens

Editor

Carlos Agustí-Brisach

MDPI • Basel • Beijing • Wuhan • Barcelona • Belgrade • Manchester • Tokyo • Cluj • Tianjin

Editor
Carlos Agustí-Brisach
University of Cordoba
Spain

Editorial Office
MDPI
St. Alban-Anlage 66
4052 Basel, Switzerland

This is a reprint of articles from the Special Issue published online in the open access journal *Plants* (ISSN 2223-7747) (available at: https://www.mdpi.com/journal/plants/special_issues/Fungal_Trunk_Dis).

For citation purposes, cite each article independently as indicated on the article page online and as indicated below:

LastName, A.A.; LastName, B.B.; LastName, C.C. Article Title. *Journal Name* **Year**, *Volume Number*, Page Range.

ISBN 978-3-0365-4411-3 (Hbk)
ISBN 978-3-0365-4412-0 (PDF)

Cover image courtesy of Carlos Agustí-Brisach

Contents

About the Editor

Carlos Agustí Brisach

Carlos Agustí-Brisach is Assistant Professor in Plant Pathology in the Department of Agronomy (Unit of Excellence María de Maeztu 2020–2023) at the University of Cordoba, Spain. He was graduated as BSc on Agricultural Engineering (2008), MSc on Plant Protection (2010), and PhD on Plant Pathology (2013) at the Polytechnic University of Valencia (Valencia, Spain). He served as postdoctoral researcher at the University of Angers, France (2013–2014), in the Kearny Agricultural Research and Extension Centre, UC Davis, Fresno, USA (4 months; 2017), and at the University of Cordoba, Spain (2016–2021). In addition, in 2015, he was recruited by IDAI Nature S.L, a private company working on biostimulation for plant protection, where he coordinated the I+D+i Department until he joined the University of Córdoba in 2016. Currently, his main research lines are 'Etiology, epidemiology and control of wood diseases in Mediterranean woody crops and 'Biocontrol of diseases in woody plants by means of antagonistic microorganisms, bio-stimulants and resistance host inducers'. The research activity of Dr. Agustí-Brisach is focused on applied plant pathology. He is the author of 57 papers in JCR scientific journals (Hi = 14), 32 outreach publications in national journals, and 75 contributions to national or international congresses.

Preface to "Decline of Mediterranean Fruit Crops and Forests Associated with Fungal Trunk Pathogens"

This book was established after closing the special issue "Decline of Mediterranean Fruit Crops and Forests Associated with Fungal Trunk Pathogens" edited by Dr. Carlos Agustí-Brisach as Guest Editor and Mr. Everett Zhu as Manager Editor.

Tree decline has been a growing syndrome in agriculture and forest ecosystems during the last decades causing major economic losses worldwide. The syndrome has been categorized as a complex disease due to the wide diversity of symptoms expression, as well as the multiple fungal species associated with the disease. Tree decline can express itself in a broad diversity of symptoms, including chlorotic and necrotic leaves, shoot blight, branch dieback, cankers in the tree trunk, crown rot, gummosis, internal wood discolouration, and/or reduction in root biomass and root necrosis. When the disease progresses, the tree shows general debilitation and eventually dies. Symptoms vary depending on the biology of the causal agent, the affected host, the environmental and agronomical conditions, and all their interactions. A broad diversity of Mediterranean fruit crops (grapevine, olive, and tree nuts) and forest trees (*Pinus* spp., *Quercus* spp., etc.), have been described as susceptible hosts. Botryosphaeriaceae, Diaporthaceae and Diatrypaceae fungi have been identified as the main causal agents of this disease syndrome. However, there are hundreds of fungal species associated with tree decline. For all these aspects, the diagnosis of the disease is difficult, and elucidating its aetiology is essential towards the establishment of effective management strategies.

In addition, elucidating the role of biotic and abiotic factors on the infection of fungal trunk pathogens as well as the interactions among fungal trunk pathogens to determine synergistic or antagonistic effects among them will provide important challenges for research.

Therefore, this book represents a collection of papers related to the etiology, epidemiology, and control of fungal trunk diseases in several Mediterranean woody crops such as almond, citrus, grapevine, or pistachio, among others. They have generated relevant knowledge on the etiology and epidemiology of the tree decline syndromes, which will be useful to build a strong foundation for developing effective management approaches to reduce the yield losses caused by these complex diseases.

This Special Issue consists of 10 research papers. In the first paper, https://www.mdpi.com/2223-7747/8/10/412, Goufo et al. demonstrated the hypothesis that invasion of grapevine wood by esca-associated fungi induces the production of defensive compounds as part of locally and systemically induced responses. Through this study, the authors concluded that the long latency period between trunk invasion by fungi and visible foliar damage and the year-to-year fluctuation in symptomatic expressions observed with "Esca complex" might be partially attributed to a better utilization of constitutive defenses.

The second paper, https://www.mdpi.com/2223-7747/9/2/134, was designed two determine the etiology of Macrophomina crown and root rot of Pistachio in California. In this study, Nouri et al. confirmed the association of *Macrophomina phaseolina* with the decline of pistachio trees, representing the first description of this fungus as a crown rot-causing agent of pistachio in California. In addition, the authors demonstrated that the widely used clonal University of California Berkeley I (UCBI) rootstock appeared highly susceptible to *M. phaseolina*, suggesting that this pathogen is an emerging threat to the production of pistachio in California.

In the third paper, https://www.mdpi.com/2223-7747/9/6/754, Espargham et al. conducted a survey of trunk disease pathogens within Citrus trees in Iran. In this study, a broad diversity of fungal species was associated with cankers and dieback symptoms in Citrus trees in Iran. Among the pathogens described in this study, *Lasiodiplodia theobromae* and *Neoscytalidium hyalinum* were reported for the first time in citrus in Iran, and several *Phaeoacremonium* species, *Stilbocrea walteri*, *Peyronellaea pinodella* and *Cadophora luteo-olivacea* were reported in citrus trees for the first time in the world.

The fourth paper, https://www.mdpi.com/2223-7747/9/9/1213, was a contribution of the Guest Editor of the Special Issue, Agustí-Brisach and co-workers. This study was designed to demonstrate that abiotic factors such as water stress may play an important role enhancing the progression of symptoms associated with almond decline under field conditions. Symptoms of branch dieback and general decline were observed over the two experimental years, mainly in the experimental plots subjected to high water deficiency, with Botryosphaeriaceae being the most consistently isolated fungi, and *Botryosphaeria dothidea* the most frequent species. In addition, this work revealed the need to elucidate the role of biotic and biotic factors that increase the rate of infection of fungal trunk pathogens.

In the fifth paper, https://www.mdpi.com/2223-7747/9/11/1427, Reis et al. suggested that early-season symptoms of Botryosphaeria dieback in grapevines may sometimes be disregarded by growers, being mistaken with symptoms from other diseases such as downy mildew or botrytis rot. To demonstrate it, grapevine samples showing necrosis on green shoots, dried inflorescences, and flowers were collected in vineyards during flowering period to conduct isolation, fundal identification and pathogenicity tests. The results of this study concluded that *Diplodia seriata* and *Neofusicoccum parvum* were the two main species apparently responsible for these symptoms.

The sixth paper, https://www.mdpi.com/2223-7747/10/4/802, aimed to investigate the role of phytotoxic metabolites (PMs) in the expression of Botryosphaeria dieback symptoms in naturally infected and artificially inoculated wood using molecular and analytical chemistry techniques. In this study, Reveglia et al., showed that (*R*)-mellein may be produced by fungal trunk pathogens during infection to break down the wood. They suggested that the foliar symptoms in vineyards may be due to a combination of PMs produced and climatic and physiological factors that require further investigation.

In the seventh paper, https://www.mdpi.com/2223-7747/10/6/1041, Calzarano et al. evaluated the factors involved on tiger-stripe foliar symptom expression of Esca of grapevine by means of macro and microelement analyses and leaf reflectance measurements on leaves of both never leaf-symptomatic vines and different categories of diseased vine shoots. Their results confirmed the strong response of the plant to symptom expression development and the possibility of limiting this response with calcium and magnesium applications carried out before the symptom onset.

The eighth paper, https://www.mdpi.com/2223-7747/10/6/1171, was conducted by Chacón-Vozmediano et al., who monitored a grapevine germplasm collection including 22 white and 25 red cultivars along three growing seasons to evaluate their susceptibility to natural infections of fungal trunk pathogens associated with grapevine trunk diseases (GTDs). The results revealed that 'Monastrell', 'Graciano', 'Cabernet Franc', 'Cabernet Sauvignon', 'Syrah', 'Moscatel de Alejandría', 'Sauvignon Blanc', and 'Airén' were highly susceptible to GTDs, whereas 'Petit Verdot', 'Pinot Noir', 'Chardonnay', and 'Riesling' were considered tolerant cultivars.

The ninth paper of this collection, https://www.mdpi.com/2223-7747/10/12/2777, aimed to elucidate the etiology of silverleaf, an important fungal trunk disease Japanese plum (*Prunus salicina*). In this study, Grinbergs et al. characterized the causal agent of the disease, *Chondrostereum purpureum*, based on morphological and molecular characters, and demonstrated its pathogenicity on healthy plum plants. In addition, the effects of the disease were also evaluated by determining the xylem water potential and fruit yield and quality in healthy and Silverleaf-diseased plum trees, showing that the water potential was altered in diseased trees, and fruit yield was reduced significantly compared to fruit from healthy trees.

Finally, the last paper of this Special Issue, https://www.mdpi.com/2223-7747/11/10/1374, was conducted by Mang and Camele, who described the etiology of apple tree dieback in Val d'Agri (Basilicata Region, Southern Italy). These authors identified *Neofusicoccum parvum*, *Diaporthe eres*, and *Trametes versicolor* as the most frequent fungi associated with the disease, among other secondary fungi such as *Pestalotiopsis funerea*, *Phomopsis* spp. and *Diaporthe foeniculina*. Pathogenicity tests were conducted on apple trees cv. Golden Delicious, with *N. parvum* being the most aggressive fungus and *Phomopsis* sp. the least.

Carlos Agustí-Brisach
Editor

plants

MDPI

Article

Exhibition of Local but Not Systemic Induced Phenolic Defenses in *Vitis vinifera* L. Affected by Brown Wood Streaking, Grapevine Leaf Stripe, and Apoplexy (Esca Complex)

Piebiep Goufo *, Ana C. Marques and Isabel Cortez

Centre for the Research and Technology of Agro-Environment and Biological Sciences, Departamento de Agronomia, Universidade de Trás-os-Montes e Alto Douro, Quinta de Prados, 5000-801 Vila Real, Portugal; anamarques@utad.pt (A.C.M.); icortez@utad.pt (I.C.)

* Correspondence: pgoufo@utad.pt

Received: 23 September 2019; Accepted: 12 October 2019; Published: 14 October 2019

Abstract: Balance between constitutive and induced responses provides plants flexibility to cope with biotic stresses. This study tested the hypothesis that invasion of grapevine wood by esca-associated fungi induces the production of defensive compounds as part of locally- and systemically-induced responses. In a vineyard, different symptomatic expressions of "Esca complex" in *Vitis vinifera* L. 'Malvasia' were evaluated in annual inspections. Then, levels of phenolics and fatty acids were determined in asymptomatic leaves of brown wood streaking (BWS) and grapevine leaf stripe (GLSD) vines, and in symptomatic leaves of GLSD and apoplectic vines; the results were compared with levels in healthy vines. In asymptomatic leaves of BWS and some GLSD vines, levels of phenolics decreased, independent of the total phenolic group. Such responses were usually associated with an increase in levels of linoleic, γ-linolenic and arachidonic acids, well-known signal transduction mediators. In symptomatic leaves, levels of phenolics increased, which is consistent with a locally-induced response; the onset of symptoms coincided with the highest increases e.g., 35% for quercetin-3-*O*-glucuronide. Thus, the long latency period between trunk invasion by fungi and visible foliar damage and the year-to-year fluctuation in symptomatic expressions observed with "Esca complex" might be partially attributed to a better utilization of constitutive defenses.

Keywords: grapevine trunk diseases; signal transduction; systemic acquired resistance; preformed defenses; symptom severity; grapevine trunk diseases

1. Introduction

In their natural environment, plants are at risks of infections by pests and pathogens. Plant resistance to such infections is attributed to multiple defenses that comprise constitutive/preformed and inducible chemical barriers. Inducible chemical barriers involve the synthesis of pathogenesis-related proteins (*PR*) and the accumulation of phytoalexins [1]. An example of a phytoalexin is tricin, which is a flavonoid that confers resistance against brown planthopper in rice [2]. Some phytoalexins are highly species-specific; for example, in pea (*Pisum sativum* L.), the synthesis and localization of pisatin are primarily associated with resistance against Fusarium wilt [3].

"Esca complex" is a widespread and destructive grapevine trunk affliction that affects grape yield and quality [4,5]. The disease is generally associated with the development of diverse wood pathogens among which the ascomycetes *Phaeomoniella chlamydospora* and *Phaeoacremonium minimun* and the basidiomycete *Fomitiporia mediterranea* are most commonly cited [6–9]. The hyphae of these fungi spread into the xylem vessel and the parenchyma cells of trunks, leading to necrosis [10]. Sometimes, these hyphae invade other woody tissues including cordons, arms, spurs, and canes [11–14].

"Esca complex" exhibits a long latency time (several years) between wood colonization and visible foliar symptoms [5] and has become increasingly frequent worldwide. A ten-year survey conducted in different vine-growing regions of France revealed that the simplification of the woody vine structure may have resulted in an increase in the incidence of the affliction in the country [15]. Some basic and practical cultural measures for preventing "Esca complex" have been proposed. For example, it was determined that increasing the length of cordons [15] and opting for a minimal pruning system instead of the standard spur-pruning [7] may help minimize the consequences of wood necroses. The foliar application of fertilizer mixtures containing calcium, magnesium, and Fucales seaweed was found effective in reducing foliar symptoms and increasing the yield and quality of berries [4,16]. The correlation between the symptomatic expression of "Esca complex" and the host physiology was highlighted by some authors [13,14,17]. This implies that characterizing the impact of the affliction on grapevine physiology could help in finding candidate biomarkers associated with disease resistance.

Several studies have indicated that phytoalexins and in particular phenolic compounds (phenolic acids, flavonoids, anthocyanins, proanthocyanidins, and stilbenes) play a role in limiting the development of "Esca complex". A typical reaction to wood colonization by esca-associated fungi is the accumulation of a mixture of polysaccharides (tyloses and gummosis) and the formation of polyphenol-rich reaction zones that obstruct the xylem to compartmentalize the fungi [18]. However, decreased levels of most phenolic compounds were observed in the xylem sap of vines with severe wood symptoms [19], as well as a decreasing trend for the levels of amino acids involved in the biosynthesis of phenolic compounds [20].

Rusjan et al. [10] found that esca-associated fungi caused the accumulation of flavonoids and stilbenes in both asymptomatic and necrotic trunks of vines. In particular, there was a high degree of flavonoid polymerization and a high level of procyanidins in the necrotic wood. However, reduction in the levels of phenolic compounds in asymptomatic wood and no effect on the levels in symptomatic wood were reported for *Pa. chlamydospora*-infected young vines [12]. Further, no difference was observed in total analyzed phenolics in asymptomatic stems of healthy and infected vines in the study by Magnin-Robert et al. [13]; however, a considerable accumulation of stilbenes—*trans*-resveratrol and *trans*-vitisin B—was observed in the affected vines.

It was also demonstrated that vines respond to "Esca complex" by accumulating stilbenes in naturally infected leaves [21] and leaves infected ex vivo with *Pa. chlamydospora* [22]. These increases were accompanied with the up-regulation of phenylalanine ammonia-lyase (*PAL*) and stilbene synthase (*StSy*)—two genes involved in the biosynthesis of polyphenolic compounds—in green [22] and dry leaves [23]. The accumulation of phenolic acids and flavonoids in symptomatic and asymptomatic leaves of field-grown vines was also reported [24]. However, in the study by Martín et al. [25], it was demonstrated that the appearance of foliar symptoms led to a decrease in the levels of flavonoids, proanthocyanidins, and hydroxycinnamic acids in the leaves of *Vitis vinifera* L. 'Tempranillo' grown under a dry and warm temperature. For the same cultivar grown under a hot and humid temperature, hydroxycinnamic acids levels increased in symptomatic leaves whereas flavonoids levels decreased. Interestingly, levels of *trans*-resveratrol in asymptomatic leaves of affected vines were slightly higher than those in healthy leaves of non-affected vines in some vineyards in Italy [26].

It is evident from the abovementioned findings that there are different responses of phenolic compounds to "Esca complex". These different results can be attributed to the types and complexities of symptomatic and asymptomatic materials studied by different authors.

Internal wood symptoms in adult vines are characterized by two diverse shapes of necrotic areas and discolorations. One shape/discoloration is caused by *F. mediterranea* and is called "white rot" or simply "esca" it is characterized by a clear/yellowish soft and spongy mass of wood usually in the center of the trunk or cordons, which can be observed alone or with dark-brown to black spots in the xylem vessels [10,11,14,19]. The second shape/discoloration refers to different types of brown wood necrosis of which "dark/brown wood streaking" (BWS) is most commonly reported; BWS consists of extended columnar strips of xylem necrosis with pink-brown to dark-brown areas or black spots around the annual growth section [13,19]. A third type of shape/discoloration (wood stripe), which is present in external vine wood, is also reported and the symptoms appear as a longitudinal and superficial yellowish-orange stripe and orange-brown discolorations of the young wood vessels located immediately below the bark [27].

Two typical severity levels of leaf symptoms are observed in esca-affected grapevines. A chronic form, characterized by tiger-striped symptoms (GLS) (also named by some authors GLSD for "Grapevine Leaf Stripe Disease"), is initially characterized by chlorosis and then light-green irregular spots and/or scorching between the main veins and/or along the leaf margins. The chlorotic and drying areas gradually expand from the basal to the distal part of the leaves, and then they coalesce to become partial necrotic stripes. As the chlorotic tissues turn yellow-brown or red-brown, the leaves exhibit a tiger stripe pattern [17,19,25–28]. GLSD symptoms are also reported in the berries and consist of tiny dark-brown or purple speckling distributed irregularly over the entire surface or scattered at the far end (termed "black measles" by some authors) and sometimes of shriveling/withering of grape bunches [18]. "Apoplexy" consists of partial or complete sudden wilting of the crown and is considered an acute form of the leaf symptomatic expression of "Esca complex" [23,27] or an acute form of GLSD by some authors [14]. BWS and GLSD vines are associated with a large procession of inhabiting fungi, although *Pa. chlamydospora* and *Pm. minimun* are most commonly found [18]. Although the percentage of necrotic areas within the wood from which pathogens can be isolated is often a key factor to determine the severity of "Esca complex," wood necrosis is not always related to the incidence of foliar symptoms [15]. Given this observation, the term "esca proper" is used by some authors to indicate the coexistence of "white rot" and GLSD in the same vine [13,14]. "Esca complex" is most commonly noted in established vineyards. In newly planted vines, scattered brown-black spots of necrotic xylem (without decay), often with a dark viscous ooze and a moderate/diffuse chlorosis of the leaves are observed, and the disease is termed "Petri disease" [6,8,12].

Studies have shown that many abiotic factors and cultural practices (alone or combined) may influence the development of "Esca complex" and the variability of its damage [15,20,26]. For example, it has been observed that heavy rainfall followed by hot winds in mid-summer favors the onset of apoplexy [18]. These observations were confirmed by other authors [25], who then reported that the biosynthesis of phenolic compounds in esca-affected leaves depended on the climate under which the vines were grown.

The above literature review shows that potential defense mechanisms developed by grapevine to resist esca-associated fungi need to be explored further. Therefore, the experiment in this study was designed to produce complementary data that would help improve the understanding of defense events occurring during an "Esca" invasion. It was hypothesized that esca-associated fungi induce the production of defensive compounds in leaves as part of both a locally- and systemically-induced defense response; local induction is defined as the enhancement of defensive traits in the organ that is attacked, while systemic induction is the enhancement of defenses in distant and undamaged organs, conferring broad-spectrum resistance throughout the plant [29,30]. To test this hypothesis, the accumulation of polyphenols in the leaves of vineyard-grown plants was monitored and levels of fatty acids were determined. Recent studies demonstrated that fatty acids play an important role in the modulation of signal transduction pathways in systemically acquired pathogen resistance. In several plants, the degree of resistance to pathogens was found to be directly correlated with the levels of C16:1 (palmitoleic acid), C18:1 (oleic and elaidic acids), C18:2 (linoleic and linolelaidic acids), C18:3 (α-linolenic and γ-linolenic acids), and C20:4 (achidonic acid) [1]. For example, rhizobacteria-induced enhanced resistance to *Botrytis cinerea* is associated with the accumulation of C18:2 and C18:3 in *Phaseolus vulgaris* L. [31], while reduction in C18:1 level induces defense responses against several pathogens by upregulating expressions of a variety of structurally diverse *R* genes in *Arabidopsis* [32]. Therefore, the levels of phenolic compounds and fatty acids in asymptomatic and symptomatic leaves of grapevine affected by BWS, GLSD and apoplexy were investigated to identify infection stages at which plant resistance mechanisms were more efficiently activated.

2. Results

2.1. *Effect of Brown Wood Streaking, Grapevine Leaf Stripe and Apoplexy on the Levels of Phenolic Compounds in Grapevine Leaves*

In this study, the amount of total phenolic compounds in the leaves of vines affected by "Esca complex" was first analyzed using colorimetric methods (Figure 1). An interesting trend emerged in that asymptomatic leaves of BWS and GLSD vines had a lower amount of TPC than that in control leaves, with a decrease of 14% in leaves of BWS vines (asymptomatic 1). The amount of TPC was particularly high in leaves exhibiting the initial foliar symptoms (GLSD stage 1) as compared to that in control leaves, and then, it decreased in proportion to the severity of chlorosis and necrosis on the leaves. Changes in the amount of TAC, TPAC, and TFC due to BWS and GLSD were similar to changes in the amount of TPC, with some exceptions: the highest amount of TAC was measured in asymptomatic leaves of GLSD vines that had both symptomatic and asymptomatic cordons (asymptomatic 2), and a 64% increase in the amount of TAC was recorded passing from chlorotic/spotting/scorching leaves (GLSD stage 2) to tiger striped (GLSD stage 3) and apoplectic leaves (Figure 1).

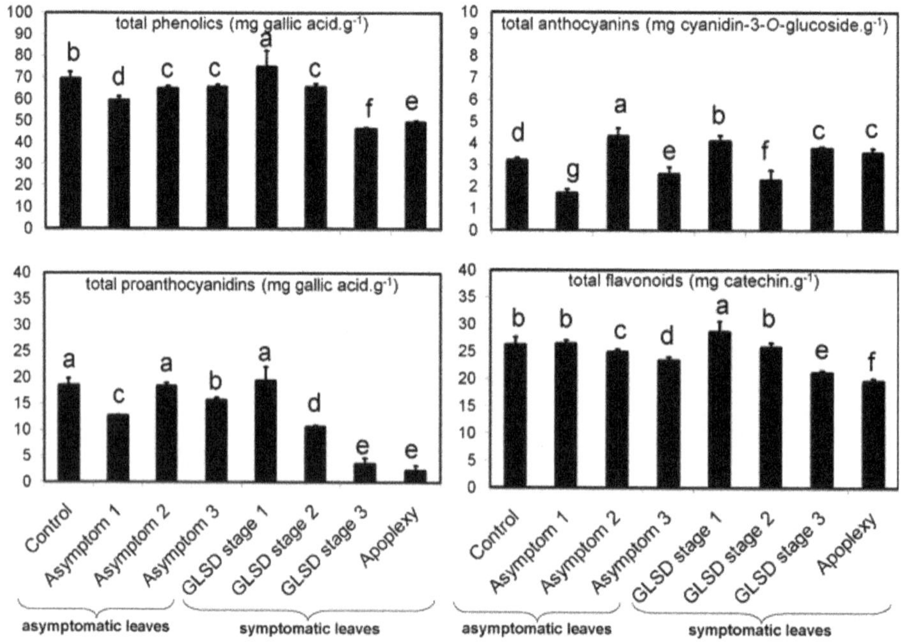

Figure 1. Total phenolic (TPC), anthocyanin (TAC), proanthocyanidin (TPAC), and flavonoid (TFC) content (dry weight basis) in asymptomatic and symptomatic leaves of vines affected by brown wood streaking, grapevine leaf stripe and apoplexy. The legend is as in Figure 6. Error bars = standard deviations (n = 4); different letters above the columns denote statistical differences (Tukey's test; $P \leq 0.05$).

The HPLC method used in this study led to the separation of 104 peaks with 95 peaks showing phenolic characteristics. Using the information provided by the detector and reports in the literature, the peaks were assigned to metabolites of the structure classes hydroxybenzoic acid (9), hydroxydiphenic acid (1), proanthocyanidin (9), stilbene (1), hydroxycinnamic acid (16), flavonoid (37), and anthocyanin (10). Eight metabolites were labeled "unknown," whereas four exhibited the characteristics of both proanthocyanidins and hydroxybenzoic acids and were labeled as "benzoic acid derivatives." On average, quercetin-3-O-glucuronide was the major phenolic compound in the leaves (2834.43 mg· kg^{-1}) followed by myricetin-3-*O*-galactoside (127.68 mg· kg^{-1}), quercetin-3-*O*-glucoside (127.32 mg· kg^{-1}), quercetin-3-*O*-galactoside (83.35 mg· kg^{-1}), caftaric acid (82.78 mg· kg^{-1}), myricetin-3-*O*-glucoside (60.82 mg· kg^{-1}), kaempferol-3-*O*-glucoside (54.01 mg· kg^{-1}), coutaric acid (49.00 mg· kg^{-1}), epicatechin (37.93 mg· kg^{-1}), quercetin-3-*O*-rutinoside (23.10 mg· kg^{-1}), kaempferol-3-*O*-rutinoside (13.72 mg· kg^{-1}), and epigallocatechin gallate (12.38 mg· kg^{-1}). The levels of the remaining compounds were below 10 mg· kg^{-1} (Table S1). In some samples, quercetin-3-*O*-glucuronide and quercetin-3-*O*-glucoside co-eluted in the chromatograms; therefore, the levels of these two compounds were summed and used in the statistical analyses; the same was true for myricetin-3-*O*-galactoside and myricetin-3-*O*-glucuronide. In *V. vinifera*, several stilbenes have been reported as stress response metabolites [22]. In this study, only one stilbene was detected, which was identified as *trans*-resveratrol, with an average content of 0.98 mg· kg^{-1} (Table S1). This value was substantially lower than 1.38–50.49 mg· kg^{-1}, which was observed in the leaves of some Italian cultivars [26]. The non-detection of stilbenes was not surprising because stilbenic compounds are usually detected by HPLC from a filtrate obtained after several solid–liquid and liquid–liquid extraction and purification steps [33]. The clean-up step used in this study was aimed at discarding chlorophylls and chromatography was optimized for the separation of flavonoids and proanthocyanidins.

The major compounds (average content ≥ 1.00 mg· kg^{-1}; Table S1) were first analyzed using ANOVA. This analysis allowed two main categories of compounds to be delineated, on the basis of similar trends in the contents observed comparing control, asymptomatic and symptomatic leaves.

The first category (Figure 2) consisted of 20 compounds that showed three characteristics. (i) The levels of these compounds were particularly high in symptomatic leaves exhibiting the initial foliar symptoms of GLSD (GLSD stage 1) compared with those in control and asymptomatic leaves. The percentage increase between control and GLSD stage 1 leaves ranged from 13% (epigallocatechin gallate) to 81% (catechin). However, there were some exceptions: the levels of myricetin-3-*O*-galactoside+myricetin-3-*O*-glucuronide and quercetin-3-*O*-rutinoside were the highest in asymptomatic leaves of BWS vines (asymptomatic 1), and the levels of epigallocatechin gallate and catechin were the highest in asymptomatic leaves of GLSD vines with berry symptoms (asymptomatic 3). (ii) The levels of these compounds progressively decreased with the increasing severity of the leaf symptom, with the lowest values usually being measured in apoplectic leaves. However, for some compounds a slight level increase was observed in apoplectic leaves as compared to tiger striped leaves (GLSD stage 3); this suggests that apoplexy might not only be a severe form of GLSD. These compounds included epicatechin, benzoic acid derivative 5, caffeic acid, hydroxycinnamic acid derivative 7, myricetin-3-*O*-galactoside+myricetin-3-*O*-glucuronide, and quercetin-3-*O*-rutinoside. (iii) The levels of these compounds were generally similar in asymptomatic leaves or lower in asymptomatic leaves of BWS and GLSD vines than those in control leaves. For example, the levels of quercetin-3-*O*-glucuronide+quercetin-3-*O*-glucoside decreased by 14, 29, and 16% in asymptomatic leaves of BWS (asymptomatic 1), GLSD foliar-symptomatic (asymptomatic 2), and GLSD berry-symptomatic (asymptomatic 3) vines, respectively. The levels of only a few compounds increased in asymptomatic leaves and that included an 18, 71, 188, 70, 99, 20, and 17 increase for hydroxycinnamic acid derivative 7, myricetin-3-*O*-galactoside+myricetin-3-*O*-glucuronide, quercetin-3-*O*-rutinoside in asymptomatic leaves of BWS vines, catechin in asymptomatic leaves of BWS vines, catechin in asymptomatic leaves of GLSD berry-symptomatic vines, epicatechin gallate in asymptomatic leaves of GLSD foliar-symptomatic vines, and epigallocatechin gallate in asymptomatic leaves of GLSD berry-symptomatic vines, respectively.

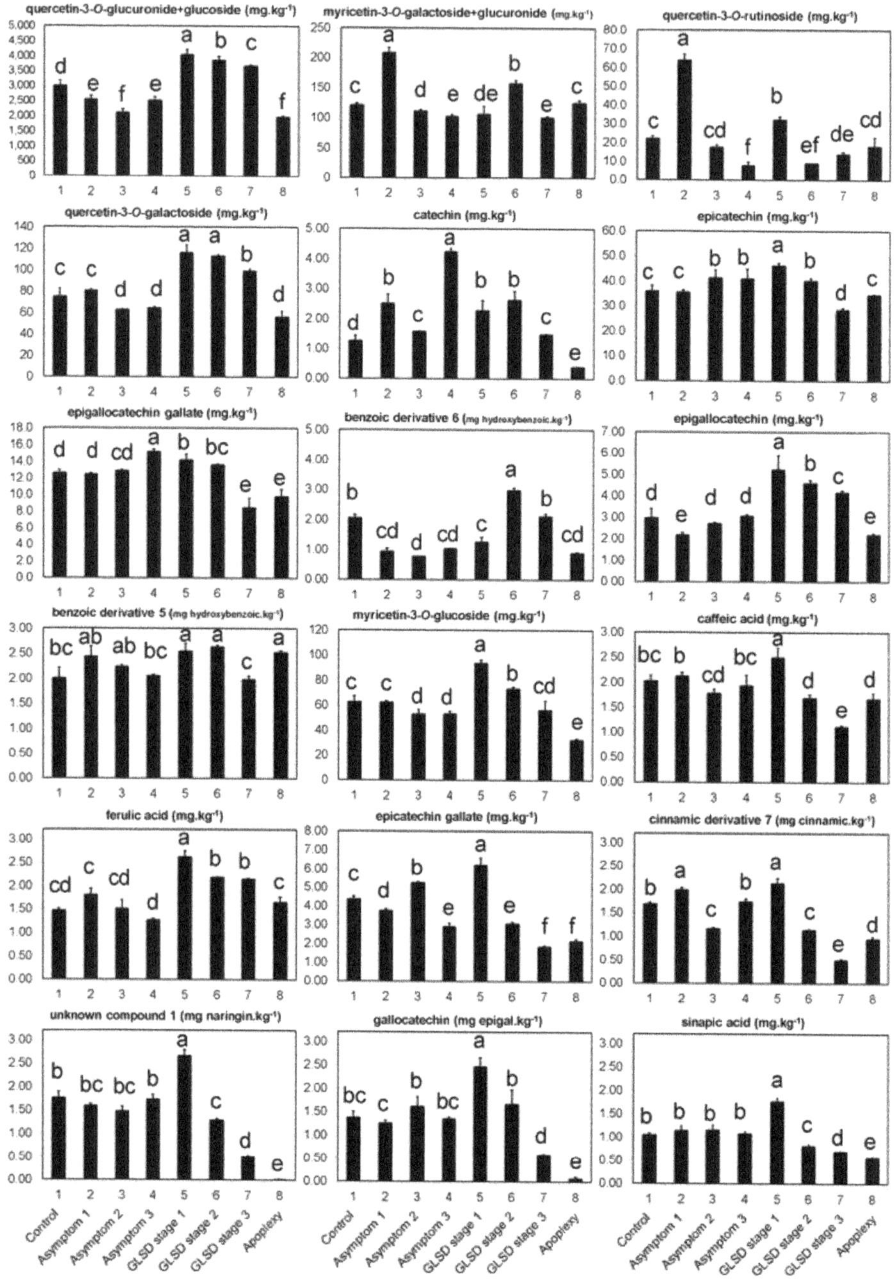

Figure 2. Phenolic compounds in asymptomatic and symptomatic leaves of vines affected by brown wood streaking, grapevine leaf stripe and apoplexy; their levels (dry weight basis) progressively decreased with increasing severity of symptoms. The legend is as in Figure 6. Error bars = standard deviations ($n = 4$); different letters above the columns denote statistical differences (Tukey's test; $P \leq 0.05$); hydroxybenzoic = hydroxybenzoic acid; cinnamic = cinnamic acid; and epigal = epigallocatechin.

The second category (Figure 3) consisted of 13 compounds whose levels were usually the highest in control leaves, confirming the general trend of decreased phenolic content in asymptomatic leaves of BWS and GLSD vines as observed in Figure 1; Figure 2. Caftaric acid—the main hydroxycinnamic acid identified in this study—belonged to that category and its levels decreased by 27% ($P \leq 0.05$), 20% ($P \leq 0.05$), and 8% ($P > 0.05$) in asymptomatic leaves of BWS, GLSD foliar-symptomatic, and GLSD berry-symptomatic vines, respectively. The levels of these compounds usually decreased in proportion to the severity of the chlorosis and necrosis on symptomatic leaves, as observed clearly for unknown compound 7 and coutaric acid. Few exceptions were kaempferol-3-*O*-rutinoside and quercetin-3-*O*-rhamnoside, whose levels tended to increase with increasing severity of symptoms.

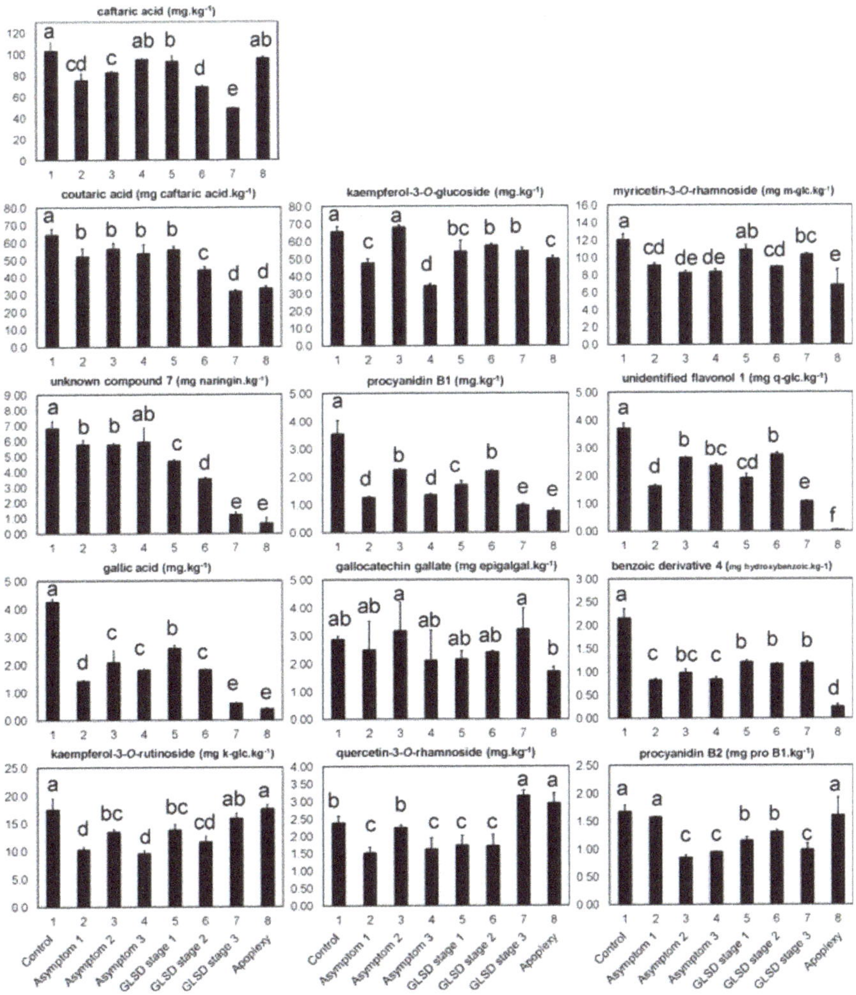

Figure 3. Phenolic compounds in asymptomatic and symptomatic leaves of vines affected by brown wood streaking, grapevine leaf stripe and apoplexy, with the highest levels (dry weight basis) in control leaves. The legend is as in Figure 6. Error bars = standard deviations ($n = 4$); different letters above the columns denote statistical differences (Tukey's test; $P \leq 0.05$); k-glc = kaempferol-3-*O*-glucoside; m-glc = myricetin-3-*O*-glucoside; epigalgal = epigallocatechin gallate; q-glc = quercetin-3-*O*-glucoside; hydroxybenzoic = hydroxybenzoic acid; and pro B1 = procyanidin B1.

A principal component analysis (PCA) was also performed to identify additional minor compounds that could help further classify the different leaf-groups. On the loading plot, 15 compounds were clearly separated from the other compounds (Figure S1). Interestingly, the levels of these compounds progressively increased with increasing symptom severity, showing strong correlations (Table S1), and reaching on average a 704% increase between control and apoplectic leaves (Figure 4). These compounds were usually undetected or detected at very low levels in asymptomatic leaves. Moreover, with the exception of quercetin, kaempferol-3-*O*-glucuronide, and kaempferol-3-*O*-galactoside, all these other compounds had average contents < 1 mg· kg^{-1}. In particular, GLSD and apoplexy stimulated the production of quinic acid, hydroxycinnamic derivative 6, and isorhamnetin-3-*O*-glucoside, which were not detected in most asymptomatic leaves.

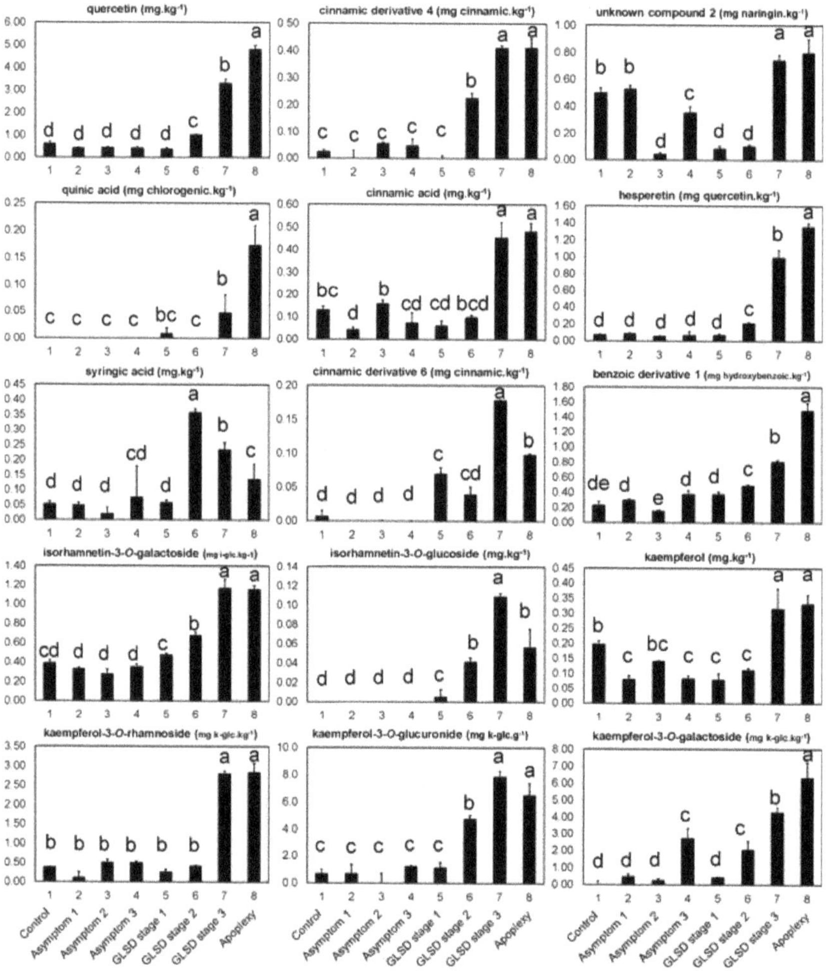

Figure 4. Phenolic compounds in asymptomatic and symptomatic leaves of vines affected by brown wood streaking, grapevine leaf stripe and apoplexy; their levels (dry weight basis) progressively increased with increasing severity of symptoms. The legend is as in Figure 6. Error bars = standard deviations (n = 4); different letters above the columns denote statistical differences (Tukey's test; $P \leq 0.05$); cinnamic = cinnamic acid; chlorogenic = chlorogenic acid; hydroxybenzoic = hydroxybenzoic acid; i-glc = isorhamnetin-3-*O*-glucoside; and k-glc = kaempferol-3-*O*-glucoside.

Overall, the levels of the remaining minor compounds were not affected in asymptomatic leaves, with the exception of four compounds that were detected primarily in these leaves (hydroxycinnamic acid derivative 2, unknown compound 4, unidentified flavonol 8, and *p*-hydroxybenzoic acid). For symptomatic leaves, the levels of some remaining minor compounds increased, while those of the others decreased with the increasing severity of symptoms (Table S2).

2.2. *Effect of Brown Wood Streaking, Grapevine Leaf Stripe and Apoplexy on the Levels of Fatty Acids in Grapevine Leaves*

In this study, 35 fatty acids present in grapevine leaves were separated by GC: 15 saturated fatty acids (SFA), 9 monounsaturated fatty acids (MUFA), and 11 polyunsaturated fatty acids (PUFA). The predominant fatty acids were γ-linolenic (C18:3n6; 34.65%), palmitic (C16:0; 15.09%), linoleic (C18:2n6c; 5.81%), elaidic (C18:1n9t; 5.49%), palmitoleic (C16:1n7; 3.71%), α-linolenic (C18:3n3; 3.35%), cis-4,7,10,13,16,19-docosahexaenoic (C22:6n3; 3.14%), caprylic (C8:0; 3.05%), arachidic (C20:0; 2.81%), and oleic (C18:1n9c; 2.22%) acids (Table S3).

The levels of most fatty acids were lower in asymptomatic leaves of BWS and GLSD vines than in control leaves (Table S4). Interestingly, the majority of C18 compounds and some other fatty acids did not seem to exhibit this decreased pattern (Figure S2; Figure 5).

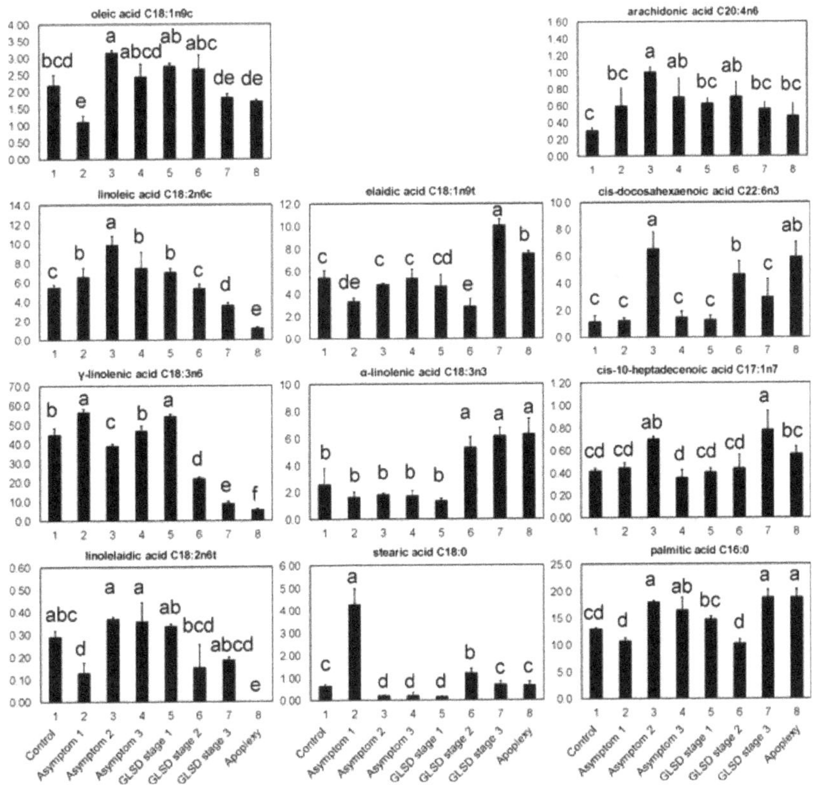

Figure 5. Fatty acids in asymptomatic and symptomatic leaves of vines affected by brown wood streaking, grapevine leaf stripe and apoplexy; their levels (%) were usually higher in asymptomatic leaves than in control leaves. The legend is as in Figure 6. Error bars = standard deviations ($n = 4$); different letters above the columns denote statistical differences (Tukey's test; $P \leq 0.05$).

In general, asymptomatic leaves of BWS vines (asymptomatic 1) had lower levels of C18:1n9c, C18:1n9t, C18:2n6t (linolelaidic acid) ($P \leq 0.05$), and C18:3n3 ($P > 0.05$) than control leaves; however, the levels of C18:2n6c, C18:0 (stearic acid), and C18:3n6 were higher in asymptomatic leaves of BWS vines than in control leaves. In particular, a 600% increase was observed for C18:0. Compared to control leaves, asymptomatic leaves of GLSD vines with both asymptomatic and symptomatic cordons (asymptomatic 2) had higher levels of C18:1n9c, C18:2n6c, and C18:2n6t, and lower levels of C18:0 and C18:3n6, while no change was recorded for C18:1n9t and C18:3n3. C18 levels in asymptomatic leaves of GLSD berry-symptomatic vines (asymptomatic 3) responded similarly to "Esca" attack as those in asymptomatic leaves of GLSD foliar-symptomatic vines, with the exception of C18:3n6, whose level remained unchanged. Substantial differences between control and asymptomatic leaves were also observed with regards to the levels of C16:0, C17:1n7 (cis-10-heptadecenoic acid), C20:4n6 (arachidonic acid), and C22:6n3. In all asymptomatic leaves, there was a strong increase in C20:4n6 levels. The levels of C17:1n7, C22:6n3, and C16:0 increased in asymptomatic leaves of GLSD foliar-symptomatic vines, while the level of C16:0 increased in asymptomatic leaves of GLSD berry-symptomatic vines (Figure 5).

In symptomatic leaves, a distinct correlation between disease symptom severity and fatty acid levels was observed (Table S3). Overall, leaf symptom severity was positively correlated with the levels of SFA (with the exception of tricosanoic acid C23:0 and heptadecanoic acid C17:0), MUFA (with the exception of C18:1n9c), and n3-PUFA (with the exception of cis-5,8,11,14,17-eicosapentaenoic acid C20:5n3), and negatively correlated with the levels of n6-PUFA (with the exception of cis-11,14-eicosadienoic acid C20:2n6) (Figure 5; Tables S3 and S4).

3. Discussion

In this study, great variability was observed in the accumulation of phenolic compounds and fatty acids in grapevine as a response to infection by esca-associated fungi, which indicated that dynamic and transient metabolic changes occur when symptoms spread from the trunk to the leaves.

3.1. *Exhibition of Locally Induced Defenses in Symptomatic Leaves*

It was clear from the data in Figures 1 and 2 that the levels of phenolic compounds increased in symptomatic leaves of GLSD vines exhibiting the first symptoms of the disease. The precocity of pathogen recognition and the velocity of the activation of defense responses are keys to enhancing the resistance of plants to infections [1,22,34]. The recognition of esca-related pathogens by grapevine plants and the formation of foliar symptoms are debated topics because propagules of *Pa. chlamydospora*, *Pm. minimun*, and *F. mediterranea* have never been detected on the leaves [18]. The most accepted interpretation is that toxic metabolites secreted by esca-associated fungi or resulting from reaction products of the infected wood are translocated from the xylem to the leaves via the transpiration/sap stream, which thus incites foliar symptom development [18]. This assumption suggests that the fungi induced local defense responses in grapevine when their metabolites reached the host leaf cells. In fact, foliar administration of calcium and subsequent accumulation of calmodulin, that mitigate the effect of the plant response, reduced GLSD leaf symptom expression [16]. The increase in the levels of phenolic compounds was the greatest when GLSD symptoms started appearing on the leaves. However, with increasing symptom severity, the levels of these compounds decreased. It is reported that the resistance of plants to infections depends partly on the balance between production/degradation of defensive compounds [19]. *Phaeomoniella chlamydospora* and *Pm. minimun* produce several enzymes that are known to travel in the plant and could reach the leaves [19]. However, the hypothesis of a phenolic decrease caused by enzymatic activities of esca-associated fungi is not tenable because these fungi lack enzymes such as ligninases, which would enable them to degrade specific phenolic bonds [9]. These decreases were also unlikely to be caused merely by chlorosis and necrosis. It is known that the development of GLSD necrotic areas in leaves leads to a decrease in photosynthetic assimilation [23]. Similarly, the expression of photosynthesis-related genes is strongly repressed in apoplectic leaves [28]. However, the reduced photosynthesis did coincide with the accumulation of hexoses and phenolic compounds in the studies by [28] and [25], respectively. It is conceivable that vines with reduced photosynthetic activity have to face with high levels of reactive oxygen species (ROS) and this can compromise the biosynthesis of primary and secondary metabolites [17,23]; this would suggest that, although the leaves initially respond to the infection with an increased production of phenolics, the vines no longer have the resources to support secondary metabolite production with increasing symptom severity, at least for many compounds detected in this study. Lambert et al. [22] also observed that the levels of *trans*-piceid and *trans*-resveratrol began to increase in grapevine leaves after 3 h of post treatment with a *Pa. chlamydospora* culture filtrate, and then decreased at 24 h. In grapevine leaf disks artificially infected with *Erysiphe necator*—a causal agent of grapevine powdery mildew—the levels of stilbenes also increased 1 to 3 days post inoculation, and then decreased with increasing disease symptom severity [34].

Interestingly, it was found that the levels of several compounds increased with increasing disease symptom severity (Figure 4). The majority of these compounds were undetected or at very low levels in asymptomatic leaves. The term "phytoalexin" has been used to describe compounds that are absent or normally present at low levels in cells, but which may increase enormously after infection and specifically inhibit the growth of a parasite [31,34]. In grapevine, the principal stress response phytoalexins studied are stilbenes [19,33]. In the study by Calzarano et al. [26], the time course of four stilbenes (*trans*-resveratrol, *trans*-ε-viniferin, *trans*-δ-viniferin, and *trans*-pterostilbene) was examined in vine leaves with different degrees of GLSD symptoms; in general, phytoalexin levels increased with increasing leaf symptom severity, although that depended on the growth stage of the plants. In this study, only *trans*-resveratrol was detected in the leaves; however, it was detected at very low amounts because of the extraction method adopted, which did not allow a clear assessment of the impact of the infection (Table S2). In a subsequent study, BWS, GLSD stage 1 and GLSD stage 2 samples were submitted to a metabolomic analysis, and ca., six stilbenes were identified in the leaves; levels of *trans*-piceid, *trans*-ε-viniferin, Ampelopsin A and *trans*-Pterostilbene increased with increasing leaf symptom severity, confirming the report in [26]. The level of a resveratrol dimer decreased, while the level of *trans*-resveratrol remained unchanged (*personal communication*). Compounds detected in Figure 4 could also act as phytoalexin in *Vitis vinifera* 'Malvasia'. The net accumulation of these compounds and stilbenes within the infected leaves may contribute to grapevine's ability to minimize the spread of the disease. A parallel could be drawn with the case of powdery mildew described by [34], where the de novo synthesis of ε-viniferin and δ-viniferin at the site of infection coincided with the interruption of the pathogen cycle. The induction of several defense genes and proteins following "Esca attack" has been reported by some authors, which strengthen the finding of this study. For example, in the study by Letousey et al. [23], the expression of the defense-related genes *StSy* (stilbene synthase), *PAL* (phenylalanine ammonialyase), *Chi4C* (class IV chitinase), *Chit1b* (class I basic chitinase), and *GST1* (glutathione-S-transferase) was strongly upregulated in dried leaves of apoplectic vines, whereas *SOD* (superoxide dismutase) was repressed. Similar inductions of genes encoding chitinases, stilbenic phytoalexins, and *PR* proteins (*Chit1b*; *CHV5*, *STS*, *GST5*, *SOD*, and *PR-6*) were observed in GLSD leaves [25]. The expression levels of PR-5 proteins, POX (peroxidase), and catechol PPO (polyphenol oxidase) were also higher in symptomatic and asymptomatic stems of grapevine affected with esca proper and apoplexy, than in the healthy vines [14].

Overall, leaf symptom severity was positively correlated with SFA, MUFA, and n3-PUFA levels, and negatively correlated with n6-PUFA levels (Figure 5; Table S4). The increased levels of SFA and MUFA observed in tiger striped and apoplectic leaves might be beneficial as they provide energy for various metabolic processes, which is particularly important for the energy-intensive processes that underlie the plant defense response. In particular, C16 and C18 fatty acids are important precursors of cuticular wax synthesis [1]. Thus, they strengthen cell membranes, provide structural integrity, and hamper the infiltration and spread of pathogens into the leaves.

3.2. Absence of Systemically Induced Defenses in Asymptomatic Leaves

Data in Figures 1 and 3 shows that, before the appearance of foliar symptoms, the presence of esca-associated fungi in the wood caused a decrease in the levels of phenolic compounds in the leaves. Only a few systemic responses for six compounds were recorded in asymptomatic leaves, including a 71% and 188% increase for myricetin-3-*O*-galactoside+myricetin-3-*O*-glucuronide and quercetin-3-*O*-rutinoside in asymptomatic leaves of BWS vines, respectively (Figure 2). These results do not support the hypothesis of a systemic induction of phenolic compounds in grapevine leaves; this was an unexpected finding given the slight increase in stilbenes in asymptomatic leaves of *V. vinifera* L. 'Trebbiano d'Abruzzo' [26] and in flavonoids and phenolic acids in asymptomatic leaves of *V. vinifera* L. 'Alvarinho' [24]. However, the findings of this study agree with some previous reports; Magnin-Robert et al. [17] found that the expression of the defense-related genes *GLUC* (β-1,3-glucanase), *GTS1, StSy, CHV5,* and *PAL* were repressed in most pre-GLSD leaves, while *Chit1b* and *Chi4C* were not affected, although the expression of stress-related genes was stimulated in vines with a decrease of net photosynthesis >75%. In most studies, *SOD* expression was found to decrease or showed a decreasing trend in both pre-apoplectic and pre-GLSD leaves [13,17,23]. In asymptomatic wood of apoplectic and esca proper vines, several genes and proteins involved in phenylpropanoid metabolism were either down- or upregulated [13,14] e.g., *IFRhom* and *IFRL4* (isoflavone reductase) and *leucoAND* (leucoanthocyanidin dioxygenase). In contrast, Valtaud et al. [28] observed an enhancement of mRNAs encoding *PR-10, Chi1b,* and *Chi3* (endochitinase 3) genes in asymptomatic leaves of GLSD vines with both symptomatic and asymptomatic cordons similar to those studied in this study. Letousey et al. [23] also reported an induction of *PAL, StSy, Chi4C, Chit1b,* and *GST1* genes in pre-apoplectic leaves.

The decline in the levels of phenolic compounds in asymptomatic leaves of BWS and GLSD vines was concomitant with the accumulation of C18:2n6c, C18:3n6, and C20:4n6, and generally a decrease in C18:1n9c levels in the leaves of BWS vines (Figure 5). These fatty acids are most prominently known for their specific signaling roles in plant defenses and they regulate ROS and nitric oxide (NO) levels by inducing specific effects on ROS- and NO-generating enzymes. For example, C18:1 in low amounts physically associates with the chloroplastic NITRIC OXIDE ASSOCIATED1 (*NOA1*) protein, inhibiting its GTPase and promoting its proteolytic turnover, which generates NO, triggering the transcriptional upregulation of NO-responsive nuclear genes, and thereby activating disease resistance [1,31,32]. Fatty acid data from this study show that systemic changes in fatty acid flux also occurred in the distal organs of grapevine. This would suggest that a mobile signal at the site of local infection is translocated to the leaves. Yet, information on the effector molecules involved in long-distance defense signaling in plants remains lacking. In tomato and other Solanaceous plants, systemic signaling appears to be mediated by systemin, an 18-amino acid peptide. Systemin is produced by wounded leaf cells, and travels to companion cells where it binds to a receptor, triggering the accumulation of jasmonic acid and fatty acids [29]. *Phaeomoniella chlamydospora, Pm. minimun,* and *F. mediterranea*, the major esca-wood-infesting fungi, were shown to produce diverse toxic metabolites detectable in the leaves. *Phaeomoniella chlamydospora* and *Pm. minimun* produce scytalone, 4-hydroxyscytalone, isosclerone [13,19,35], and pullulan [19], among other compounds. In the case of *F. mediterranea,* metabolites secreted that can be considered as toxic include 4-hydroxybenzaldehyde, dihydroactinolide, and 6-methoxymellein [35]. These toxic compounds have been identified in higher amounts in symptomatic than in asymptomatic leaves of affected vines [19]. Further, they may function as both pathogenic and virulence factors, thus representing the specific signals sensed by grapevine distal organs that result in fatty acid accumulation in asymptomatic leaves.

Specific changes in the levels of these fatty acids indicate that asymptomatic leaves are mounting a defense response in time to cope with the infection. That assumption is in good agreement with previous studies that report biochemical and physiological changes in grapevine leaves before the appearance of visible symptoms. For example, downregulation of *SOD* [23], low abundance of the *SODCP* protein *s6205* [14], upregulation of *GST1* [23], high abundance of GSTU1 and GSTF2 proteins, enhanced activity of *GST5* [28], and a decrease in the number and size of starch grains [28] have been reported in asymptomatic leaves of esca-affected grapevine as an early response of cells distant from the damaged wood. Within the week preceding leaf symptoms, drastic physiological alterations of photosynthesis were also registered in pre-apoplectic and pre-GLSD leaves, as revealed by a decrease in CO_2 assimilation, chlorophyll *a* fluorescence, and the repression of photosynthesis-related genes *psbP1, rbcL, rbcS, SBP* [17,23], *PRK*, and *Lhca3* [17] probably due to a lower activity of Rubisco or carbonic anhydrase [18]. Other metabolic alterations detected in asymptomatic leaves of esca-affected vines include a slight upregulation of the aquaporin-encoding water-stress-related gene *TIP1* [23]. In this study, resistance mechanisms activated in asymptomatic leaves seem to first involve the use of conserved antimicrobial compounds by the vines to respond to infection rapidly, as revealed by the decreased levels of phenolic compounds (Figures 1 and 3).

It is known that esca-associated fungi require several years of wood colonization to establish infection in the leaves [19]. Moreover, "Esca complex" is characterized by partial remission or total disappearance of foliar symptoms on plants in some years [12]. The year-to-year fluctuation in symptom expression has been attributed to the combination of optimal circumstances which include rainfall [5], and the occurrence of abiotic stresses such as drought [20]. It has also been hypothesized that each season, the newly formed vessels redefining the vascular system of the vine may affect the appearance of foliar symptoms [27]. Another hypothesis is that local defense reaction is the consequence of the development of drying zones and discolorations after a sudden sap disruption after or along the apparition and development of leaf stripe symptoms [27]. A complementary hypothesis could be that symptoms sometimes do not appear because of the activation of constitutive defenses. These pre-formed compounds could be effective in restraining the propagation of fungi in the wood and the translocation of their metabolites to the leaves, depending on the years and environmental conditions. The induced production of phenolic compounds would only occur after the metabolites have reached the leaves at doses sufficient to incite symptoms. This assumption is strengthened by the finding that ex vivo, stilbenes do not influence the damaging effects of *Pa. chlamydospora* on healthy leaves [26], which indicates that phytoalexins are synthesized in the leaves not before, but after, the apparition GLSD symptoms.

4. Materials and Methods

4.1. Site Characterization: Cultivar, Location, and Weather

Experiments were performed on *V. vinifera* L. 'Malvasia' composed of 21–24-year-old plants in the vineyard of Quinta de Nossa Senhora de Loures (465 m, 41° 17.12′ 31″ N, 7° 44.07′ 22″ W) in Vila Real, Portugal. The vineyard has 1247 vines grafted on 196-17-Castel rootstock and trained to a bilateral cordon according to the royal-type trellis system, on an area of 0.27 ha. The vines were planted at a distance of 1.80 × 1.20 m in 22 longitudinal rows. The climatic conditions in the Quinta are characterized by an average annual air temperature of 14.35 °C (2.04/29.23 °C day/night) and 814 mm annual precipitation, with 75% relative humidity and a 16-h photoperiod (1350 $\mu mol \cdot m^{-2} \cdot s^{-1}$). The vineyard is located on Anthrosol (62% sand, 25% silt, 13% clay; pH 4.2) and the vines are managed without irrigation. Pruning, fertilization, and plant protection practices are undertaken annually according to local practices. The position of the vineyard in the Quinta allows all vines to be grown under the same soil and climatic conditions. The vineyard is naturally infected with esca-associated fungi, and since 2010, research at the experimental field focused on the prevention and cure of "Esca complex."

4.2. Sampling Procedure

A characteristic trait of "Esca complex" is the unpredictable year-to-year discontinuity in foliar symptomatic expression [5,15,19,25]. Thus, to assess the incidence of the affliction, vines were inspected over four years through visual observations of leaf and berry symptoms and internal observations of wood symptoms by destructive means.

Several vines that did not show external symptoms since 2010 when work started at the experimental site were inspected during a four-year study period for the presence of discolorations associated with "Esca complex;" these vines were characterized as "apparently healthy" by several authors [5,11,13,14,17,23,26,28]. In this study, it was decided that an internal inspection of the wood was necessary before selecting "apparently healthy" vines. Therefore, wood cores were retrieved with a sterilized Pressler increment borer at 30 and 110 cm above the ground from the trunk of the vines, as described in [19]. Based on the analysis of wood cores, the vines were categorized into two groups. The first group consisted of vines that did not exhibit symptoms either in the trunk or in the leaves; these vines were presumed healthy and considered as "controls," as suggested in several papers [7,10,19,25]. Woods cores were subsequently subjected to fungal isolation and identification as described in [6]; *Pa. chlamydospora*, *Pm. minimun*, and *F. mediterranea* were usually not identified in these wood cores. The second group consisted of vines with brown necrosis and dark streaking of the xylem vessels, or BWS vines. These vines did not exhibit visible leaf or berry symptoms during the four-year survey. *Phaeomoniella chlamydospora* and *Pm. minimun* were identified in these wood cores, along with some *Phaeoacremonium*, *Botryospaeriaceae*, and other species (data not shown). The wood deterioration characteristic of "white rot" was not observed. GLSD was the prevalent form of "Esca complex" in the vineyard. Some GLSD vines had both symptomatic and asymptomatic shoots (one cordon symptomatic and one cordon asymptomatic), and they were selected for the study; such vines were also studied by several authors [4,14,17,24,28]. Other vines that showed GLSD leaf symptoms in a one or more inspection years and in some years only berry symptoms were also studied; however, this was a rare observation in the vineyard.

GLSD leaf symptoms at different degrees of severity were easily identifiable in the field. Leaf symptoms appeared between late June and early August, and although they usually increased in severity with plant growth, this increase was highly variable. In order to understand the biosynthesis of phenolic compounds by symptomatic expression, rather than selecting leaves with different degrees of symptom severity, vines with the majority of their leaves showing the same degree of symptom severity at harvest were targeted. In some vines, small chloroses characteristic of GLSD appeared, but did not evolve rapidly into spotting/scorching or tiger stripes. At the time of berry harvest, the surface of most leaves on these vines was still covered with discolorations, although some leaves started producing spotting/scorching or assuming the "tiger stripes" pattern (GLSD severity stage 1). At harvest, GLSD symptoms appeared in some vines as mainly chlorotic/spotting/scorching zones scattered over the leaf lamina (GLSD severity stage 2) or mainly tiger striped leaves (GLSD severity stage 3). An attempt was made to group vines exhibiting apoplectic symptoms; these symptoms appeared in a highly discontinuous manner in time (usually between early August and early September) and space in the vineyard. All selected vines were numbered and marked according to their place in the lines and rows.

4.3. Sample Collection

The occurrence of symptoms in the vineyard allowed the collection of different sets of leaves, which were divided into eight groups (Figure 6): (1) Asymptomatic leaves from apparently healthy vines (control); (2) asymptomatic leaves from BWS vines (asymptomatic 1), to analyze the systemic effects of trunk-localized fungi attack; (3) asymptomatic leaves from asymptomatic cordons on GLSD vines (asymptomatic 2), to assess whether the biosynthesis of defensive compounds was similar in symptomatic and asymptomatic parts of the same vine; (4) asymptomatic leaves from GLSD vines with berry symptoms (asymptomatic 3), to analyze the systemic effects of berry-localized infection; (5) symptomatic leaves from vines with initial symptoms of GLSD i.e., chlorotic leaves (GLSD stage 1); (6) symptomatic leaves from vines with moderate symptoms of GLSD i.e., chlorotic/spotting/scorching leaves (GLSD stage 2); (7) symptomatic leaves from vines with advanced symptoms of GLSD i.e., tiger striped leaves (GLSD stage 3); (8) symptomatic leaves from apoplectic vines (apoplexy). In the field, apoplexy appeared quickly, affecting the entire vine with total wilt and immediate drying caused by the hot weather (an average of 32 °C day temperature during apoplexy expression); thus, apoplectic leaves were harvested and studied already dried as in [23]. All symptomatic leaves were collected to study locally induced defenses.

All samples were collected mid-September, one day prior to berry harvesting. This ensured that the leaves were at the same stage of maturity. For each leaf-group, four vines were used for sampling and were considered as replicates. Six to twelve leaves of the same size from different parts of a vine were selected. Only two vines exhibited berry symptoms at harvest; hence, two sets of leaves were harvested from each vine to make four replicates, allowing for statistical comparisons. Leaves were immediately frozen in the field with liquid nitrogen to halt enzymatic activities and stored at −80 °C. Prior to use, the leaves were lyophilized, finely powdered with a hand blender, and sieved (0.2-mm mesh).

Figure 6. Description of the sampling procedure: A view of the foliar morphology of asymptomatic and symptomatic leaves of *Vitis vinifera* L. 'Malvasia' affected by brown wood streaking, grapevine leaf stripe (GLSD) and apoplexy.

4.4. Determination of Total Amounts of Phenolic Compounds

Phenolic compounds were extracted using an optimized laboratory protocol. After defatting with 1 mL hexane for 16 h, 0.2 g samples were extracted using 1 mL 70% methanol added with 10 μL naringin as an internal standard, during ultrasonication in ice water for 20 min. The extract was centrifuged at 13,000× *g* for 15 min (25 °C), and the extraction was repeated using the pellet. The combined supernatants were pre-purified on a Sep-Pak C18 cartridge (Waters, Milford, MA, USA) to remove chlorophylls, and then filtered through a Spartan 13/0.2 RC filter (Whatman, Dassel, Germany). The filtrate was used for the determination of total phenolic content (TPC) in mg gallic acid equivalent [GAE]·g^{-1} using the Folin–Ciocalteu method as described in [36]; total flavonoid content (TFC) in mg catechin equivalent [CAE]·g^{-1} using aluminum chloride as described in [36]; and total proanthocyanidin content (TPAC) in mg [GAE]·g^{-1} using polyvinylpyrrolidone, as described in [37]. The total anthocyanin content (TAC) was estimated using the pH differential assay [38], and the results were expressed in mg cyanidin 3-*O*-glucoside [CGE]·g^{-1}.

4.5. Chromatographic Separation and Identification of Phenolic Compounds

The quantitative analysis of individual phenolic compounds was carried out on a Gilson (Villers-le-bel, France) high-performance liquid chromatography (HPLC) instrument consisting of an autosampler, binary pump, column compartment, and a Finnigan photodiode array detector (DAD 81401; Thermo Electron, San Jose, CA, USA). Chromatography was performed on 10 μL samples of the phenolic filtrate injected into the HPLC onto a C18 column (5 μm, 250 × 4.5 mm i.d.) supplied from Sigma/Aldrich (Steinheim, Germany), and maintained at 25 °C. The solvent system consisted of 0.1% trifluoroacetic acid in water (mobile phase A) and 0.1% trifluoroacetic acid in acetonitrile (mobile phase B). Elution was performed at a constant flow rate of 1 mL.min^{-1} using a linear gradient program starting with 100% mobile phase A for 5 min, decreasing to 80% at 15 min, 50% at 30 min, 0% at 45 min, and then reverting to 100% at 55 until reaching 60 min.

The detection of compounds by DAD was conducted by scanning between 210–520 nm, with a resolution of 1.2 nm. Eluting peaks were monitored at 280, 320, 360, and 520 nm for hydroxybenzoic acids and other low molecular weight compounds, hydroxycinnamic acids and stilbenes, flavonoids, and anthocyanins, respectively, using the software Excalibur 2.0, which generated a three-dimensional dataset (absorbance, retention time, and wavelength). Eluting peaks at 450 nm were also monitored because two peaks were consistently observed with large areas at that wavelength. The peaks were selected using both the Gensis and the ICIS detection algorithms of Xcalibur. The threshold for quantification by peak areas was 5000 μAU·min^{-1}, and compounds whose peak areas were below this value were considered "non-detected."

For identification, 38 reference compounds previously reported in grapevine leaves [10,12,13,19,22,25,26,33], and representatives of the chemical classes under study were purchased (Table S1); they were also separated by HPLC. Peaks were identified with "some certainty" to compounds by matching UV/vis spectra and retention times with those of the reference compounds. The remaining peaks were putatively identified by comparison with UV/vis bibliographic data. Some peaks could not match to any compounds or phenolic group and were labeled as "unknown." Compounds were quantified by dividing their peak areas with that of the internal standard (naringin) and the results were converted to mg· kg^{-1} after correction by the peak area of the reference, its response factor, and the amount of biomass extracted. For compounds identified putatively, quantification was carried out using reference compounds with similar chemical characteristics as shown in Table S1.

4.6. Extraction, Separation, and Identification of Fatty Acids

The extraction of lipids was based on the method presented in [39]. Leaf samples (5 mg) were added with 0.8 mL water and 2 mL methanol in a DSR-2800V rotary shaker (Digisystem Laboratory Instruments Inc, Taipei, Taiwan) at room temperature; after continuous shaking for 5 min, 1 mL chloroform was added, and it was followed by agitation for 5 min. The mixture was centrifuged for 5 min at 2000× *g* (25 °C). The supernatant was collected and 2 mL chloroform/water (1/1, *v*/*v*) and five drops of 100 mM KCl were added. After vortexing, the mixture was centrifuged for 5 min at 2000× *g* (25 °C). The lipid fraction in the bottom layer was collected and the chloroform phase was evaporated to dryness under nitrogen. The dried extract was then transesterified with 5 mL 14% boron trifluoride in methanol under nitrogen at 70 °C for 60 min. Transesterified lipids were extracted by adding 5 mL hexane, followed by 3 min of vortexing. The upper phase, constituting fatty acid methyl esters (FAME), was collected and 1 g Na_2SO_4 was added to remove water.

FAME were separated via capillary gas chromatography (CG) using Shimadzu GC-2010 Plus (Shimadzu, Kyoto, Japan) equipped with an autosampler and an automatic split/splitless injector. Exactly 1 μL of FAME extract was injected into the GC at an inlet temperature of 270 °C and a split ratio of 5:1; compounds were separated on a 30 m long, 0.25-μm-thick-film DB-225MS column with a 0.25 mm i.d. (Agilent, Wilmington, DE, USA). The flow rate of the carrier gas (helium) was maintained at a constant value of 1 mL·min^{-1} at an inlet pressure of 200 kPa. The column temperature was maintained at 200 °C for 10 min, and it was then increased to 220 °C at a rate of 5 °C·min^{-1}.

The resolved compounds were detected using a flame ionization detector (FID-2010 Plus) set at 270 °C; the compounds were identified by comparing their retention times to those of a standard FAME mixture (FAME 37, Supelco, Bellefonte, PA, USA) run under the same conditions. Quantification was achieved by integrating the peaks with the Lab Solution 5.71 software, setting the minimum peak area/height at 2000 count. The amount of each FAME was expressed as a weight percentage of the total FAMEs represented in the chromatogram.

4.7. Statistical analyses

All data from four replications were subjected to an analysis of variance (ANOVA) using SPSS 16.0 (SPSS Inc., Chicago, IL, USA). In the case of TPC, TPAC, TFC and TAC, each replicate was analyzed two times and the average values used in statistical analyses. The Tukey's test was applied for assessing the mean differences, and a P value of ≤ 0.05 was considered as meaning statistical difference between the leaf groups.

Supplementary Materials: The following are available online at http://www.mdpi.com/2223-7747/8/10/412/s1, Figure S1: Principal component analysis (PCA) score and loading plots of phenolic compounds in leaves of vines infected by brown wood streaking, grapevine leaf stripe and apoplexy (Esca complex), Figure S2: Principal component analysis (PCA) score and loading plots of fatty acids in leaves of vines infected by brown wood streaking, grapevine leaf stripe and apoplexy (Esca complex), Table S1: Phenolic compounds identified in *Vitis vinifera* L. cv. Malvasia leaves from healthy, brown wood streaking, grapevine leaf stripe and apoplexy-infected vines and listed in decreasing order based on their average contents (dry weight basis), Table S2: Effect of brown wood streaking, grapevine leaf stripe and apoplexy (Esca complex) on minor phenolic compounds (average content < 1.00 mg·kg^{-1}, dry weight basis) in *Vitis vinifera* L. cv. Malvasia leaves, Table S3: Fatty acids identified in *Vitis vinifera* l. cv. Malvasia leaves from healthy, brown wood streaking, grapevine leaf stripe and apoplexy-infected vines and listed n decreasing order based on their average contents (dry weight basis), Table S4: Effect of brown wood streaking, grapevine leaf stripe and apoplexy (Esca complex) on the levels (%, dry weight basis) of 24 fatty acids in *Vitis vinifera* L. cv. Malvasia leaves.

Author Contributions: Conceptualization, P.G.; and I.C.; methodology, P.G.; I.C.; and A.C.M.; software, P.G.; validation, P.G.; I.C.; and A.C.M.; formal analysis, P.G.; investigation, P.G.; and A.C.M..; resources, I.C.; data curation, P.G.; writing—original draft preparation, P.G.; writing—review and editing, P.G.; I.C.; and A.C.M.; visualization, P.G.; supervision, I.C.; project administration, I.C.; funding acquisition, I.C.

Funding: This research was funded by the R&D project "INNOVINE & WINE –Vineyard and Wine Innovation Platform (NORTE-01-0145-FEDER-000038)", co-supported by Fundo Europeu de Desenvolvimento Regional (FEDER) and Norte 2020 (Programa Operacional Regional do Norte 2014/2020), and the project UID/AGR/04033/2019 supported by FCT – Fundação para a Ciência e a Tecnologia.

Conflicts of Interest: The authors declare no conflict of interest. The funders had no role in the design of the study; in the collection, analyses, or interpretation of data; in the writing of the manuscript, or in the decision to publish the results.

References

1. Lim, G.H.; Singhal, R.; Kachroo, A.; Kachroo, P. Fatty acid- and lipid-mediated signaling in plant defense. *Ann. Rev. Phytopathol.* **2017**, *55*, 505–536. [CrossRef] [PubMed]
2. Zhang, Z.; Cui, B.; Zhang, Y. Electrical penetration graphs indicate that tricin is a key secondary metabolite of rice, inhibiting phloem feeding of brown planthopper, *Nilaparvata lugens*. *Entomol. Exp. Appl.* **2015**, *156*, 14–27. [CrossRef]
3. Bani, M.; Cimmino, A.; Evidente, A.; Rubiales, D.; Rispail, N. Pisatin involvement in the variation of inhibition of *Fusarium oxysporum* f. sp. *pisi* spore germination by root exudates of *Pisum* spp. Germplasm. *Plant Pathol.* **2018**, *67*, 1046–1054.
4. Calzarano, F.; Di Marco, S. Further evidence that calcium, magnesium and seaweed mixtures reduce grapevine leaf stripe symptoms and increase grape yields. *Phytopathol. Mediterr.* **2018**, *57*, 459–471.
5. Calzarano, F.; Osti, F.; Baránek, M.; Di Marco, S. Rainfall and temperature influence expression of foliar symptoms of grapevine leaf stripe disease (esca complex) in vineyards. *Phytopathol. Mediterr.* **2018**, *57*, 488–505.

6. Aroca, A.; Raposo, R. PCR-based strategy to detect and identify species of *Phaeoacremonium* causing grapevine diseases. *Appl. Environ. Microbiol.* **2007**, *73*, 2911–2918. [CrossRef] [PubMed]
7. Travadon, R.; Lecomte, P.; Diarra, B.; Lawrence, P.D.; Renault, D.; Ojeda, H.; Rey, P.; Baumgartner, K. Grapevine pruning systems and cultivars influence the diversity of wood-colonizing fungi. *Fungal Ecol.* **2016**, *24*, 82–93. [CrossRef]
8. Surico, G. Towards a redefinition of the diseases within the esca complex of grapevine. *Phytopathol. Mediterr.* **2009**, *48*, 5–10.
9. Valtaud, C.; Larignon, P.; Roblin, G.; Fleurat-Lessard, P. Developmental and ultrastructural features of *Phaeomoniella chlamydospora* and *Phaeoacremonium aleophilum* in relation to xylem degradation in esca disease of the grapevine. *J. Plant Pathol.* **2009**, *91*, 37–51.
10. Rusjan, D.; Persic, M.; Likar, M.; Biniari, K.; Mikulic-Petkovsek, M. Phenolic responses to esca-associated fungi in differently decayed grapevine woods from different trunk parts of 'Cabernet Sauvignon'. *J. Agric. Food Chem.* **2017**, *65*, 6615–6624. [CrossRef]
11. Bruez, E.; Vallance, J.; Gerbore, J.; Lecomte, P.; Da Costa, J.P.; Guerin-Dubrana, L.; Rey, P. Analyses of the temporal dynamics of fungal communities colonizing the healthy wood tissues of esca leaf-symptomatic and asymptomatic vines. *PLoS ONE* **2014**, *9*, e95928. [CrossRef] [PubMed]
12. Martin, N.; Vesentini, D.; Rego, C.; Monteiro, S.; Oliveira, H.; Ferreira, B.R. *Phaeomoniella chlamydospora* infection induces changes in phenolic compounds content in *Vitis vinifera*. *Phytopathol. Mediterr.* **2009**, *48*, 101–116.
13. Magnin-Robert, M.; Spagnolo, A.; Boulanger, A.; Joyeux, C.; Clément, C.; Abou-Mansour, E.; Fontaine, F. Changes in plant metabolism and accumulation of fungal metabolites in response to Esca proper and apoplexy expression in the whole grapevine. *Phytopathology* **2016**, *106*, 541–553. [CrossRef] [PubMed]
14. Spagnolo, A.; Magnin-Robert, M.; Alayi, D.T.; Cilindre, C.; Mercier, L.; Schaeffer-Reiss, C.; Dorsselaer, V.A.; Clement, C.; Fontaine, F. Physiological changes in green stems of *Vitis vinifera* L. cv. chardonnay in response to esca proper and apoplexy revealed by proteomic and transcriptomic analyses. *J. Proteome Res.* **2012**, *11*, 461–475. [CrossRef] [PubMed]
15. Lecomte, P.; Diarra, B.; Carbonneau, A.; Rey, P.; Chevrier, C. Esca of grapevine and training practices in France: Results of a 10-year survey. *Phytopathol. Mediterr.* **2018**, *57*, 472–487.
16. Calzarano, F.; Osti, F.; D'agostino, V.; Pepe, A.; Di Marco, S. Mixture of calcium, magnesium and seaweed affects leaf phytoalexin contents and grape ripening on vines with grapevine leaf stripe disease. *Phytopathol. Mediterr.* **2017**, *56*, 445–457.
17. Magnin-Robert, M.; Letousey, P.; Spagnolo, A.; Rabenoelina, F.; Jacquens, L.; Mercier, L.; Clément, C.; Fontaine, F. Leaf stripe form of esca induces alteration of photosynthesis and defense reactions in presymptomatic leaves. *Funct. Plant Biol.* **2011**, *38*, 856–866. [CrossRef]
18. Fontaine, F.; Pinto, C.; Vallet, J.; Clément, C.; Gomes, C.A.; Spagnolo, A. The effects of grapevine trunk diseases (GTDs) on vine physiology. *Eur. J. Plant Patholol.* **2016**, *144*, 707–721. [CrossRef]
19. Bruno, G.; Sparapano, L. Effects of three esca-associated fungi on Vitis vinifera L.: II. Characterization of biomolecules in xylem sap and leaves of healthy and diseased vines. *Physiol. Mol. Plant Pathol.* **2006**, *69*, 195–208. [CrossRef]
20. Lima, R.M.M.; Machado, F.A.; Gubler, D.W. Metabolomic study of chardonnay grapevines double stressed with esca-associated fungi and drought. *Phytopathology* **2017**, *107*, 669–680. [CrossRef]
21. Calzarano, F.; D'agostino, V.; Pepe, A.; Osti, F.; Pelle, D.F.; De Rosso, M.; Flamini, R.; Di Marco, S. Patterns of phytoalexins in the grapevine leaf stripe disease (esca complex)/grapevine pathosystem. *Phytopathol. Mediterr.* **2016**, *55*, 410–426.
22. Lambert, C.; Khiook, I.L.K.; Lucas, S.; Télef-Micouleau, N.; Mérillon, J.-M.; Cluzet, S. A faster and a stronger defense response: One of the key elements in grapevine explaining its lower level of susceptibility to Esca? *Phytopathology* **2013**, *103*, 1028–1034. [CrossRef] [PubMed]
23. Letousey, P.; Baillieul, F.; Perrot, G.; Rabenoelina, F.; Boulay, M.; Vaillant-Gaveau, N.; Clément, C.; Fontaine, F. Early events prior to visual symptoms in the apoplectic form of grapevine esca disease. *Phytopathology* **2010**, *100*, 424–431. [CrossRef] [PubMed]
24. Lima, R.M.M.; Felgueiras, M.L.; Cunha, A.; Chicau, G.; Ferreres, F.; Dias, C.P.A. Differential phenolic production in leaves of Vitis vinifera cv. Alvarinho affected with esca disease. *Plant Physiol. Biochem.* **2017**, *112*, 45–52. [CrossRef] [PubMed]

25. Martín, L.; Fontaine, F.; Castañoa, J.F.; Songy, A.; Roda, R.; Vallet, J.; Ferrer-Gallego, R. Specific profile of Tempranillo grapevines related to Esca-leaf symptoms and climate conditions. *Plant Physiol. Biochem.* **2019**, *135*, 575–587. [CrossRef]
26. Calzarano, F.; Osti, F.; D'agostino, V.; Pepe, A.; Della Pelle, F.; De rosso, M.; Flamini, R.; Di Marco, S. Levels of phytoalexins in vine leaves with different degrees of grapevine leaf stripe disease symptoms (Esca complex of diseases). *Phytopathol. Mediterr.* **2017**, *56*, 494–501.
27. Lecomte, P.; Darrieutort, G.; Limiñana, J.M.; Comont, A.; Muruamendiaraz, A.; Legorburu, F.-J.; Choueiri, E.; Jreijiri, F.; El Amil, R.; Fermaud, M. New insights into Esca of grapevine: The development of foliar symptoms and their association with xylem discoloration. *Plant Dis.* **2012**, *96*, 924–934. [CrossRef]
28. Valtaud, C.; Foyer, H.C.; Fleurat-Lessard, P.; Bourbouloux, A. Systemic effects on leaf glutathione metabolism and defence protein expression caused by esca infection in grapevines. *Funct. Plant Biol.* **2009**, *36*, 260–279. [CrossRef]
29. Pearce, G. Systemin, hydroxyproline-rich systemin, and the induction of protease inhibitors. *Curr. Prot Pep. Sci.* **2011**, *12*, 399–408. [CrossRef]
30. Tsunoda, T.; Grosser, K.; van Dam, M.N. Locally and systemically induced glucosinolates follow optimal defence allocation theory upon root herbivory. *Funct. Ecol.* **2018**, *32*, 2127–2137. [CrossRef]
31. Ongena, M.; Duby, F.; Rossignol, F.; Fauconnier, M.L.; Dommes, J.; Thonart, P. Stimulation of the lipoxygenase pathway is associated with systemic resistance induced in bean by a nonpathogenic *Pseudomonas* strain. *Mol. Plant-Microbe Interact.* **2004**, *17*, 1009–1018. [CrossRef] [PubMed]
32. Mandal, M.K.; Chandra-Shekara, A.C.; Jeong, R.D.; Yu, K.; Zhu, S.; Chanda, B.; Navarre, D.; Kachroo, A.; Kachroo, P. Oleic acid–dependent modulation of NITRIC OXIDE ASSOCIATED1 protein levels regulates nitric oxide–mediated defense signaling in Arabidopsis. *Plant Cell* **2012**, *24*, 1654–1674. [CrossRef] [PubMed]
33. Gabaston, J.; Cantos-Villar, E.; Biais, B.; Waffo-Teguo, P.; Renouf, E.; Corio-Costet, M.; Richard, T.; Mérillon, J. Stilbenes from *Vitis vinifera* L. waste: A sustainable tool for controlling *Plasmopara Viticola*. *J. Agric. Food Chem.* **2017**, *65*, 2711–2718. [CrossRef] [PubMed]
34. Schnee, S.; Viret, O.; Gindro, K. Role of stilbenes in the resistance of grapevine to powdery mildew. *Physiol. Mol. Plant Pathol.* **2008**, *72*, 128–133. [CrossRef]
35. Tabacchi, R.; Fkeyrat, A.; Poliart, C.; Dubin, G.-M. Phytotoxins from fungi of esca of grapevine. *Phytopathol. Mediterr.* **2000**, *39*, 156–161.
36. Goufo, P.; Pereira, J.; Figueiredo, N.; Oliveira, M.B.P.P.; Carranca, C.; Rosa, E.A.S.; Trindade, H. Effect of elevated carbon dioxide (CO_2) on phenolic acids, flavonoids, tocopherols, tocotrienols, γ-oryzanol and antioxidant capacities of rice (*Oryza sativa L.*). *J. Cereal Sci.* **2014**, *59*, 15–24. [CrossRef]
37. Queiroz, C.R.A.A.; Morais, S.A.L.; Nascimento, E.A. Characterization of aroeira-preta (*Myracrodruon urundeuva*) wood tannins. *Rev. Árvore* **2002**, *26*, 485–492. [CrossRef]
38. Giusti, M.M.; Wrolstad, E.R. Characterization and measurement of anthocyanins by UV-visible spectroscopy. *Curr. Prot. Food Anal. Chem.* **2001**, F1.2.1–F1.2.13. [CrossRef]
39. Bligh, E.G.; Dyer, W.J. A lipid method of total lipid extraction and purification. *Can. J. Biochem.* **1959**, *37*, 911–917.

plants

MDPI

Article

Macrophomina Crown and Root Rot of Pistachio in California

Mohamed T. Nouri [1,2], Daniel P. Lawrence [2], Craig E. Kallsen [3] and Florent P. Trouillas [1,2,*]

1 Kearney Agricultural Research and Extension Center, Parlier, CA 93648, USA; mnouri@ucanr.edu
2 Department of Plant Pathology, University of California, Davis, CA 95616, USA; dlawrence@ucdavis.edu
3 University of California Cooperative Extension Kern County, Bakersfield, CA 93307, USA; cekallsen@ucanr.edu
* Correspondence: flotrouillas@ucdavis.edu; Tel.: +1-559-646-6566

Received: 3 January 2020; Accepted: 20 January 2020; Published: 21 January 2020

Abstract: In this study, declining pistachio rootstocks were detected in newly planted commercial pistachio orchards in Kern County, California. Symptoms were characterized by wilted foliage combined with crown rot in the rootstock. From diseased trees, 42 isolates were obtained, and all had similar cultural and morphological characteristics of *Macrophomina phaseolina*. Analyses of nucleotide sequences of three gene fragments, the internal transcribed spacer region (ITS1–5.8S–ITS2), partial sequences of β-tubulin, and translation elongation factor 1-α (*TEF1*) confirmed this identification, and 20 representative isolates are presented in the phylogenetic study. Testing of Koch's postulates showed that *M. phaseolina*, when inoculated to stems and roots of the pistachio rootstocks using mycelial plugs or a microsclerotial suspension, is indeed pathogenic to this host. The widely used clonal University of California Berkeley I (UCBI) rootstock appeared highly susceptible to *M. phaseolina*, suggesting that this pathogen is an emerging threat to the production of pistachio in California. This study confirmed the association of *M. phaseolina* with the decline of pistachio trees and represents the first description of this fungus as a crown rot-causing agent of pistachio in California.

Keywords: pistachio; crown rot; root rot; *Macrophomina phaseolina*; pathogenicity

---

1. Introduction

Pistachio (*Pistacia vera* L.) is one of California's many important, high-value nut crops. In 2017, California accounted for more than 99.1% of the United States pistachio crop, with approximately 101,327 hectares of bearing trees with a value of approximately $3.6 billion (http://www.acpistachios.org). Although pistachio cultivation in California is relatively new, this industry has had a record of steady expansion. As pistachio demand and acreage continues to grow, fungal pathogens pose a continually evolving challenge for the industry.

Panicle and shoot blight of pistachio has been one of the main fungal diseases affecting pistachio in California, with early reports dating back to the 1980s [1]. In California, at least eight species of Botryosphaeriaceae fungi have been associated with this disease. Botrytis blossom and shoot blight caused by *Botrytis cinerea* Pers. [2] and Alternaria late blight caused by multiple species of *Alternaria* Nees [3] are additional important diseases of pistachio in California. Management of these various diseases has mainly relied upon the use of synthetic fungicides. Additional above-ground disease of pistachio includes Cytospora canker caused by several *Cytospora* spp. and Botryosphaeria canker caused by *Neofusicoccum mediterraneum* Crous, M.J. Wingf. and A.J.L. Phillips [4,5].

Verticillium wilt caused by *Verticillium dahliae* Kleb. has been the most important soil-borne disease affecting pistachio trees in California [6]. *Verticillium dahliae* is found worldwide in all types of soils and has an extremely wide host range, affecting more than 400 different plant species, including vegetables, flowers, fruit crops, ornamentals, and perennial agronomic crops [7–9]. Verticillium wilt of pistachio mostly occurred in orchards planted after crops such as cotton and tomato, which are highly susceptible to *V. dahliae*, or in orchards established adjacent to fields with susceptible crops [10]. The fungus can survive in the soil for many years due to extremely persistent resting structures, the microsclerotia. Accordingly, control of Verticillium wilt begins before the trees are planted and the site location is an important consideration. Former cotton, tomato, and alfalfa fields should be avoided, as they increase the risk of Verticillium wilt. The most effective control measures to combat Verticillium wilt is the use of resistant rootstocks. In California, commercial pistachio nut production began in the 1970s with *Pistacia atlantica* Desf. as the main pistachio rootstock [11]. However, *P. atlantica* is highly susceptible to infection by *V. dahliae*, and the fungus is widespread in soils of the San Joaquin Valley. The development and widespread planting of resistant University of California Berkeley I (UCBI) rootstocks has diminished the Verticillium wilt problem of pistachio in California. Symptoms of Verticillium wilt are now only observed sporadically in the southern half of the San Joaquin Valley, where the *P. atlantica* rootstock was initially planted.

California pistachio is also subject to root and crown rots caused by oomycetes such as *Phytophthora* and *Phytopythium* [12–14]. Disease development with oomycetes is generally enhanced in poorly drained soils, where orchards receive long durations of flood irrigation, or in trees in lower areas or along creeks and natural drainage creeks in the orchard [6]. When the soil remains saturated for a long time, these pathogens are able to infect susceptible roots.

Crespo et al. (2019) recently revealed species of *Fusarium* and *Neocosmospora* from declining pistachio rootstocks and stem cankers in the southern San Joaquin Valley of California that were pathogenic to this host [15]. Three *Fusarium* (*Fusarium equiseti* (Corda) Sacc., *F. oxysporum* Schltdl., and *F. proliferatum* (Matsush.) Nirenberg), and two *Neocosmospora* species (*Neocosmospora falciformis* (Carrión) L. Lombard and Crous (syn: *Fusarium falciforme* (Carrión) Summerb. and Schroers) and *N. solani* (Mart.) L. Lombard and Crous (syn: *Fusarium solani* (Mart.) Sacc.)) were found associated with crown rot symptoms and vascular discoloration in stems of clonal UCBI rootstocks in California [15]. Necrotic root lesions and black discoloration of the root cortex, epidermis, and vascular tissues associated with cylindrocarpon-like fungi included species in *Dactylonectria*, *Neonectria*, and *Thelonectria*, which were also detected in multiple counties in California [16].

Other disease symptoms observed in recent surveys of pistachio orchards have included young pistachio trees that were collapsing, showing crown rot with black discolorations in the rootstocks. Isolation from these young declining pistachio rootstocks revealed the occurrence of a single putative pathogen tentatively identified as a member of the Botryosphaeriaceae. The aims of this study were to (1) identify and characterize the putative pathogen associated with pistachio rootstock decline based on molecular and morphological methods; (2) test the pathogenicity of the putative pathogen on the commonly planted UCBI pistachio rootstock in California.

2. Results

2.1. Field Surveys and Collection of Fungal Isolates

Forty-two fungal isolates resembling a member of the Botryosphaeriaceae were isolated from the rootstock of young declining pistachio trees showing root crown and lower trunk rot symptoms (Figure 1). Affected rootstocks mainly included the clonal UCBI rootstocks. All pistachio isolates were obtained from orchards in Kern County, California. Orchards characteristics included heavy clay soils and history of vegetable crops or cotton. In addition to pistachio, 21 isolates with typical characteristics of the putative pathogen were also collected: 8 from sweet cherry in Fresno and San Joaquin Counties and 13 from grapevine in Fresno County.

Figure 1. Symptoms in pistachio rootstocks associated with *Macrophomina phaseolina*; (**A**) wilting and death of unbudded University of California Berkeley I (UCBI) pistachio rootstocks; (**B**,**C**) associated crown rot symptom; (**D**,**E**) pistachio tree showing crown rot and black discoloration.

2.2. Phylogenetic Analyses

PCR amplification of the ITS region, *TEF1*, and *TUB2* generated 452–498, 196–213, and 373–380 bp fragments, respectively. For ML analysis, the best-fit model of nucleotide evolution was K2+G for each dataset. The three-gene (ITS+*TEF1*+*TUB2*) 49 sequence dataset consisted of 1161 characters (643 characters were constant, 83 characters were parsimony-uninformative, and 435 characters were parsimony informative). MP analysis produced 30 equally most parsimonious trees of 996 steps and a consistency index (CI), retention index (RI), and rescaled consistency index (RC) of 0.776, 0.890, and 0.690, respectively. MP and ML analyses revealed that 30 Californian fungal isolates, isolated in this study, strongly clustered (100%/100% MP and ML bootstraps, respectively) with the type specimen of *Macrophomina phaseolina* (Tassi) Goid. isolate CBS 227.33 (Figure 2).

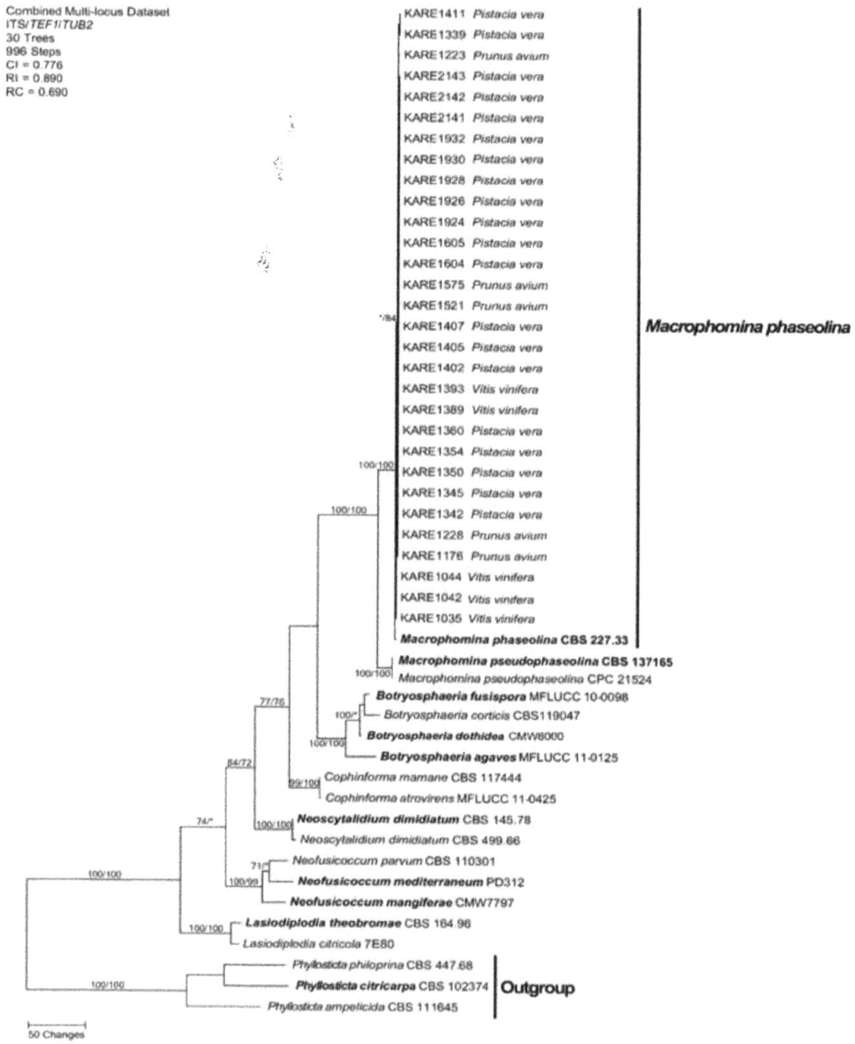

Figure 2. One of 30 equally most parsimonious trees resulting from the analysis of the three-gene combined dataset (ITS, *TUB2*, and *TEF1*). Numbers in front and after the slash represent maximum parsimony and maximum likelihood bootstrap values, respectively. Values represented by an asterisk were less than 70%. Scale bar represents the number of nucleotide changes.

2.3. Morphological Characterization

Colonies in culture ranged in color from light to dark gray and became black with age (Figure 3A). The average growth rate was 4 cm per day, and most colonies reached the edge of an 85 mm potato dextrose agar (PDA) dish in 48 h. Aerial mycelia generally did not develop on PDA. Hyphae were septate, initially subhyaline turning dark-brown with time. Abundant microsclerotia developed in water agar (WA) and PDA and were black, spherical to oblong, and averaged 84.5 (Length) × 53.5 (Width) µm (n = 40) with a length-width ratio of 1.57 at maturity (Figure 3B,C). Pycnidia, produced on pistachio leaf agar (PLA) medium after 2 weeks incubation, were black, subglobose to lageniform, solitary or gregarious, and ranged from 100 to 200 µm in diameter. Conidia were ellipsoid to obovoid and averaged 24.5 × 11.0 μm with a length-width ratio of 2.28. Immature conidia possessed apical mucoid appendages (Figure 3D). Morphological features of the isolates were typical of those of *M. phaseolina* [17]. The optimal growth temperature was 30 °C for all three isolates tested (Figure 4).

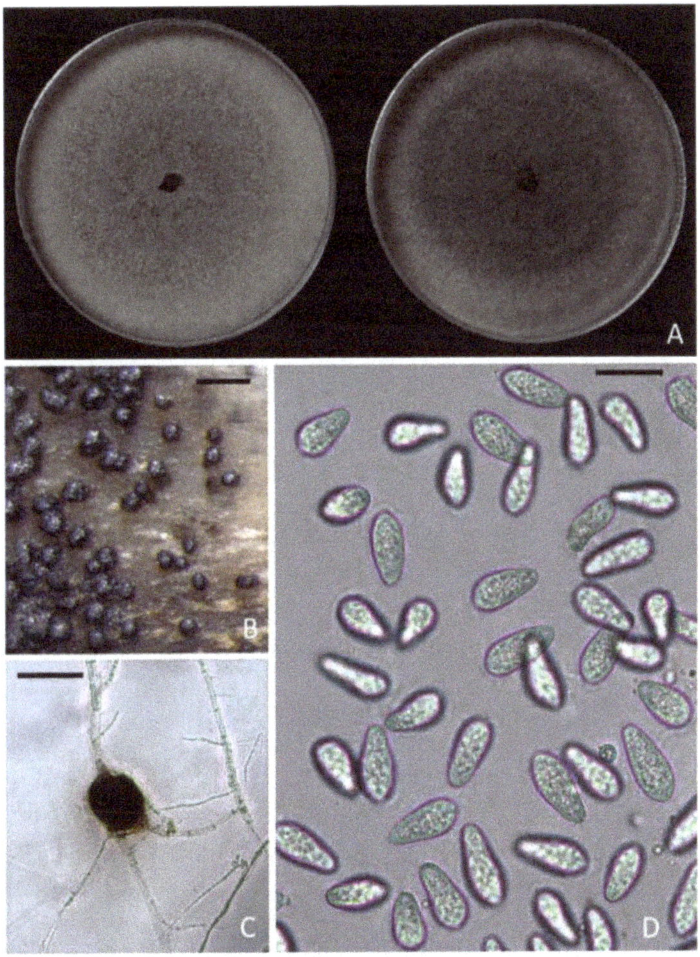

Figure 3. Morphological characteristics of *Macrophomina phaseolina*; (**A**) colonies on potato dextrose agar after 10 days of incubation at 25 ± 1 °C in darkness; (**B**,**C**) microsclerotia on pistachio wood and close up on a microsclerotium produced on water agar; (**D**) conidia produced from mature pycnidia forming on pistachio leaf agar medium. Scale bar in B = 300 μm; in C = 60 μm; and in D = 20 μm.

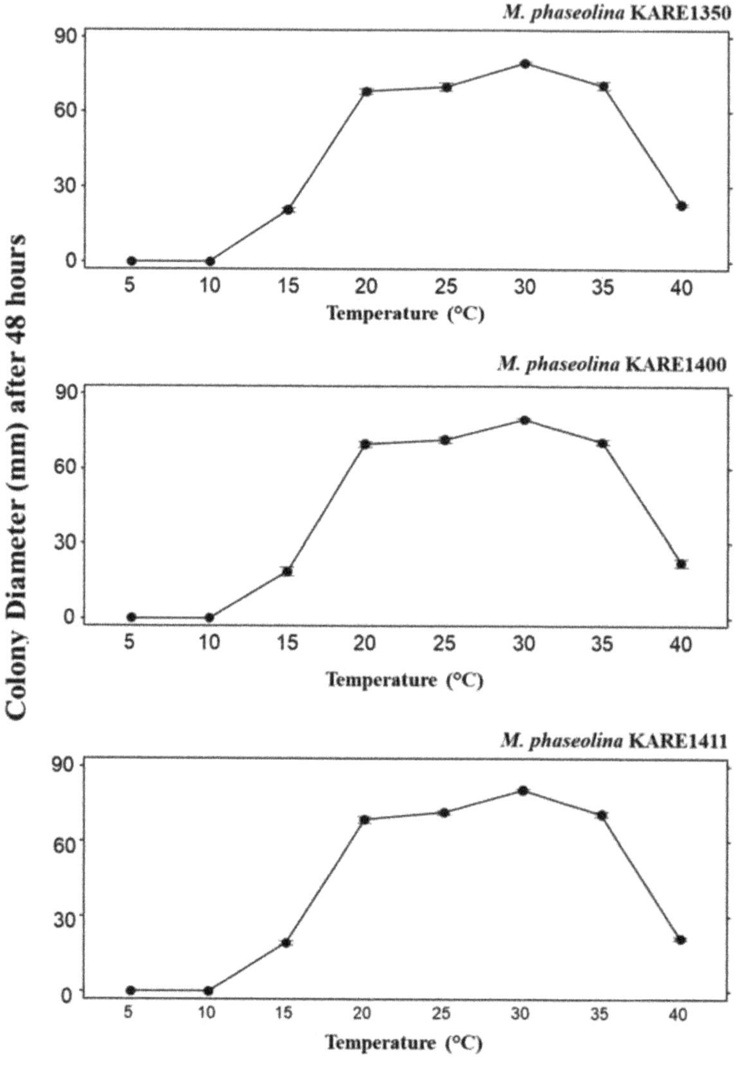

Figure 4. Effect of the temperature on the mycelial growth of three isolates of *Macrophomina phaseolina*: KARE1350, KARE1400 and KARE1411 on potato dextrose agar after 48 h of incubation. Error bars represent the standard error of the mean.

2.4. Pathogenicity Tests

2.4.1. Stem Inoculation of Pistachio Rootstocks with Mycelium Plugs

Ten months after inoculation, lesion and rot symptoms observed from the inoculated stems were similar to those observed in the field. For the first pathogenicity test conducted in September 2016, the two fungal isolates tested produced dark vascular discoloration in the wood that ranged from 4 to 4.76 cm in length (Figure 5). Both isolates tested produced significantly longer ($p = 0.0005$) lesion lengths on pistachio stems compared to the control treatment (1.2 cm).

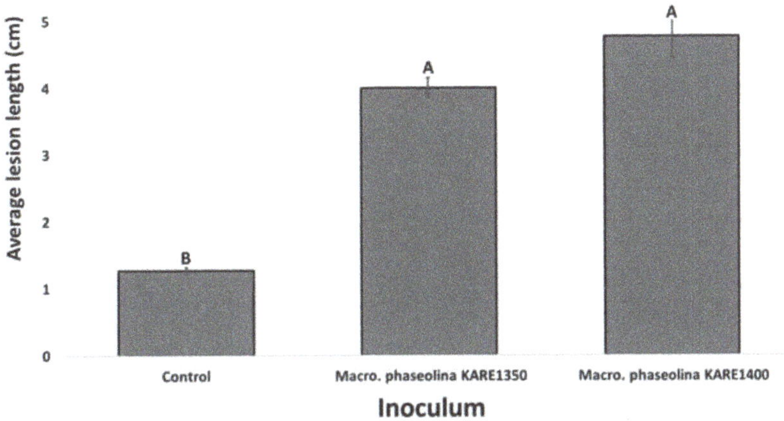

Figure 5. Average lesion lengths (cm) on pistachio clonal UCBI rootstocks 10 months after inoculation with mycelial plugs of *Macrophomina phaseolina* isolates KARE1350 and KARE1400 (September 2016 inoculations). Columns with different letters indicate treatment means that are significantly different ($p = 0.0005$).

In the second experiment, conducted in July 2017, all three isolates tested produced substantial lesions to the stems of pistachio clonal UCB1 rootstocks (Figure 6). Mean lesion length varied between 16.15 and 16.83 cm and was significantly longer ($p < 0.00001$) as compared to the control plants (2 cm) (Figure 7). Each fungal isolate was successfully recovered from inoculated plants; pathogen recovery varied between 66.6% and 100% among the three isolates, and morphologically matched the inoculated fungus, thereby fulfilling Koch's postulates of pathogenic organisms. The controls showed no disease symptoms and no pathogen was isolated from them.

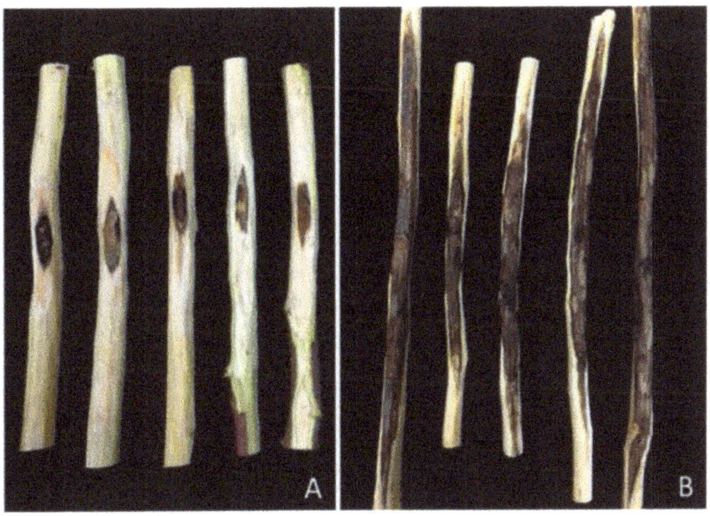

Figure 6. Lesions produced in 2-year-old potted clonal UCBI rootstocks, 10 months after inoculation: (**A**) Control; (**B**) *Macrophomina phaseolina* isolate KARE1400 (July 2017 inoculations).

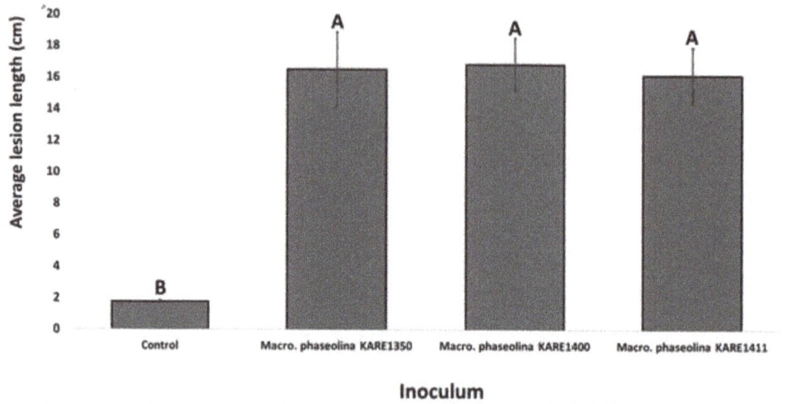

Figure 7. Average lesion lengths (cm) on pistachio clonal UCBI rootstocks 10 months after inoculation with mycelial plugs of *Macrophomina phaseolina* isolates KARE1350, KARE1400, and KARE1411 (July 2017 inoculations). Columns with different letters indicate treatment means that are significantly different ($p < 0.00001$).

2.4.2. Root Inoculation of Pistachio Rootstocks with Microsclerotial Suspension

Three weeks after inoculation, all isolates tested had caused severe wilting or death of all the inoculated plantlets of clonal UCBI rootstocks (Figure 8). Isolates were recovered consistently from resulting root or crown lesions thus, fulfilling Koch's postulates. *Macrophomina phaseolina* was not isolated from control plants, which remained asymptomatic.

Figure 8. Symptoms produced on 4-month-old plantlets of clonal UCBI rootstocks after 22 days incubation and following root inoculation using a microsclerotia suspension with control plants (left) and inoculated plants (right).

3. Discussion

This is the first study to report *M. phaseolina* associated with pistachio rootstock decline in California. *Macrophomina phaseolina* was isolated from black lesions developing at the crown or basal stem of rapidly declining young pistachio trees in Kern County California. The identification of *M. phaseolina* was supported by morphological examinations of the unique colony characteristics, including fast growth and other anamorphic morphologies typical of the species including obtuse conidia with apical mucoid appendages and the production of microsclerotia in culture [17,18].

Pathogen identification was confirmed by phylogenetic analyses of ITS, *TEF1*, and *TUB2* DNA sequence data. These analyses revealed that isolates collected from different symptomatic parts in pistachio rootstocks (crown rot and root rot) represented a single species. Results of phylogenetic analyses also indicated that *M. phaseolina* isolates from pistachio were genetically similar to other isolates isolated from symptomatic grapevine and sweet cherry in Fresno and San Joaquin Counties, respectively.

Macrophomina phaseolina is an important phytopathogenic fungus, infecting more than 750 plant species [17,19]. The pathogen incites a stem canker disease in many crops that is often referred to as charcoal rot disease, due to the charcoal type coloration imparted to the symptomatic plant tissues. Charcoal rot disease affects many field crops including soybean (*Glycine max* (L.) Merr.) [20], chickpea (*Cicer arietinum* L.) [21], common bean (*Phaseolus vulgaris* L.) [22], sunflower (*Helianthus annuus* L.) [23], and sorghum (*Sorghum bicolor* (L.) Moench.) [24]. In California, charcoal rot has emerged as a serious concern for strawberry cultivation [25]. By 2014, the pathogen was confirmed in all major coastal strawberry counties in California [26]. *Macrophomina phaseolina* has been reported only sporadically as a pathogen of perennial woody crops. It is known to affect grapevine in Australia, Iran, South Africa, Spain, and California [27–29] and olive in Australia [30]. In California, *M. phaseolina* has also been reported from almond cankers [31].

Pathogenicity tests in potted 2-year-old clonal UCBI pistachio rootstocks using mycelium plugs showed that *M. phaseolina* was highly aggressive following July infections, causing dark and elongated lesions in the stems of the pistachio trees. In the microsclerotial inoculum assay, *M. phaseolina* caused the death of clonal UCBI pistachio plantlets, inducing both root and crown rots. These results indicate that clonal UCBI pistachio rootstock most likely do not possess genetic resistance against this aggressive plant pathogen. To our knowledge, this work is the first to report *M. phaseolina* associated with the decline of pistachio trees worldwide.

Macrophomina phaseolina causes important annual losses to its host crop and can survive in the soil for many years, mainly as microsclerotia that germinate repeatedly during the growing season. The pathogen generally attacks young plants when their growth is retarded due to unfavorable conditions including low-water potentials that occur during periods of severe drought [32]. In the present study, *M. phaseolina* was mostly isolated from young rootstocks that were newly planted in the field and were subjected to stressful environmental conditions such as high temperature, poor soil fertility, or trees that had suffered bark damage due to gophers.

In recent years, California has been severely affected by drought and increased mean annual temperatures, which may contribute to fluxes in the plant environment and increase plant stress. Water shortage combined with high temperatures, the intensification and expansion of pistachio cultivation, the aging of orchards, and the planting of new varieties are also factors that could contribute to disease emergence and spread. Recently, multiple *Cytospora* spp., *Neofusicoccum mediterraneum*, *Colletotrichum karsti* You L. Yang, Zuo Y. Liu, K.D. Hyde and L. Cai, *Diaporthe ambigua* Nitschke, and *Didymella glomerata* (Corda) Qian Chen and L. Cai, were reported as new pathogens of pistachio in California, causing cankers in the trunks and branches of mature trees [4,5].

In conclusion, this study has provided new information regarding the identity and pathogenicity of *M. phaseolina* associated with pistachio rootstock decline in California, and this pathogen represents a new and emerging threat to the industry. The occurrence of this fungal pathogen, in pistachio but also sweet cherry and grapevine in California, has become of increasing concern in recent years. Research needs to be continued to investigate the disease epidemiology as well as develop efficient control strategies to mitigate the impact of Macrophomina crown and root rot disease of pistachio in California.

4. Materials and Methods

4.1. Field Surveys and Collection of Fungal Isolates

Isolates used in this study were isolated from declining pistachio trees and rootstocks expressing crown and root rot symptoms in Kern County California from 2016 to 2018. In total, 22 symptomatic samples were collected from 8 orchards ranging from 1- to 5-years-old. Samples of phloem (live bark) and vascular-cambium tissue from the lower trunk/crown of young trees showing crown rot and associated decline symptoms were collected and taken to the laboratory for detailed examination and isolation of putative pathogens. Symptomatic bark and root tissues were surface disinfested by submerging them in a 0.5% sodium hypochlorite for 2 min and rinsed twice with sterile water. Subsequently, the bark and root tissues were placed in Petri dishes containing potato dextrose agar (PDA) acidified with lactic acid (2.5 mL of 25% [vol/vol] per liter of medium) (APDA). Cultures were incubated at ambient laboratory light and temperature conditions (24 ± 2 °C) with approximately 12 h of daylight and 12 h of darkness until fungal colonies were observed. The most prevalent fungal taxa growing from the symptomatic tissues were then individually transferred to fresh APDA Petri dishes. To obtain pure cultures, single hyphal tips from colonies with typical growth characteristics of the Botryosphaeriaceae (initially white, fast-growing, and darkening with age) were transferred to fresh PDA and incubated as described above for 5 to 7 days. Similar isolates were also isolated from symptomatic grapevine and sweet cherry wood tissues and were included in the phylogenetic analyses. Isolates collected in the present study are summarized in Table 1 and maintained in the culture collection of the Department of Plant Pathology of the University of California, Davis at the Kearney Agricultural Research and Extension (KARE) Center, Parlier, CA, USA.

Table 1. Fungal isolates recovered in this study.

Species	Isolate [a]	Host	Substrate	Location	GenBank Accession Number [b]		
					ITS	TEF1	TUB2
Macrophomina phaseolina	**KARE1339**	***Pistacia vera***	**Crown**	Kern Co., CA	MN097202	MN106057	MN106087
Macrophomina phaseolina	KARE1342	*Pistacia vera*	Crown	Kern Co., CA	MN097203	MN106058	MN106088
Macrophomina phaseolina	KARE1345	*Pistacia vera*	Root	Kern Co., CA	MN097204	MN106059	MN106089
Macrophomina phaseolina	KARE1350	*Pistacia vera*	Crown	Kern Co., CA	MN097205	MN106060	MN106090
Macrophomina phaseolina	KARE1354	*Pistacia vera*	Crown	Kern Co., CA	MN097206	MN106061	MN106091
Macrophomina phaseolina	KARE1360	*Pistacia vera*	Root	Kern Co., CA	MN097207	MN106062	MN106092
Macrophomina phaseolina	KARE1402	*Pistacia vera*	Crown	Kern Co., CA	MN097210	MN106065	MN106095
Macrophomina phaseolina	KARE1405	*Pistacia vera*	Crown	Kern Co., CA	MN097211	MN106066	MN106096
Macrophomina phaseolina	KARE1407	*Pistacia vera*	Crown	Kern Co., CA	MN097212	MN106067	MN106097
Macrophomina phaseolina	KARE1411	*Pistacia vera*	Root	Kern Co., CA	MN097213	MN106068	MN106098
Macrophomina phaseolina	KARE1604	*Pistacia vera*	Crown	Kern Co., CA	MN097216	MN106069	MN106101
Macrophomina phaseolina	KARE1605	*Pistacia vera*	Crown	Kern Co., CA	MN097217	MN106070	MN106102
Macrophomina phaseolina	KARE1924	*Pistacia vera*	Crown	Kern Co., CA	MN097218	MN106073	MN106103
Macrophomina phaseolina	KARE1926	*Pistacia vera*	Crown	Kern Co., CA	>MN097219	MN106074	MN106104
Macrophomina phaseolina	KARE1928	*Pistacia vera*	Crown	Kern Co., CA	MN097220	MN106075	MN106105
Macrophomina phaseolina	KARE1930	*Pistacia vera*	Crown	Kern Co., CA	MN097221	MN106076	MN106106
Macrophomina phaseolina	KARE1932	*Pistacia vera*	Crown	Kern Co., CA	MN097222	MN106077	MN106107
Macrophomina phaseolina	KARE2141	*Pistacia vera*	Crown	Kern Co., CA	MN097223	MN106078	MN106108
Macrophomina phaseolina	KARE2142	*Pistacia vera*	Crown	Kern Co., CA	MN097224	MN106079	MN106109
Macrophomina phaseolina	KARE2143	*Pistacia vera*	Crown	Kern Co., CA	MN097225	MN106080	MN106110
Macrophomina phaseolina	KARE1176	*Prunus avium*	Root	San Joaquin Co., CA	MN097199	MN106054	MN106084
Macrophomina phaseolina	KARE1223	*Prunus avium*	Root	San Joaquin Co., CA	MN097200	MN106055	MN106085
Macrophomina phaseolina	KARE1228	*Prunus avium*	Root	San Joaquin Co., CA	MN097201	MN106056	MN106086
Macrophomina phaseolina	KARE1521	*Prunus avium*	Root	Fresno Co., CA	MN097214	MN106071	MN106099
Macrophomina phaseolina	KARE1575	*Prunus avium*	Root	Fresno Co., CA	MN097215	MN106072	MN106100
Macrophomina phaseolina	KARE1035	*Vitis vinifera*	Trunk	Fresno Co., CA	MN097196	MN106051	MN106081
Macrophomina phaseolina	KARE1042	*Vitis vinifera*	Cordon	Fresno Co., CA	MN097197	MN106052	MN106082
Macrophomina phaseolina	KARE1044	*Vitis vinifera*	Cordon	Fresno Co., CA	MN097198	MN106053	MN106083
Macrophomina phaseolina	KARE1389	*Vitis vinifera*	Trunk	Fresno Co., CA	MN097208	MN106063	MN106093
Macrophomina phaseolina	KARE1393	*Vitis vinifera*	Trunk	Fresno Co., CA	MN097209	MN106064	MN106094

[a] KARE = Kearney Agricultural Research and Extension. [b] ITS = internal transcribed spacer, *TEF1* = translation elongation factor 1α, and *TUB2* = beta-tubulin.

4.2. DNA Extraction, Sequencing, and Phylogenetic Analyses

Total genomic DNA was isolated from fungal mycelium scraped with a sterile scalpel from the surface of 7-day-old PDA cultures using the DNeasy Plant Kit (Qiagen, Valencia, CA, USA), following the manufacturer's instructions. All PCR reactions utilized AccuPower™ PCR Premix (Bioneer, Alameda, CA, USA), following the manufacturer's instructions. Amplification of rDNA, including the intervening ITS regions and 5.8S rDNA (ITS1–5.8S–ITS2), using the primer set ITS1 and ITS4 followed the protocol of White et al. (1990) [33]. Amplification of translation elongation factor 1-α (*TEF1*) fragments utilized the primer set EF1-728F and EF1-986R [34], and beta-tubulin (*TUB2*) utilized primers Bt2a and Bt2b [35]. PCR products were visualized on 1.5% agarose gels (120 V for 25 min) stained with GelRed® (Biotium, Fremont, CA, USA), following the manufacturer's instructions, to confirm presence and size of PCR amplicons, purified via Exonuclease I and recombinant Shrimp Alkaline Phosphatase (Affymetrix, Santa Clara, CA, USA), and sequenced in both directions utilizing the primers above on an ABI 3730 Capillary Electrophoresis Genetic Analyzer (College of Biological Sciences Sequencing Facility, University of California, Davis, CA, USA).

Forward and reverse DNA sequences were assembled, edited, and proofread in Sequencher v. 5 (Gene Codes Corporation, Ann Arbor, MI, USA) and deposited in GenBank (Table 1). Homologous sequences with high similarity from ex-type and non-type Botryosphaeriaceae species were included for phylogenetic reference utilizing the BLASTn function in NCBI and literature review (Table 2). Multiple sequence alignments were conducted in MEGA v. 6 [36] and manually adjusted where necessary in Mesquite v. 3.10 [37]. Alignments were submitted to TreeBASE under accession number S24559. The three-gene dataset was analyzed using two different optimality search criteria, maximum parsimony (MP) and maximum likelihood (ML), in PAUP * v. 4.0a164 and GARLI v. 0.951 [38,39], respectively. For the MP analysis, a heuristic search with 1000 random sequence additions was implemented with the Tree-Bisection-Reconnection algorithm and gaps were treated as missing data. Bootstrap analysis with 1000 replicates was used to estimate branch support. For the ML analysis, MEGA was used to infer a model of nucleotide substitution for each dataset, using the Akaike Information Criterion (AIC). ML analyses were conducted according to the best fit model of nucleotide substitution for each dataset using default parameters in GARLI, and branch support was determined by 1000 bootstrap replicates. Sequences of *Phyllosticta* Pers. (Botryosphaeriales, Phyllostictaceae) served as the outgroup taxon in the phylogenetic analyses.

Table 2. Fungal isolates retrieved from Genbank for phylogenetic reference.

Species	Isolate [a]	Host/Substrate	Location	GenBank Accession Number [b]		
				ITS	*TEF1*	*TUB2*
Botryospheria agaves	**MFLUCC 11-0125**	***Agaves*** **sp.**	Thailand	JX646791	JX646856	JX646841
Botryospheria corticis	**CBS 119047**	*Vaccinium corymbosum*	NJ, USA	DQ299245	EU017539	EU673107
Botryospheria dothidea	**CMW8000**	*Prunus* sp.	Switzerland	AY236949	AY236898	AY236927
Botryospheria fusispora	**MFLUCC 10-0098**	*Caryota* sp.	Thailand	JX646789	JX646854	JX646839
Cophinforma eucalypti	MFLUCC 11-0425	*Eucalyptus* sp.	Thailand	JX646800	JX646865	JX646848
Cophinforma mamane	CBS 117444	*Eucalyptus* sp.	Venezuela	KF531822	KF531801	KF531802
Lasiodiplodia theobromae	**CBS 164.96**	Fruit on coral reef coast	New Guinea	AY640255	AY640258	EU673110
Lasiodiplodia citricola	7-E80	*Juglans regia*	CA, USA	KC357300	KC357312	KC357306
Macrophomina phaseolina	**CBS 227.33**	*Zea mays*	Palestine	KF531825	KF531804	KF531806
Macrophomina pseudophaseolina	**CBS 137165**	*Arachis hypogaea*	Senegal	KF951791	KF952153	KF952233
Macrophomina pseudophaseolina	CPC 21524	*Hibiscus sabdarifa*	Senegal	KF951799	KF952161	KF952240
Neofusicoccum mangiferae	CBS 118532	*Mangifera indica*	Australia	AY615186	DQ093220	AY615173
Neofusicoccum mediterraneum	**PD312**	*Eucalyptus* sp.	Greece	GU251176	GU251308	GU251836
Neofusicoccum parvum	CBS 110301	*Vitis vinifera*	Portugal	AY259098	AY573221	EU673095
Neoscytalidium dimidiatum	**CBS 145.78**	*Homo sapiens*	United Kingdom	KF531816	KF531795	KF531796
Neoscytalidium dimidiatum	CBS 499.66	*Mangifera indica*	Mali	KF531820	KF531798	KF531800
Phyllosticta ampelicida	CBS 111645	*Parthenocissus quinquefolia*	MO, USA	FJ824766	FJ824772	FJ824777
Phyllosticta citricarpa	**CBS 102374**	*Citrus aurantium*	Brazil	FJ824767	FJ538371	FJ824778
Phyllosticta philoprina	CBS 447.68	*Taxus baccata*	United States	FJ824768	FJ824773	FJ824779

[a] Isolates in bold represent type specimens. [b] ITS = internal transcribed spacer, *TEF1* = translation elongation factor 1α, and *TUB2* = beta-tubulin.

4.3. Morphological Characterization

Three representative isolates (KARE1045, KARE1350, and KARE1400) were selected to study their cultural and conidial features. For each selected isolate, 5 mm mycelial plugs from developing colonies were removed and transferred to the center of new 85 mm diameter PDA Petri dishes. Cultures were incubated in the dark at temperatures ranging from 5 to 40 °C at 5-degree intervals. At each temperature (5°, 10°, 15°, 20°, 25°, 30°, 35°, and 40 °C), 3 replicate dishes of each isolate were used. Measurements of colony diameter were taken each 24 h until the fastest-growing colony had reached the edge of the dish. For each of the 8 temperatures, the average colony diameter per 48 h of growth was calculated. Colony characters and pigment production were noted after 48 h of incubation on PDA at 30 °C in darkness. The experiment was repeated once.

The representative isolates, as mentioned above, were further used to study conidial morphology. Pycnidia were induced to form by incubating colonies on pistachio leaf agar (PLA) medium [40]. Pistachio leaves were autoclaved twice at 120 °C for 20 min and then placed into Petri dishes containing PDA (5 g agar, 5 g potato dextrose broth and 500 mL water). Cultures were then incubated under near UV irradiation (12 h light/12 h dark) at 25 ± 1 °C for 2 weeks. Pycnidia containing conidia were mounted in sterile water on glass slides, and conidial dimensions included length and width of 40 conidia for each isolate were measured at ×400 magnification using a Leica compound microscope (Leica DM2000 LED Microscope, Wetzlar, Germany). Microsclerotia were also induced to form by incubating isolates on WA and characterized as described above.

4.4. Pathogenicity Tests

4.4.1. Stem Inoculation of Clonal UCBI Pistachio Rootstocks with Mycelium Plugs

Potted 2-year-old clonal UCBI pistachio rootstocks showing no apparent decline symptoms, or weak growth, were used for the pathogenicity tests. Two experiments were conducted using the same methodology. The first pathogenicity test was conducted in September 2016. Four tree replicates were inoculated per combination of 3 inoculation treatments (isolates KARE1350, KARE1400, and the control). Plants were arranged in a completely randomized design and maintained in a greenhouse with a natural photoperiod and temperatures ranging from 22 to 30 °C for 10 months. Plants were watered once weekly during the course of the experiment. The second pathogenicity test was conducted in July 2017. Eight replicate trees were inoculated per combination of 4 inoculation treatments (isolates KARE1350, KARE1400, KARE1411, and the control) and arranged in a completely randomized design in an outdoor gravel bed for 10 months. Inoculations were conducted by using a 5 mm-diameter mycelium plug from a 7-day-old culture. The bark surface was disinfected at the point of inoculation with 70% ethanol. Stems of the potted clonal UCBI rootstocks were inoculated 4 cm above the soil line, wounds to the stem/crown were produced with a flame-sterilized 5 mm cork borer followed by placing a 5 mm agar plug bearing aerial mycelia, upper surface facing inward, into the fresh wound, which was then sealed with petroleum jelly, and wrapped with Parafilm. Control plants were treated with sterile PDA agar plugs followed by sealing and wrapping as above. Plants were watered twice weekly.

Infection data were recorded 10 months after inoculation. The bark was removed, and the length of wood discoloration (LWD) upwards and downwards from the point of inoculation was measured using a digital caliper. In an attempt to fulfill Koch's postulates, small pieces of necrotic tissue from the edge of each lesion were surface disinfested as described above and plated on APDA filled Petri dishes to recover the inoculated fungus. Emerging fungal colonies were recorded and identified based on morphology as described above. Data for lesion length were tested for normality and homogeneity of variance using Shapiro-Wilk's and Bartlett's tests, respectively. ANOVAs were done in Statistix 10 (Analytical Software, Tallahassee, FL, USA) to evaluate differences in the LWD between the control and fungal treatments. Differences in the means between fungal treatments and the mock-inoculated controls were assessed using Tukey's honestly significant difference (HSD) test at $p = 0.05$.

4.4.2. Root Inoculation of UCBI Pistachio Rootstocks with Microsclerotial Suspensions

Clonal UCBI pistachio saplings were root-inoculated using a microsclerotial suspension. The experiment was conducted using 3 isolates (KARE1350, KARE1400, and KARE1411) of *M. phaseolina* inoculated to 4-month-old plantlets of clonal UCBI rootstocks. The inoculum, consisting of mycelium and microsclerotia (1×10^5 propagules per mL), was prepared by scraping the surface of 12-day-old PDA cultures containing microsclerotia of the pathogen with a sterile scalpel blade and homogenizing in sterile water using a kitchen blender on high speed for 30 s. Inoculations were performed by dipping the bare root system of each rootstock into the inoculum slurry for 5 min, and subsequently mixing the microsclerotia suspension homogeneously with potting soil in 10 × 24 cm tree pots. In total, 20 clonal UCBI plantlet rootstocks were used. These included 5 replicates for each fungal isolate and the control. The 5 mock-inoculated controls were subjected to the same process as described above but were treated with a mixture of distilled sterile water and potted and incubated as above. Disease development and symptoms were assessed every 4 days for 1 month.

Author Contributions: Conceptualization, F.P.T. and M.T.N.; methodology, M.T.N., D.P.L., and F.P.T.; software, D.P.L. and M.T.N.; formal analysis, D.P.L. and M.T.N.; surveys, F.P.T., C.E.K. and M.T.N.; writing—original draft preparation, M.T.N.; writing—review and editing, F.P.T. and D.P.L.; supervision, F.P.T.; funding acquisition, F.P.T. All authors have read and agreed to the published version of the manuscript.

Funding: This research was funded by the California Pistachio Research Board (CPRB).

Conflicts of Interest: The authors declare no conflict of interest.

References

1. Michailides, T.J. Pathogenicity, distribution, sources of inoculum, and infection courts of Botryosphaeria dothidea on pistachio. *Phytopathology* **1991**, *81*, 566–573. [CrossRef]
2. Michailides, T.J. Pest, Disease, and Physiological Disorders Management: Above Ground Fungal Diseases. 2005. Available online: http://fruitsandnuts.ucdavis.edu/files/73707.pdf (accessed on 14 August 2019).
3. Pryor, B.M.; Michailides, T.J. Morphological, Pathogenic, and Molecular Characterization of *Alternaria* isolates associated with Alternaria late blight of pistachio. *Phytopathology* **2002**, *92*, 406–416. [CrossRef]
4. Lawrence, D.P.; Holland, L.A.; Nouri, M.T.; Travadon, R.; Abramians, A.; Michailides, T.J.; Trouillas, F.P. Molecular phylogeny of *Cytospora* species associated with canker diseases of fruit and nut crops in California, with the descriptions of ten new species and one new combination. *IMA Fungus* **2018**, *9*, 333–370. [CrossRef] [PubMed]
5. Nouri, M.T.; Lawrence, D.P.; Holland, L.A.; Doll, D.A.; Kallsen, C.E.; Culumber, C.M.; Trouillas, F.P. Identification and pathogenicity of fungal species associated with canker diseases of pistachio in California. *Plant Dis.* **2019**, *103*, 2397–2411. [CrossRef] [PubMed]
6. Holtz, B.A.; Teviotdale, B.L. Soil-borne diseases. In *Pistachio Production Manual*; Ferguson, L., Haviland, D., Eds.; University of California Agriculture and Natural Resources: Oakland, CA, USA, 2016; pp. 257–264.
7. Fradin, E.F.; Thomma, B.P.H.J. Physiology and molecular aspects of Verticillium wilt diseases caused by *V. dahliae* and *V. albo-atrum*. *Mol. Plant Pathol.* **2006**, *7*, 71–86. [CrossRef] [PubMed]
8. Pegg, G.F.; Brady, B.L. *Verticillium Wilts*; CAB International: Oxon, UK, 2002.
9. Klosterman, S.J.; Atallah, Z.K.; Vallad, G.E.; Subbarao, K.V. Diversity, pathogenicity and management of *Verticillium* species. *Annu. Rev. Phytopathol.* **2009**, *47*, 39–62. [CrossRef]
10. Epstein, L.; Beede, R.; Kaur, S.; Ferguson, L. Rootstock effects on pistachio trees grown in *Verticillium dahliae*-infested soil. *Phytopathology* **2004**, *94*, 388–395. [CrossRef]
11. Ogawa, J.M.; English, H. *Diseases of Temperate Zone Tree Fruit and Nut Crops*; UC ANR Publications: Oakland, CA, USA, 1991.
12. Fichtner, E.J.; Kallsen, C.E.; Blomquist, C.L. First report of crown rot caused by *Phytophthora parsiana* on pistachio in the Southern San Joaquin Valley, California. *Plant Dis.* **2016**, *100*, 1795. [CrossRef]
13. Fichtner, E.J.; Browne, G.T.; Mortaz, M.; Ferguson, L.; Blomquist, C.L. First report of root rot caused by *Phytopythium helicoides* on pistachio rootstock in California. *Plant Dis.* **2016**, *100*, 2337. [CrossRef]

14. MacDonald, J.D.; Banihashemi, Z.; Mircetich, S.M.; Browne, G.; Bolkan, L. Trunk and branch canker of pistachio caused by *Phytophthora* spp. *Phytopathology* **1992**, *82*, 1089.
15. Crespo, M.; Lawrence, D.P.; Nouri, M.T.; Doll, D.A.; Trouillas, F.P. Characterization of *Fusarium* and *Neocosmospora* species associated with crown rot and stem canker of pistachio rootstocks in California. *Plant Dis.* **2019**, *103*, 1931–1939. [CrossRef] [PubMed]
16. Lawrence, D.P.; Nouri, M.T.; Trouillas, F.P. Taxonomy and multi-locus phylogeny of cylindrocarpon-like species associated with diseased roots of grapevine and other fruit and nut crops in California. *FUSE* **2019**, *4*, 59–75. [CrossRef]
17. Sarr, M.P.; Ndiaye, M.B.; Groenewald, J.Z.; Crous, P.W. Genetic diversity in *Macrophomina phaseolina*, the causal agent of charcoal rot. *Phytopathol. Mediterr.* **2014**, *53*, 250–268.
18. Crous, P.W.; Slippers, B.; Wingfield, M.J.; Rheeder, J.; Marasas, W.F.O.; Philips, A.J.L.; Alves, A.; Burgess, T.; Barber, P.; Groenewald, J.Z. Phylogenetic lineages in the Botryosphaeriaceae. *Stud. Mycol.* **2006**, *55*, 235–253. [CrossRef]
19. Baird, R.E.; Watson, C.E.; Scruggs, M. Relative longevity of *Macrophomina phaseolina* and associated mycobiota on residual soybean roots in soil. *Plant Dis.* **2003**, *87*, 563–566. [CrossRef]
20. Mengistu, A.; Wrather, A.; Rupe, J.C. Charcoal rot. In *Compendium of Soybean Diseases and Pests*, 5th ed.; Hartman, G.L., Rupe, J.C., Sikora, E.J., Domier, L.L., Davies, J.A., Steffey, L.K., Eds.; APS Press: Saint Paul, MN, USA, 2015; pp. 67–69.
21. Srivastava, A.K.; Singh, T.; Jana, T.K.; Arora, D.K. Microbial colonization of *Macrophomina phaseolina* and suppression of charcoal rot of chickpea. In *Microbes and Plants*; Sinha, A., Ed.; Vedamse Books (P) Ltd.: New Delhi, India, 2001; pp. 269–319.
22. Mayék-Pérez, N.; Garcia-Espinosa, R.; López-Castañeda, C.; Acosta-Gallegos, J.A.; Simpson, J. Water relations, histopathology and growth of common bean (*Phaseolus vulgaris* L.) during pathogenesis of *Macrophomina phaseolina* under drought stress. *Physiol. Mol. Plant Pathol.* **2002**, *60*, 185–195. [CrossRef]
23. Bokor, P. *Macrophomina phaseolina* causing a charcoal rot of sunflower through Slovakia. *Biologia* **2007**, *62*, 136–138. [CrossRef]
24. Sharma, I.; Kumari, N.; Sharma, V. Defense gene expression in Sorghum bicolor against *Macrophomina phaseolina* in leaves and roots of susceptible and resistant cultivars. *J. Plant Interact.* **2014**, *9*, 315–323. [CrossRef]
25. Koike, S.T. Crown rot of strawberry caused by *Macrophomina phaseolina* in California. *Plant Dis.* **2008**, *92*, 1253. [CrossRef]
26. Koike, S.T.; Arias, R.S.; Hogan, C.S.; Martin, F.N.; Gordond, T.R. Status of *Macrophomina phaseolina* on strawberry in California and preliminary characterization of the pathogen. *Int. J. Fruit Sci.* **2016**, *16*, 148–159. [CrossRef]
27. Abed-Ashtiani, F.; Narmani, A.; Arzanlou, M. *Macrophomina phaseolina* associated with grapevine decline in Iran. *Phytopathol. Mediterr.* **2018**, *57*, 107–111. [CrossRef]
28. Gonzalez, V.; Tello, M.L. The endophytic mycota associated with *Vitis vinifera* in central Spain. *Fungal Divers.* **2011**, *47*, 29–42. [CrossRef]
29. Nouri, M.T.; Zhuang, G.; Culumber, C.M.; Trouillas, F.P. First Report of *Macrophomina phaseolina* Causing trunk and cordon canker disease of grapevine in the United States. *Plant Dis.* **2018**, *103*, 579. [CrossRef]
30. Sergeeva, V.; Tesoriero, L.; Spooner-Hart, R.; Nair, N. First report of *Macrophomina phaseolina* on olives (*Olea europaea*) in Australia. *Australas. Plant Pathol.* **2005**, *34*, 273–274. [CrossRef]
31. Inderbitzin, P.; Bostock, R.M.; Trouillas, F.P.; Michailides, T.J. A six locus phylogeny reveals high species diversity in Botryosphaeriaceae from California almond. *Mycologia* **2010**, *102*, 1350–1368. [CrossRef]
32. Diourte, M.; Starr, J.L.; Jeger, M.J.; Stack, J.P.; Rosenow, D.T. 1995. Charcoal rot (*Macrophomina phaseolina*) resistance and the effects of water stress on disease development in sorghum. *Plant Pathol.* **1995**, *44*, 196–202. [CrossRef]
33. White, T.J.; Bruns, T.; Lee, S.; Taylor, J. Amplification and direct sequencing of fungal ribosomal RNA genes for phylogenetics. In *PCR Protocols: A Guide to Methods and Applications*; Innis, M.A., Gelfand, D.H., Snisky, J.J., White, T.J., Eds.; Academic: San Diego, CA, USA, 1990; pp. 315–322.
34. Carbone, I.; Kohn, L.M. A method for designing primer sets for speciation studies in filamentous ascomycetes. *Mycologia* **1999**, *3*, 553–556. [CrossRef]

35. Glass, N.L.; Donaldson, G.C. Development of primer sets designed for use with the PCR to amplify conserved genes from filamentous Ascomycetes. *Appl. Environ. Microbiol.* **1995**, *61*, 1323–1330. [CrossRef]
36. Tamura, K.; Stecher, G.; Peterson, D.; Filipski, A.; Kumar, S. MEGA6: Molecular evolutionary genetics analysis version 6.0. *Mol. Biol. Evol.* **2013**, *30*, 2725–2729. [CrossRef]
37. Maddison, W.P.; Maddison, D.R. Mesquite: A Modular System for Evolutionary Analysis. Version 3.10. 2016. Available online: http://mesquiteproject.org (accessed on 19 July 2019).
38. Swofford, D. *PAUP 4: Phylogenetic Analysis Using Parsimony (*and Other Methods)*; Sinauer Associates: Sunderland, MA, USA, 2002.
39. Zwickl, D.J. Genetic Algorithm Approaches for the Phylogenetic Analysis of Large Biological Datasets under the Maximum Likelihood Criterion. Ph.D. Thesis, Department of Integrative Biology, University of Texas at Austin, Austin, TX, USA, 2006.
40. Chen, S.F.; Morgan, D.P.; Hasey, J.K.; Anderson, K.; Michailides, T.J. Phylogeny, morphology, distribution, and pathogenicity of *Botryosphaeriaceae* and *Diaporthaceae* from English walnut in California. *Plant Dis.* **2014**, *98*, 636–652. [CrossRef]

plants

MDPI

Article

A Survey of Trunk Disease Pathogens within Citrus Trees in Iran

Nahid Espargham [1], Hamid Mohammadi [1,*] and David Gramaje [2,*]

1 Department of Plant Protection, Faculty of Agriculture, Shahid Bahonar University of Kerman, Kerman 7616914111, Iran; hmho279@yahoo.com
2 Instituto de Ciencias de la Vid y del Vino (ICVV), Consejo Superior de Investigaciones Científicas, Universidad de la Rioja, Gobierno de La Rioja, 26007 Logroño, Spain
* Correspondence: hmohammadi@uk.ac.ir (H.M.); david.gramaje@icvv.es (D.G.); Tel.: +98-34-3132-2682 (H.M.); +34-94-1899-4980 (D.G.)

Received: 4 May 2020; Accepted: 12 June 2020; Published: 16 June 2020

Abstract: Citrus trees with cankers and dieback symptoms were observed in Bushehr (Bushehr province, Iran). Isolations were made from diseased cankers and branches. Recovered fungal isolates were identified using cultural and morphological characteristics, as well as comparisons of DNA sequence data of the nuclear ribosomal DNA-internal transcribed spacer region, translation elongation factor 1α, β-tubulin, and actin gene regions. *Dothiorella viticola*, *Lasiodiplodia theobromae*, *Neoscytalidium hyalinum*, *Phaeoacremonium (P.) parasiticum*, *P. italicum*, *P. iranianum*, *P. rubrigenum*, *P. minimum*, *P. croatiense*, *P. fraxinopensylvanicum*, *Phaeoacremonium* sp., *Cadophora luteo-olivacea*, *Biscogniauxia* (*B.*) *mediterranea*, *Colletotrichum gloeosporioides*, *C. boninense*, *Peyronellaea* (*Pa.*) *pinodella*, *Stilbocrea* (*S.*) *walteri*, and several isolates of *Phoma*, *Pestalotiopsis*, and *Fusarium* species were obtained from diseased trees. The pathogenicity tests were conducted by artificial inoculation of excised shoots of healthy acid lime trees (*Citrus aurantifolia*) under controlled conditions. *Lasiodiplodia theobromae* was the most virulent and caused the longest lesions within 40 days of inoculation. According to literature reviews, this is the first report of *L. theobromae* and *N. hyalinum* on citrus in Iran. Additionally, we report several *Phaeoacremonium* species, *S. walteri*, *Pa. pinodella* and *C. luteo-olivacea* on citrus trees for the first time in the world.

Keywords: bscogniauxia; botryosphaeriaceae; cadophora; citrus dieback; colletotrichum; phaeoacremonium

---

1. Introduction

Iran is the sixth largest Citrus producer, accounting for 3.3% of the world's Citrus production, which yielded 4.1 million tons in 2016 [1]. A total of 276,000 ha of various *Citrus* species are cultivated in Iran, including sweet orange (*Citrus sinensis* L.), acid lime (*C. aurantifolia* (Christm.) Swingle), sour orange (*C. aurantium* L.), mandarin (*C. reticulata* Blanco), lemon (*C. limon* (L.) Osbeck), and grapefruit (*C. paradisi* Macfad). The most important producing regions in Iran are Mazandaran, Fars, Hormozgan, Giroft, and Kahnouj.

Fungal trunk diseases have been studied in detail in grapevine, which are the main biotic factor limiting vineyard productivity and longevity [2]. However, recent findings of high incidence in stone and pome fruits, small fruits, nut crops, citrus, and olive worldwide highlight the need for a focus on this novel group of hosts [3]. Trunk diseases are caused by a broad range of taxonomically unrelated fungi that primarily infect wood hosts through winter pruning wounds, thus colonizing the vascular tissues. Members of the families Botryosphaeriaceae, Togninaceae, Diatrypaceae, Diaporthaceae, as well as several basidiomycetes are included in this group of fungi. Members of Diatrypaceae (Xylariales) can often be observed on dead wood and bark of a wide range of plant species around the world. Nevertheless, some species of this family are reported as putative plant pathogens on fruit, ornamental, and forest trees [4–8]. Some species of *Eutypella* have been previously isolated from citrus species, including *Citrus limon*, *C. paradisi*, *C. maxima*, and *C. aurantium* in Australia, Argentina, Brazil, Coted'Ivoire, Philippines, and USA [9–14]. Diatrypaceae spp. were also isolated from citrus trees in Australia [15].

Species of Botryosphaeriaceae have a cosmopolitan distribution and have been associated with numerous plant species worldwide [16–18]. Many species of the genera *Lasiodiplodia* [19–23], *Diplodia*, *Dothiorella*, *Neofusicoccum* [20], and *Neoscytalidium* [14,20] have been previously reported to affect citrus trees. Togniniaceae (Togniniales), with the well-known asexual morph genus *Phaeoacremonium*, is another family of fungi traditionally associated with dieback, canker, and yellowing of various fruit, forest, and ornamental trees, worldwide [24–31]. To date, 56 *Phaeoacremonium* species have been identified from woody hosts [32]. The most prevalent *Phaeoacremonium* species isolated from woody hosts are *P. minimum*, followed by *P. parasiticum* [33]. Dieback and related disease symptoms have been achieved by inoculating *Phaeoacremonium* species onto several hosts such as *Prunus* spp., kiwifruit, oak, and grapevine [33,34]. *Phaeoacremonium* species isolated from grapevine have been intensively studied because of their involvement in two trunk diseases, Petri disease in young vines and esca in mature vines [32,35]. To our knowledge, there are no reports of *Phaeoacremonium* species affecting *Citrus* spp.

In spring 2014, a severe decline of citrus trees was noticed in some orchards in Bushehr (Bushehr province, Iran). External disease symptoms included chlorosis of leaves, defoliation, branch and shoot cankers, and dieback. Internal wood symptoms ranged from brown to black wood streaking and black spots to wedge-shaped necrosis, irregular wood discoloration, central necrosis, and arch-shaped necrosis. Many fungi associated with trunk diseases have been isolated from several woody hosts in Iran, including grapevine [36,37], pome and stone fruit trees [28,38], and ornamental and forest trees [29,31,39,40]. However, little information is presently available on the causal agents of the severe decline of citrus trees in Iran. Therefore, the aim of this study was to investigate the etiology of fungal trunk diseases associated with wood necrosis of citrus trees in Iran and to determine their pathogenicity.

2. Results

2.1. Field Survey and Diversity of Disease Symptoms

In this study, wood samples were collected from lime (46 trees), sweet lemon (23 trees), sweet orange (22 trees), mandarin (eight trees), sour orange (four trees) and lemon (three trees). Citrus trees showed various external disease symptoms, including yellowing, canker, defoliation, dieback, cracking of the bark associated with gumming, and sooty cankers (form a black powder underneath the bark). Examination of infected branches from symptomatic trees revealed different types of wood discoloration in cross-sections, black to brown streaking in the wood, wedge-shaped necrosis, black spots, irregular wood discoloration, central necrosis, and arch-shaped necrosis (Figure 1).

Figure 1. Diversity of external (**A–D**) and internal (**E–I**) trunk disease symptoms on Citrus species in Iran (**A**) a severe dieback on *Citrus sinensis*; (**B**) two cankers on a trunk of a *C. sinensis* tree indicated by arrows; (**C**) an extended canker on the branch of *Citrus aurantifolia* indicated by arrow; (**D**) gummosis on *Citrus limetta*; (**E**) cross-section of a healthy branch of *C. aurantifolia*; (**F**) central necrosis on *C. sinensis*; (**G**) Co-occurrence of brown wood streaking (black arrow), wedge-shaped necrosis (white arrow) and irregular wood necrosis (red arrow) on *C. sinensis*; (**H**) Arch-shaped necrosis on *C. aurantifolia*; (**I**) a young wedge shaped necrosis on *Citrus reticulata*; (**J**) Co-occurrence of wedge-shaped necrosis is indicated by the white arrow and black spots are indicated by the black arrow on the *C. sinensis*.

2.2. Fungal Isolation and Morphological Identification

In this survey, 326 fungal isolates were collected from citrus trees (Table 1). According to colony appearance, culture characteristics, and microscopic structures, the main fungal isolates were classified as *Phaeoacremonium* spp., Botryosphaeriaceae spp. *Cadophora* sp., *Colletotrichum* spp., Peyronellaea sp., *Phoma* spp., and *Biscogniauxia mediterranea*. Thirty-nine isolates (11.96% of total isolates) were identified as *Phaeoacremonium* species and characterized by beige to medium brown flat slow-growing cultures on potato dextrose agar (PDA; Merck, Darmstadt, Germany) and on malt extract agar (MEA; 2% malt extract, Merck, Darmstadt, Germany). Septate hyphae were single or fasciculate, and three types of phialides, variable in shape and size (I, II, and III types), were recorded in these isolates [41]. Morphological features of 49 isolates (15.03%) were consistent with the description of species of Botryosphaeriaceae [16,17,42]. These isolates were characterized by dark green to gray or fast-growing gray mycelium on the PDA. All isolates produced fruit bodies, pycnidia, on pine needles within 15–35 days. Conidia were pigmented or hyaline. These isolates belonged to the genera *Lasiodiplodia*, *Neoscytalidium*, and *Dothiorella*. Twelve isolates of the phialides fungus were identified as *Cadophora* sp. These isolates formed flat, felty, and black-olivaceous and white to gray colonies on PDA, and their conidia were ellipsoid or elongate. Cultural and morphological characteristics observed were similar with the description of the *Cadophora* spp. [43,44]. Based on morphological characteristics, the remaining isolates were classified to *Colletotrichum*, *Peyronellea*, *Pestalotiopsis*, *Fusarium*, *Microsphaeropsis*, *Alternaria*, *Trichoderma*, *Paecilomyces*, *Aspergillus*, *Penicillium*, *Phoma*, *Biscogniauxia*, and *Stilbocrea* genera.

Table 1. Fungal species isolated from *Citrus* species in Iran.

Fungal Species	*Citrus* Species						Total Isolates
	C. sinensis	*C. aurantifolia*	*C. reticulata*	*C. limetta*	*C. aurantium*	*C. limon*	
Phaeoacremonium parasiticum	1	11	0	0	0	0	12
P. rubrigenum	0	4	0	0	0	0	4
P. minimum	0	8	0	0	0	0	8
P. italicum	0	5	0	0	0	0	5
P. croatiense	2	0	0	2	0	0	4
P. iranianum	0	0	2	0	0	0	2
P. fraxinopennsylvanicum	0	0	0	0	1	1	2
Phaeoacremonium sp.	0	2	0	0	0	0	2
Cadophora luteo-olivacea	0	0	6	6	0	0	12
Biscogniauxia mediterranea	2	0	0	0	0	0	2
Neoscytalidium hyalinum	0	7	0	7	0	0	14
Dothiorella viticola	7	7	0	0	6	0	20
Lasiodiplodia theobromae	7	8	0	0	0	0	15
Colletotrichum gleoesporioides	8	0	0	10	0	0	18
Colletotrichum boninense	0	0	0	7	0	0	7
Peyronellea pinodella	5	6	0	0	0	0	11
Phoma herbarum	0	6	0	5	0	0	11
Phoma fungicola	5	0	5	4	0	0	14
Microsphaeropsis olivacea	6	0	0	0	0	0	6
Stilbocrea walteri	0	3	0	0	1	4	8
Pestalotiopsis sp.	4	0	0	0	0	0	4
Fusarium spp.	2	3	0	1	1	5	12
Paecilomyces spp.	2	3	5	1	3	1	15
Phoma spp.	3	0	0	0	0	0	3
Penicillium spp.	7	4	9	0	4	5	29
Aspergillus spp.	11	10	8	8	3	8	48
Trichoderma spp.	5	0	2	5	0	1	13
Alternaria spp.	3	7	3	10	0	2	25
Total fungal isolates	80	94	40	66	19	27	326
Total number of trees surveyed	39	57	19	43	18	12	188

No association was found between wood symptoms and fungal species. Dual infections by trunk disease fungi in a single tree occurred. *Phaeoacremonium parasiticum* and *P. italicum* were isolated from one tree of *C. aurantifolia*; *P. parasiticum*, *P. croatiense*, and *Do. viticola* from one tree of *C. sinensis*, *L. theobromae* and *Neoscytalidium hyalinum* from one tree of *C. aurantifolia*, and *C. luteo-olivacea* and *P. croatiense* from one tree of *C. limetta*. In addition, some fungal species grew from an individual wood segment, such as *Stilbocrea walteri* and *P. fraxinopennsylvanicum* from *C. limon*.

2.3. Molecular Characterization and Phylogenetic Analyses

BLASTn searches in GenBank showed that the nuclear ribosomal DNA-internal transcribed spacer region (ITS) and translation elongation factor 1α (tef1-α) sequences of Botryosphaeriaceae isolates had 99–100% identity with isolates of *Lasiodiplodia theobromae* (strain CBS559.70), *Neoscytalidium hyalinum* (strain CBS 145.78), and *Dothiorella viticola* (strain CBS 117006). The ITS sequences of the *Cadophora* isolates had 99–100 % identity with isolates previously identified as *Cadophora luteo-olivacea* in GenBank (strain CBS 855.69). ITS and β-tubulin (BT) sequences of *Colletotrichum* isolates were identical to isolates previously reported as *Colletotrichum gloeosporioides* (ITS: strain CBS 132465; BT: strain CBS 100471) and *Colletotrichum boninense* (ITS and BT: strain CBS:123755) in GenBank. ITS and BT sequences of *Peyronellea* isolates in our study showed 99–100 % identity with those isolates previously submitted as *Didymella pinodella* (strain CBS 531.66) in GenBank. ITS and tef1-α sequences of *Stilbocrea* isolates from citrus trees had 99–100% identity with *Stilbocrea walteri* (strain NQI). Regarding the *Biscogniauxia* isolates, ITS of our isolates had 99–100% identity with isolates previously identified as *Biscogniauxia mediterranea* (strain CBS 129072).

Datasets of the BT and actin (ACT) alignments of *Phaeoacremonium* were congruent and could be combined ($p = 0.225$). The Hasegawa–Kishino–Yano model (HKY) with gamma distributed with invariant sites rates (G+I) was identified as the BIC best-fit nucleotide substitution model by the jModelTest for the *Phaeoacremonium* multi-locus analysis. Maximum likelihood (ML) of the combined ACT-BT regions provided a phylogeny with 98 to 100% ML bootstrap support for all species-level clades, with the exception of *P. alvesii* (paraphyletic, 87% bootstrap support), *P. griseorubrum* (paraphyletic, 66% bootstrap support), *P. roseum* (89% bootstrap support), and *P. viticola* (paraphyletic with regard to *P. roseum* and *P. angustius*) (Figure 2). The 39 strains from Iran clustered in eight clades (*P. italicum*, *Phaeoacremonium* sp., *P. rubrigenum*, P. parasiticum, *P. minimum*, *P. iranianum*, *P. fraxynopennsylvanicum*, and *P. croatiense*). The isolates of the clade 2 grouped together in a polyphyletic clade with 100% bootstrap support with the *P. italicum* as a closely related species. The BT and ACT sequences of the second clade of *Phaeoacremonium* isolates were 98% (BT) and 98.77% (ACT) identical to those of *P. italicum* CBS 137763 (GenBank KJ534074, KJ534046). Three nucleotides varied in the ACT region and ten nucleotides in the BT region between the second clade of *Phaeoacremonium* isolates and the *P. italicum* CBS 137763 sequences.

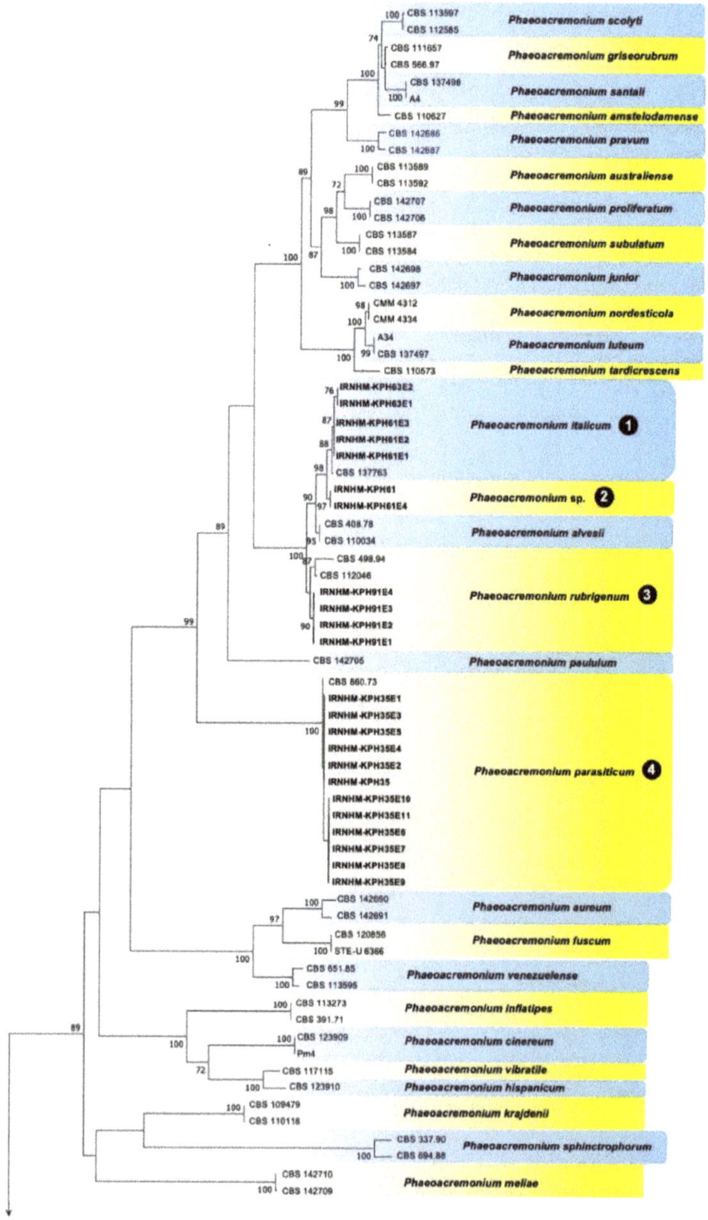

Figure 2. *Cont.*

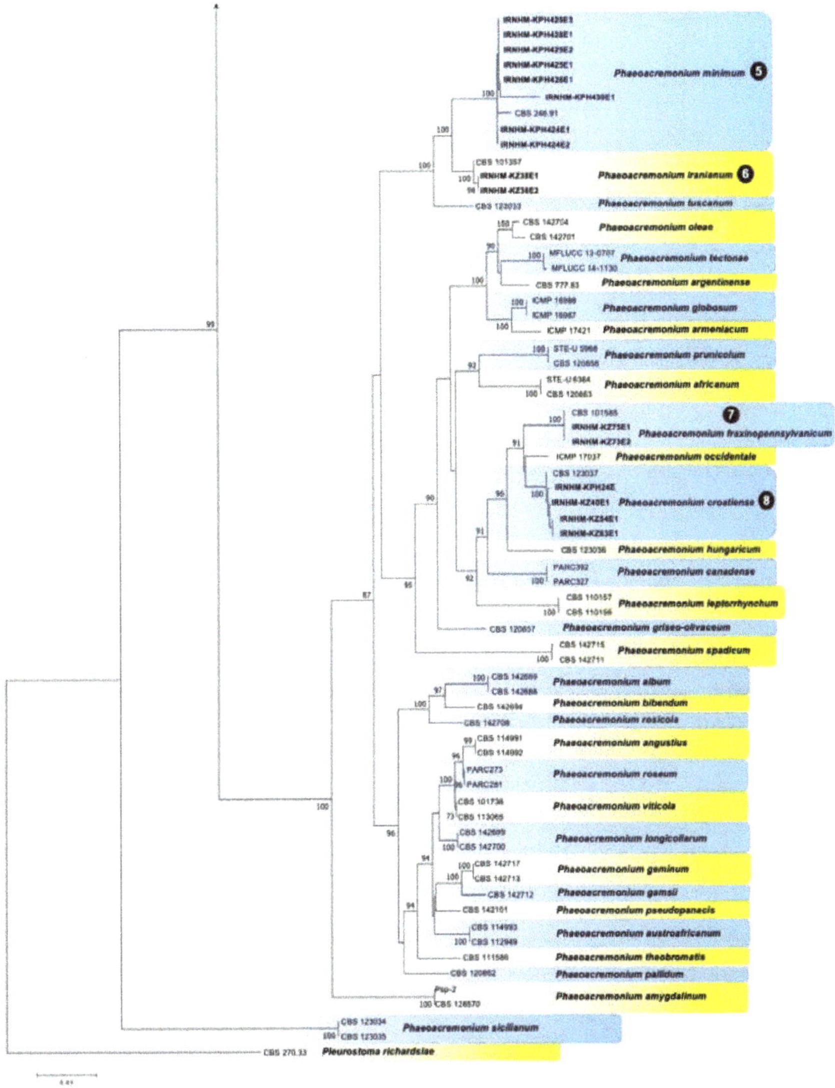

Figure 2. Maximum likelihood phylogeny of *Phaeoacremonium* spp. according to concatenated alignments of the actin (ACT) and beta-tubulin (BT) gene regions. Support values less than 70% bootstrap were omitted. Maximum likelihood bootstrap percentages are indicated at the nodes. Isolates obtained in this study are indicated in bold. The eight clades associated with the *Phaeoacremonium* spp. obtained in this study are indicated by numbers.

2.4. Pathogenicity Test

Mean lengths of wood discolorations caused by inoculated isolates obtained from Citrus species on the detached shoots of *C. aurantifolia* are shown in Figures 3 and 4. Our results showed a variation in the total (Figure 4a), and both the upward and downward lesion lengths (Figure 4b) from the point of inoculation and re-isolation frequencies of inoculated isolates on lime shoots. *L. theobromae* was the most aggressive fungal species and produced the longest necrotic lesions (57.67 mm) on the inoculated shoots followed by *Do. viticola* (38.17 mm) and *P. parasiticum* (34.33 mm) (Figure 4a). In contrast, two species of *S. walteri* (6.33 mm) and *P. rubrigenum* (6.00 mm) produced the smallest wood lesions on the inoculated shoots, and no significant differences were observed between these species and the control treatments (3.67 mm).

Figure 3. Pathogenicity tests of fungal species inoculated onto Citrus aurantifolia detached shoots, 40 days after inoculation: (**A**) control; (**B,C**) Lasiodiplodia theobromae; (**D**); Dothiorella viticola; (**E**) Neoscytalidium hyalinum; (**F,G**) Phaeoacremonium parasiticum; (**H**) Cadophora luteo-olivacea; (**I**) Phaeoaremonium minimum; (**J**) Phaeoacremonium iranianum; (**K**) Colletotrichum gloeosporoides; (**L**) Phaeoacremonium fraxinopennsylvanicum; (**M**) Phaeoacremonium italicum; (**N**) Stilbocrea walteri; (**O**) Phaeoacremonium rubrigenum; (white arrows show the point of inoculation, and the red dashed lines indicate the lesion length caused by each isolate).

All inoculated fungi caused longer basipetal than acropetal lesions on the lime shoots (Figure 4b). Of the isolates inoculated, 10 species caused downward and upward wood lesions that were significantly different to those in the control ($p < 0.05$). *L. theobromae* also produced the longest wood lesion lengths both in upward (22.34 mm) and in downward (35.33) directions, while *S. walteri* (upward = 2.5, downward = 3.83 mm) and *P. rubrigenum* (upward = 2.5, downward = 3.50 mm) did not cause any significant necrotic lesion lengths both in the downward and in the upward directions compared to the control treatments (upward = 1.34, downward = 2.33 mm) on the inoculated shoots. Re-isolation percentages were between 40.0% (*C. luteo-olivacea*) and 100% (*L. theobromae* and *N. hyalinum*) on the inoculated lime shoots, and no fungal isolates were recovered from control treatments.

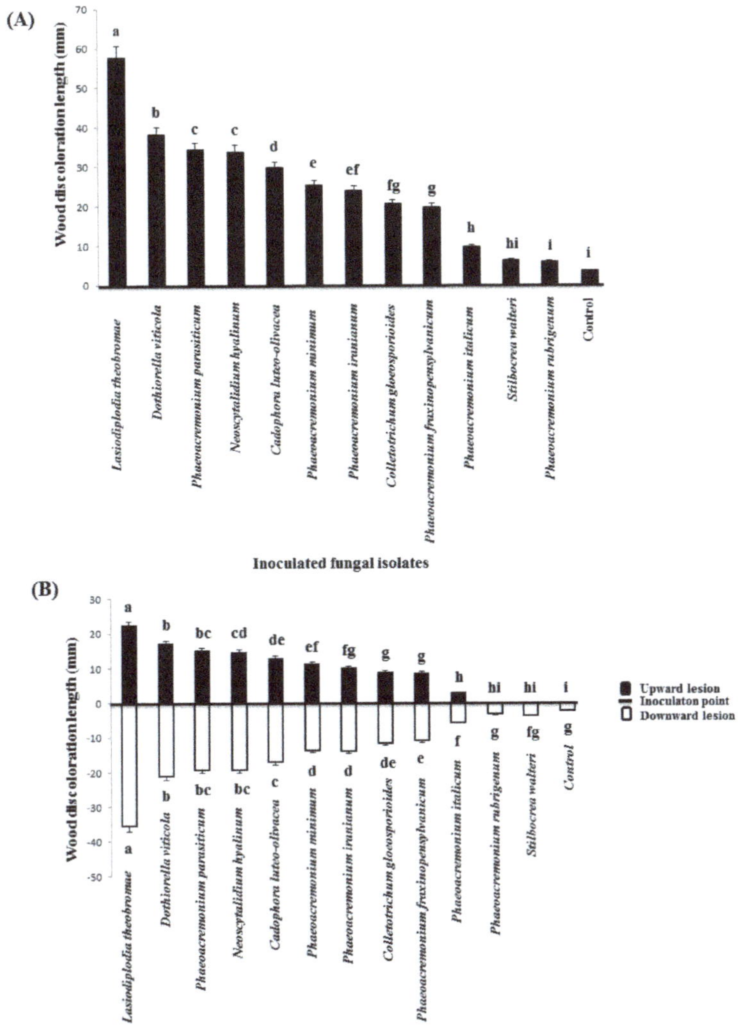

Figure 4. Total (**A**) and upward and downward (**B**) wood lesion size produced by inoculated isolates onto *Citrus aurantifolia* detached shoots, 40 days after inoculation. Different letters in boldface indicate significant differences at $p = 0.05$. Bars represent standard error of the means.

3. Discussion

This study shows the high incidence and severity of fungal trunk pathogens associated with wood decay symptoms of six *Citrus* species (*C. sinensis*, *C. aurantifolia*, *C. reticulate*, *C. limetta*, *C. aurntium*, and *C. limon*) in Iran. During the last decade, extensive studies have been done on fungal trunk pathogens of fruit trees, including grapevine [36,37], stone [38], and pome fruit trees [28,30], pistachio [45], almond [46–48], walnut [49,50], pomegranate, and fig trees [51] in Iran. The current study shows that *Citrus* also represents a rich catch host for fungi associated with trunk diseases in this country. Different trunk disease fungi often co-occurred in the same tree and even in the same type of symptom, thus showing the complexity of the etiology of wood symptoms observed. The co-infection of several trunk disease fungi on woody crops could lead to an increase in disease severity compared to the single occurrence of a fungal species, as it has been previously demonstrated on grapevine with Botryosphaeriaceae and *Ilyonectria* spp. [52].

Morphological comparisons of trunk disease fungi often reveal an overlap between species in several characters [13,16,41]. In our study, the use of these characters to distinguish fungal species within a genus or family was inadequate, thus highlighting the convenience of DNA-based methods for such purposes. This is particularly important for species of the genus *Phaeoacremonium* [41]. Throughout this survey, seven *Phaeoacremonium* species, including *P. parasiticum, P. minimum, P. rubrigenum, P. italicum, P. iranianum, P. croatiense, P. fraxinopennsylvanicum,* and an unidentified species of *Phaeoacremonium* were recovered from *Citrus* spp. showing a decline in symptoms. All *Phaeoacremonium* species reported herein have been found associated with grapevine [41,53–55]. *P. parasiticum* was the dominant *Phaeoacremonium* species in this study, with 12 isolates collected from *C. sinensis* and *C. aurantifolia*. This fungus has previously been reported from *C. reticulata* in Iran [56], and from various fruit trees, such as grapevine [41], *Actinidia chinensis* [57], *Cydonia oblonga, Ficus carica* [58], *Olea europaea* [58,59], *Malus domestica* [28,58], *Prunus armeniaca* [24], *Prunus avium* [60], *Punica granatum* [58], and *Pyrus communis* [28] worlwide. In our study, *P. minimum, P. rubrigenum* and *P. italicum* were isolated only from *C. aurantifolia*. Similar to *P. parasiticum, P. minimum* was also reported from a wide range of fruit trees, including *A. chinensis* [61], *A. deliciosa* [62], *C. oblonga* [28,58], *M. domestica* [25,28,63], *O. europaea* [64], *P. armeniaca* [24], *Prunus dulcis* [57], *P. pennsylvanica* [65], *Prunus salicina* [24], *P. granatum* [58], *P. communis* [25,28], in Iran and other parts of the world. *P. rubrigenum* has previously been reported from *C. oblonga* [28], *O. europaea* [59], and *P. communis* [28]. More recently, fruit tree infections by *P. italicum* have also been reported from South Africa and this fungus has been isolated from *C. oblonga, Ficus carica, M. domestica, O. europaea, P. persica,* and *P. granatum* in this country [58]. *P. croatiense* was isolated from *C. sinensis* and *C. limetta*, while *P. fraxinopennsylvanicum* was isolated form *C. aurantium*. Related to fruit trees, *P. fraxinopennsylvanicum* was previously reported to affect *A. deliciosa* [62], *M. domestica* [28,58], *P. salicina* [24], and *Pyrus communis* [25], while *P. croatiense* was only reported from grapevine [54]. Our research confirms the broad distribution of *Phaeoacremonium* spp. affecting woody crops, and provides their first record on citrus trees in the world.

Three species of Botryosphaeriaceae, namely *N. hyalinum, Do. viticola* and *L. theobromae* were obtained from citrus trees in this study. *Neoscytalidium hyalinum* was isolated from *C. aurantifolia* and *C. limetta, Do. viticola* was recovered from *C. sinensis, C. aurantifolia,* and *C. aurantium*, and *L. theobromae* was associated with *C. sinensis* and *C. aurantifolia*. Several species of Botryosphaeriaceae are known to dieback and branch cankers in *Citrus* spp. worldwide [14,20,22,23,66–70]. *Dothiorella viticola* has been previously reported to cause gummosis in citrus in California [20] and Tunisia [71]. This fungus has also been reported from cultivar Parent Washington on sour orange rootstock [68], *C. sinensis* and *C. latifolia* Tan. in California [20], and *C. sinensis* in New Zealand [72]. Abdollahzadeh et al. reported this species from *Citrus* sp. in Guilan province of Iran [73]. Our study provides the first report of this fungus from *C. aurantifolia* and *C. aurantium*.

Neoscytalidium hyalinum has been reported as the most prevalent Botryosphaeriaceae species associated with citrus branch cankers in the desert regions of southern California [14]. This fungus has been recovered from *C. paradise* showing gummosis in California [20] and also from *C. sinensis* in Italy [74]. Therefore, our work is the first report of *N. hyalinum* from two *Citrus* species, *C. aurantifolia* and *C. limetta*. *L. theobromae* has been previously reported from some *Citrus* species, including *C. limon* in Chile [23] and Persian lime (*Citrus latifolia*) trees in Mexico [70]. Our study represents the first report of this species on *C. sinensis* and *C. aurantifolia*.

In the current study, 12 isolates of *Cadophora luteo-olivacea* were obtained from *C. reticulata* and *C. limetta*. *C. luteo-olivacea* has previously been reported with black vascular streaking and a decline in the symptoms characteristic of Petri disease on grapevine [44,54,75,76], bark cracks of kiwifruit [62], and from pear fruits showing dark-brown and slightly sunken spots [77]. Aside from these reports, little is known regarding the role of *Cadophora* species involved in trunk diseases of trees. This is the first time that *C. luteo-olivacea* has been found on *Citrus* spp.

Most species of the genus *Biscogniauxia* are reported from forest trees, mainly from *Quercus* spp. [78–80]. Some species of this genus have also been found associated with fruit trees such as *B. pruni* and *B. granmoi* on *Prunus padus* [81,82], *B. marginata* on *M. communis* [83], *B. rosacearum* on *P. communis*, *C. oblonga* and *Prunus domestica* [84], and *B. capnodes* on *Averrhoa carambola* [85]. *Biscogniauxia mediterranea* is known to be the causal agent of charcoal cankers on a wide range of trees worldwide, in particular *Quercus* spp. [79,80,86]. In Iran, this pathogen was already reported from *C. sinensis* [87], along with other woody hosts, such as *Quercus castaneifolia* [88], *Zelkova carpinifolia* [89], *Q. brantii* [89], and *Amygdalus scoparia* [90].

In the current study, eight isolates of *Stilbocrea walteri* were isolated from *C. aurantifolia, C. aurntium,* and *C. limon*. This species was originally reported from dead corticated branches of *Quercus ilex* in Portugal [91], and to our knowledge, it has not been reported from necrotic wood tissues of trees. Therefore, this study is the first report of this species in Iran and on *Citrus* species worldwide.

Peyronellea pinodella (Didymellaceae) is a destructive necrotrophic pathogen on some plant families, including *Fabaceae, Amaranthaceae, Asteraceae, Amaryllidaceae, Appiaceae Rubiaceae, Malvaceae, Poaceae,* and *Polemoniaceae* [92]. To date, there is no report on the occurrence of *P. pinodella* on *Citrus* species and this is the first data on the occurrence of this species on *C. sinensis* and *C. aurantifolia*.

Two species of *Colletotrichum* were found to be associated with trunk diseases of citrus trees in this work, *C. gleoesporioides* on *C. sinensis* and *C. limetta* and *C. boninense* on *C. limetta*. Several species of *Colletotrichum* are associated with fruit and leaf anthracnose diseases of *Citrus* species; however, other diseases such as twig and shoot dieback caused by *Colletotrichum* spp. have been documented on citrus trees [14,93]. *Colletotrichum gloeosporioides* has been reported from a wide range of fruit trees such as strawberry, olive, almond, mango, apple, avocado, and citrus [94]. This fungus was found to be associated with twig dieback of lemon trees in Portugal [93]. *C. boninense* has been associated with fruit and leaf anthracnose on citrus trees [95,96]. According to a recent study, some *Colletotrichum* species have been isolated and reported from stems of citrus trees in Iran. These included *C. karstii* from *C. aurantifolia* and *C. sinensis* and four species, *C. gloeosporioides, C. novae–zelandiae, C. siamense,* and *C. fructicola* from *C. sinensis* [97]. Therefore, our study represents the first report of *C. gleoesporioides* and *C. boninense* from branches of *C. limetta*.

In our work, six isolates of *M. olivacea* were obtained from sweet orange. This fungus has been reported from various plant species worldwide. This taxon has previously been isolated as an endophytic species from *P. persica* [98], from xylem and stems of *Pinus sylvestris* [99] and Chilean gymnosperms [100]. Carlucci et al. isolated this species from internal wood discoloration of olive trees in Italy [101]. *Microsphaeropsis olivacea* has also been isolated and reported from some woody plants, such as *Prunus cerasus, P. avium* [102], and Persian oak (*Quercus brantii*) [103] in Iran. To our knowledge, this is the first report of *M. olivacea* on citrus trees. Several isolates of *Fusarium, Pestalotiopsis, Phoma, Penicillium, Aspergillus, Trichoderma,* and *Alternaria* species were also obtained from *Citrus* species in this study. Therefore, more studies are needed on these taxa in order to elucidate their potential impact on citrus trunk diseases.

Pathogenicity of selected fungal species in detached shoots of lime tree were confirmed in the current study. Results revealed that *L. theobromae* was more virulent on lime shoots than other species. In contrast to our results, Bautista-Cruz et al. reported that *L. theobromae* was the least virulent species when inoculated in Persian lime branches [70]. Several factors differed from the study carried out by Bautista-Cruz et al. and might have contributed to the discrepancy between the experiments, including the type of planting material inoculated, the environmental conditions for disease development, the time for virulence assessment, and the fungal strain used in the pathogenicity test. *L. theobromae* has been considered the most aggressive species on *Eucalyptus* [104,105], grapevine [42,106], and pistachio trees [107]. *Lasiodiplodia theobromae* was considered an important pathogen on greengage, sour cherry, peach, apricot, cherry [38], and willow trees [29] in Iran. Our study improved the knowledge on the occurrence of fungal trunk pathogens on *Citrus* species showing a decline in symptoms. Further investigations are needed throughout the citrus orchards to determine the potential impact of these fungi on citrus decline.

4. Materials and Methods

4.1. Tree Sampling and Fungal Isolation

During 2014 and 2015, several field surveys were performed in important citrus-producing regions of Bushehr province, Tallhe and Tang Eram. This province is located in the south of Iran, within 28.7621° N latitude and 51.5150° E longitude. Symptomatic wood samples were collected from various species of citrus trees including, acid lime, sweet orange, mandarin, sour orange, sweet lemon (*C. limetta*), and lemon showing yellowing, defoliation, canker, dieback, and gummosis. In total, 325 wood samples were collected from branches of 106 symptomatic trees (15- to 35-year-old) in 27 orchards. A map with the point locations of the sampled orchards is shown in Figure 5. Collected samples were brought to the laboratory and inspected for internal wood lesions and fungal isolation. Small fragments (4 × 4 mm) of symptomatic wood tissues were cut from the edges of wood lesions, surface-sterilized in sodium hypochlorite solution (1.5%) for 60 s, and rinsed three times in sterilized water. Wood chips were dried in sterilized filter paper and placed on PDA amended with 90 to 100 mg/L streptomycin sulfate (PDAS). For each branch sampled, three to five Petri dishes were obtained. All Petri dishes were incubated at 25 °C until fungal colonies were observed. Pure cultures of the fungal isolates were obtained by hyphal-tipping or transferring single conidia to fresh PDA.

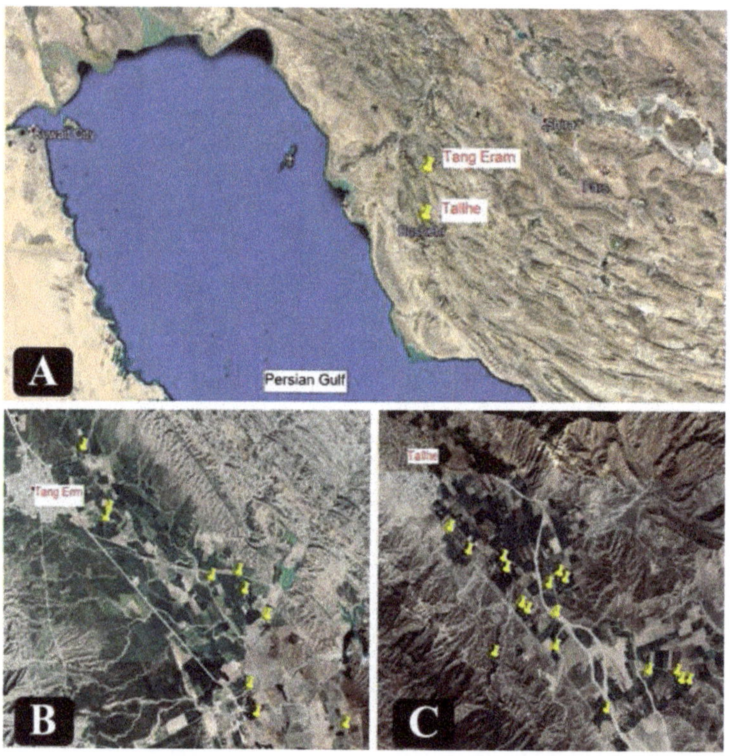

Figure 5. (**A**) Location of the Bushehr province. Two Citrus growing regions were surveyed, Tallhe and Tang Eram. (**B**) Location of surveyed orchards in the Tallhe region. (**C**) Location of surveyed orchards in the Tang Eram region.

4.2. Morphological Identification

All fungal isolates were identified initially to the genus level based on colony morphology and main microscopic structures using published articles and descriptions. Botryosphaeriaceae isolates were identified based on colony appearance and conidial morphology [16,108]. To induce sporulation, three to five mycelial plugs from each isolate were placed on 2% water agar (WA; Biokar-Diagnostics) plates amended with sterilized pine needles and incubated at 25 °C under near-ultraviolet light for 15–45 days [42]. Conidial characteristics (size, shape, color, and presence or absence of septa) were recorded for all isolates. *Phaeoacremonium* isolates were grouped based on colony appearance, pigment production on MEA, PDA and oatmeal agar (OA; 60 g oatmeal; 12.5 g agar; Difco, France) and the main microscopic structures (phialide shape and type, conidiophore morphology, size of hyphal warts, and conidial shape and size) [41,57,109]. Identification of *Cadophora* isolates was based on the colony and micro-morphological structures, such as conidiogenous cell size and shape, and conidia. The remaining fungal isolates were identified based on available identification keys and published papers [91,110–114].

4.3. DNA Extraction, Amplification, and Sequencing

Identities of representative isolates were confirmed using molecular data. Fungi selected for molecular studies were grown on PDA for 10 to 15 days at 25 °C in the dark. DNA was extracted using an AccuPrep®Genomic DNA Extraction Kit (Bioneer, South Korea) following the instructions of the manufacturer. Four primer sets, ITS1/ITS4 [115], EF1-728F/EF1-986R [116], T1/Bt2b [117,118], and ACT-512F/ACT-783R [116] were used to amplify the ITS region ITS1-5.8S-ITS2, portions of the *tef1-α*, BT and ACT genes, respectively. The identification of Botryosphaeriaceae isolates was confirmed by the sequencing of ITS and a partial sequence of *tef-1a*. For *Phaeoacremonium* isolates, a partial sequence of BT and ACT genes were amplified and sequenced. Molecular identifications of other isolates were confirmed by sequence analysis of ITS (*Cadophora*, *Colletotrichum*, *Peyronellea*, *Stilbocrea*, and *Biscogniauxia* isolates), BT (*Colletotrichum* and *Peyronellea* isolates), or *tef1-α* (*Stilbocrea* isolates). The polymerase chain reaction (PCR) was performed in a Techne TC-312 Thermal Cycler (Techne, Cambridge, UK), as described by Hashemi and Mohammadi [29]. For each isolate, 3–4 μL of PCR product was separated by electrophoresis on a 1% agarose gel (UltraPureTM Agarose, Invitrogen) containing ethidium bromide and visualized under UV illumination. The size of the products was evaluated using a 100 bp ladder (Gene Ruler, TMDNA Ladder Mix, Fermentas). PCR products were submitted to Bioneer Corporation (Daejeon, South Korea) for sequencing. MegaBLAST approach of the NCBI database (https://www.ncbi.nlm.nih.gov/) was initially used to identify fungal species.

4.4. Phylogenetic Analysis

Due to the broad range of *Phaeoacremonium* spp. obtained in this study, a phylogenetic analysis was carried out for the *Phaeoacremonium* spp. isolates. Sequences from citrus in Iran were aligned with sequences available in GenBank/NCBI. These were compared using MAFFT sequence alignment program v. 6 [119] with ex-type specimens from different hosts. Alignments were inspected in Sequence Alignment Editor v. 2.0a11 [120]. PAUP version 4.0 b 10 [121] was used to perform a partition homogeneity test. The congruence between the ACT and BT datasets was tested at 1000 replicates, and the maximum likelihood (ML) was carried out on the concatenated alignment. The MEGA version 7 software [122] was used for ML analysis. Bayesian information criterion in jModelTest 2.1.10 [123] was used to estimate the best fit model. Single and concatenated datasets were tested for branch support (1000 bootstrap replicates). We included sequences published by Spies et al. as reference sequences [58]. *Pleurostoma richardsiae* CBS 270.33 was included as an outgroup. *Phaeoacremonium* sequences obtained in this study were submitted to GenBank/NCBI (Table 2) and the sequence alignments were deposited in TreeBASE under study number 26006 (http://treebase.org).

Table 2. Host, origin, and GenBank accession numbers of *Phaeoacremonium* isolates obtained from *Citrus* spp. in Iran (used in phylogenetic studies).

Fungal Isolates		*Citrus* spp.	GenBank Accession Number	
Phaeoacremonium Species	Code		b-Tubulin	Actin
P. parasiticum	IRNHM-KPH35*	*C. aurantifolia*	KU737504	MT127573
	IRNHM-KPH35E1	*C. sinensis*	MT122909	MT127574
	IRNHM-KPH35E2	*C. aurantifolia*	MT122910	MT127575
	IRNHM-KPH35E3	*C. aurantifolia*	MT122911	MT127576
	IRNHM-KPH35E4	*C. aurantifolia*	MT122912	MT127577
	IRNHM-KPH35E5	*C. aurantifolia*	MT122913	MT127578
	RNHM-KPH35E6	*C. aurantifolia*	MT122914	MT127579
	IRNHM-KPH35E7	*C. aurantifolia*	MT122915	MT127580
	IRNHM-KPH35E8	*C. aurantifolia*	MT122916	MT127581
	IRNHM-KPH35E9	*C. aurantifolia*	MT122917	MT127582
	IRNHM-KPH35E10	*C. aurantifolia*	MT122918	MT127583
	IRNHM-KPH35E11	*C. aurantifolia*	MT122919	MT127584
Phaeoacremonium sp.	IRNHM-KPH61	*C. aurantifolia*	KU737517	MT127585
	IRNHM-KPH61E4	*C. aurantifolia*	MT122920	MT127586
P. italicum	IRNHM-KPH61E1	*C. aurantifolia*	MT122921	MT127587
	IRNHM-KPH61E2*	*C. aurantifolia*	MT122922	MT127588
	IRNHM-KPH61E3	*C. aurantifolia*	MT122923	MT127589
	IRNHM-KPH63E1	*C. aurantifolia*	MT122924	MT127590
	IRNHM-KPH63E2	*C. aurantifolia*	MT122925	MT127591
P. rubrigenum	IRNHM-KPH91E1	*C. aurantifolia*	MT122926	MT127592
	IRNHM-KPH91E2	*C. aurantifolia*	MT122927	MT127593
	IRNHM-KPH91E3	*C. aurantifolia*	MT122928	MT127594
	IRNHM-KPH91E4*	*C. aurantifolia*	MT122929	MT127595
	IRNHM-KPH424E1	*C. aurantifolia*	MT122930	MT127596
	IRNHM-KPH424E2	*C. aurantifolia*	MT122931	MT127597
P. minimum	IRNHM-KPH425E1*	*C. aurantifolia*	MT122932	MT127598
	IRNHM-KPH425E2	*C. aurantifolia*	MT122933	MT127599
	IRNHM-KPH425E3	*C. aurantifolia*	MT122934	MT127600
	IRNHM-KPH426E1	*C. aurantifolia*	MT122935	MT127601
	IRNHM-KPH428E1	*C. aurantifolia*	MT122936	MT127602
	IRNHM-KPH430E1	*C. aurantifolia*	MT122937	MT127603
P. iranianum	IRNHM-KZ38E1	*C. reticulata*	MT122938	MT127604
	IRNHM-KZ38E2*	*C. reticulata*	MT122939	MT127605
P. croatiense	IRNHM-KPH24E	*C. sinensis*	MT122940	MT127606
	IRNHM-KZ40E1	*C. limetta*	MT122941	MT127607
	IRNHM-KZ54E1	*C. limetta*	MT122942	MT127608
	IRNHM-KZ63E1	*C. sinensis*	MT122943	MT127609
P. fraxinopennsylvanicum	IRNHM-KZ73E1*	*C. limon*	MT122944	MT127610
	IRNHM-KZ73E2	*C. aurantium*	MT122945	MT127611

Isolates used for pathogenicity tests on detached shoots of *C. aurantifolia*.

4.5. Pathogenicity Tests

Pathogenicity tests were carried out with 12 species on detached shoots of *C. aurantifolia* under controlled conditions. These include *Do. viticola, P. italicum, P. minimum, P. rubrigenum,* and *P. parasiticum* isolated from *C. aurantifolia, L. theobromae,* and *Col. gloeosporioides* obtained from *C. sinensis, C. luteo-olivacea,* and *N. hyalinum* recovered from *C. limetta, P. fraxinopensylvanicum* from *C. limon, P. iranianum* from *C. reticulata* and *S. walteri* isolated from *C. aurantium.* The shoots (38–40 cm in length and 2–2.5 cm in diameter) were surface-disinfected with alcohol (96%) and then were wounded at the uppermost internode with a 4-mm cork borer. To assess pathogenicity, wounds were inoculated with a 4-mm colonized PDA agar from 14-days-old cultures. All inoculated sites first were covered by moist cotton and then were wrapped with a strip of Parafilm (Pechiney Plastic Packaging, Menasha, USA). Six shoots per fungal isolate were used, and an equal number of shoots were also inoculated with 4-mm non-colonized PDA agar plugs for negative controls. Inoculated shoots were arranged at random, including the six inoculated shoots per isolate. Inoculated shoots were placed in moist chambers and incubated at 25 °C. The total, upward, and downward lesion length data were evaluated individually, 40 days after inoculation. Recorded data were checked for normality of distribution by means of the Shapiro–Wilk and Kolmogorov–Smirnov tests. The data were subjected to analysis of variance (one-way ANOVA) using SAS v 9.1 (SAS Institute, Cary, NC, USA) (Dataset S1; Dataset S2). The least significant difference (LSD) test was used for comparison of treatment means at $p < 0.05$. Fungal re-isolations were made from the edges of the lesions on the test and control shoots and placed on PDA. The identity of the re-isolated fungi was confirmed based on morphological characteristics and molecular analysis in order to complete Koch's postulates. The pathogenicity of other species was not tested in this work because they were identified after the pathogenicity trials had begun on the detached shoots of *C. aurantifolia*.

Supplementary Materials: The following are available online at http://www.mdpi.com/2223-7747/9/6/754/s1, Dataset S1: SAS code, Dataset S2: Lesion length data.

Author Contributions: Conceived the idea of the research, designed the scientific experiments and methodology, H.M.; field surveys, sample preparation, laboratory works, sequencing of fungal isolates, pathogenicity tests, N.E.; software, H.M. and D.G.; writing—original draft preparation, H.M.; writing—review and editing, H.M. and D.G.; supervision, H.M.; phylogenetic analysis, D.G. All authors have read and agreed to the published version of the manuscript.

Funding: This research received no external funding.

Conflicts of Interest: The authors declare no conflict of interest.

References

1. FAO. Citrus Fruit Fresh and Processed. Statistical Bulletin. 2016. Available online: http://fao.org/3/a-i8092e.pdf (accessed on 15 May 2020).
2. Gramaje, D.; Úrbez-Torres, J.R.; Sosnowski, M.R. Managing grapevine trunk diseases with respect to etiology and epidemiology: Current strategies and future prospects. *Plant Dis.* **2018**, *102*, 12–39. [CrossRef] [PubMed]
3. Gramaje, D.; Baumgartner, K.; Halleen, F.; Mostert, L.; Sosnowski, M.R.; Úrbez-Torres, J.R.; Armengol, J. Fungal trunk diseases: A problem beyond grapevines? *Plant Pathol.* **2016**, *65*, 355–356. [CrossRef]
4. Carter, M.V. *The Status of Eutypa Lata as a Pathogen. Monograph, Phytopathological Paper No 32*; Commonwealth Agricultural Bureau, International Mycological Institute: Surrey, UK, 1991.
5. Munkvold, G.P.; Marois, J.J. Eutypa dieback of sweet cherry and occurrence of *Eutypa lata* perithecia in the central valley of California. *Plant Dis.* **1994**, *78*, 200–207. [CrossRef]
6. Sinclair, W.A.; Lyon, H.H. *Diseases of Trees and Shrubs*, 2nd ed.; Cornell University Press: Ithaca, NY, USA, 2005; p. 659.
7. Jurc, D.; Ogris, N.; Slippers, B.; Stenlid, J. First report of Eutypella canker of *Acer pseudoplatanus* in Europe. *Plant Pathol.* **2006**, *55*, 577. [CrossRef]
8. Úrbez-Torres, J.R.; Adams, P.; Kama, J.; Gubler, W.D. Identification, incidence and pathogenicity of fungal species associated with grapevine dieback in Texas. *Am. J. Enol. Vitic.* **2009**, *60*, 497–507.
9. Reinking, O.A. Philippine economic plant diseases. *Philipp. J. Sci.* **1918**, *13*, 165–274.
10. Rappaz, F. Taxonomie et nomenclature des Diatrypacees a asques octosporees. *Mycol. Helv.* **1987**, *2*, 285–648.
11. Resplandy, R.; Chevaugeon, J.; Delassus, M.; Luc, M. Premiere liste annotee de champignons parasites de plantes cultivees en Cote d'Ivoire. *Ann. Epiphyt.* **1954**, *1*, 1–61.
12. Hanlin, R.T. Index to genera and species of ascomycetes described by A. P. Viegas. *Mycotaxon* **1992**, *43*, 207–230.
13. Trouillas, F.P.; Pitt, W.M.; Sosnowski, M.R.; Huang, R.; Peduto, F.; Loschiavo, A.; Savocchia, S.; Scott, E.S.; Gubler, W.D. Taxonomy and DNA phylogeny of Diatrypaceae associated with Vitis vinifera and other woody plants in Australia. *Fungal Divers.* **2011**, *49*, 203–223. [CrossRef]
14. Mayorquin, J.S.; Wang, D.H.; Twizeyimana, M.; Eskalen, A. Identification, distribution, and pathogenicity of Diatrypaceae and Botryosphaeriaceae associated with citrus branch canker in the southern California desert. *Plant Dis.* **2016**, *2016. 100*, 2402–2413. [CrossRef]

15. Trouillas, F.P.; Urbez-Torres, J.R.; Gubler, W.D. Diversity of Diatrypaceous fungi associated with grapevine canker diseases in California. *Mycologia* **2010**, *102*, 319–336. [CrossRef] [PubMed]
16. Phillips, A.J.L.; Alves, A.; Abdollahzadeh, J.; Slippers, B.; Wingfield, M.J.; Groenewald, J.Z.; Crous, P.W. The Botryosphaeriaceae: Genera and species known from culture. *Stud. Mycol.* **2013**, *76*, 51–167. [CrossRef] [PubMed]
17. Slippers, B.; Boissin, E.; Phillips, A.J.; Groenewald, J.Z.; Lombard, L.; Wingfield, M.J.; Postma, A.; Burgess, T.; Crous, P.W. Phylogenetic lineages in the Botryosphaeriales: A systematic and evolutionary framework. *Stud. Mycol.* **2013**, *76*, 31–49. [CrossRef] [PubMed]
18. Dissanayake, A.J.; Phillips, A.J.L.; Li, X.H.; Hyde, K.D. Botryosphaeriaceae: Current status of genera and species. *Mycosphere* **2016**, *7*, 1001–1073. [CrossRef]
19. Abdollahzadeh, J.; Javadi, A.; Mohammadi Goltapeh, E.; Zare, R.; Phillips, A.J.L. Phylogeny and morphology of four new species of *Lasiodiplodia* from Iran. *Persoonia* **2010**, *25*, 1–10. [CrossRef]
20. Adesemoye, A.O.; Mayorquin, J.S.; Wang, D.H.; Twizeyimana, M.; Lynch, S.C.; Eskalen, A. Identification of species of *Botryosphaeriaceae* causing bot gummosis in citrus in California. *Plant Dis.* **2014**, *98*, 55–61. [CrossRef]
21. Linaldeddu, B.T.; Dedda, A.; Scanu, B.; Franceschini, A.; Serra, S.; Berraf-Tebbal, A.; Zouaoui Boutiti, M.; Ben Jamâa, M.L.; Phillips, A.J.L. Diversity of Botryosphaeriaceae species associated with grapevine and other woody hosts in Italy, Algeria and Tunisia, with descriptions of *Lasiodiplodia exigua* and *Lasiodiplodia mediterranea* sp. nov. *Fungal Divers.* **2015**, *71*, 201–214. [CrossRef]
22. Coutinho, I.B.L.; Freire, F.C.O.; Lima, C.S.; Lima, J.S.; Gonçalves, F.J.T.; Machado, A.R.; Silva, A.M.S.; Cardoso, J.E. Diversity of genus *Lasiodiplodia* associated with perennial tropical fruit plants in northeastern Brazil. *Plant Pathol.* **2017**, *66*, 90–104. [CrossRef]
23. Guajardo, J.; Riquelme, N.; Tapia, L.; Larach, A.; Torres, C.; Camps, R.; Besoain, X. First report of *Lasiodiplodia theobromae* causing bot gummosis in *Citrus limon* in Chile. *Plant Dis.* **2018**, *102*, 818. [CrossRef]
24. Damm, U.; Mostert, L.; Crous, P.W.; Fourie, P.H. Novel *Phaeoacremonium* species associated with necrotic wood of *Prunus* trees. *Persoonia* **2008**, *20*, 87–102. [CrossRef]
25. Cloete, M.; Fourie, P.H.; Damm, U.; Crous, P.W.; Mostert, L. Fungi associated with die-back symptoms of apple and pear trees, a possible inoculum source of grapevine trunk disease pathogens. *Phytopathol. Mediterr.* **2011**, *50*, S176–S190.
26. Gramaje, D.; Agustí-Brisach, C.; Pérez-Sierra, A.; Moralejo, E.; Olmo, D.; Mostert, L.; Damm, U.; Armengol, J. Fungal trunk pathogens associated with wood decay of almond trees on Mallorca (Spain). *Persoonia* **2012**, *28*, 1–13. [CrossRef] [PubMed]
27. Olmo, D.; Gramaje, D.; Agustí-Brisach, C.; Leon, M.; Armengol, J. First report of *Phaeoacremonium venezuelense* associated with decay of apricot trees in Spain. *Plant Dis.* **2014**, *98*, 1001. [CrossRef] [PubMed]
28. Sami, S.; Mohammadi, H.; Heydarnejad, J. *Phaeoacremonium* species associated with necrotic wood of pome fruit trees in Iran. *J. Plant Pathol.* **2014**, *96*, 487–495.
29. Hashemi, H.; Mohammadi, H. Identification and characterization of fungi associated with internal wood lesions and decline disease of willow and poplar trees in Iran. *For. Pathol.* **2016**, *46*, 341–352. [CrossRef]
30. Mohammadi, H.; Sharifi, S. Association of Botryosphaeriaceae and *Phaeoacremonium* species with insect-damaged quince shoots. *J. Plant Pathol.* **2016**, *98*, 35–42.
31. Kazemzadeh Chakusary, M.; Mohammadi, H.; Khodaparast, A. Decline-associated *Phaeoacremonium* spp. occurring on forest trees in the north of Iran. *For. Pathol.* **2017**, *47*, e12368. [CrossRef]
32. Marin-Felix, Y.; Hernández-Restrepo, M.; Wingfield, M.J.; Akulov, A.; Carnegie, A.J.; Cheewangkoon, R.; Gramaje, D.; Groenewald, J.Z.; Guarnaccia, V.; Halleen, F.; et al. Genera of phytopathogenic fungi: GOPHY 2. *Stud. Mycol.* **2018**, *92*, 47–133. [CrossRef]
33. Gramaje, D.; Mostert, L.; Groenewald, J.Z.; Crous, P.W. *Phaeoacremonium*: From esca disease to phaeohyphomycosis. *Fungal Biol.* **2015**, *119*, 759–783. [CrossRef]
34. Baloyi, M.A.; Mostert, L.; Halleen, F. Pathogenicity of ten *Phaeoacremonium* species associated with esca and Petri disease of grapevine. *Phytopathol. Mediterr.* **2018**, *57*, 538–546.
35. Mugnai, L.; Graniti, A.; Surico, G. Esca (black measles) and Brown wood-streaking: Two old and elusive diseases of grapevines. *Plant Dis.* **1999**, *83*, 404–416. [CrossRef] [PubMed]
36. Mohammadi, H.; Banihashemi, Z.; Gramaje, D.; Armengol, J. Fungal pathogens associated with grapevine trunk diseases in Iran. *J. Agric. Sci. Technol.* **2013**, *15*, 137–150.

37. Mohammadi, H.; Banihashemi, Z.; Gramaje, D.; Armengol, J. Characterization of *Diplodia seriata* and *Neofusicoccum parvum* associated with grapevine decline in Iran. *J. Agric. Sci. Technol.* **2013**, *15*, 603–616.
38. Soltaninejad, N.; Mohammadi, H.; Massumi, H. Isolation, identification and pathogenicity of Botryosphaeriaceae and *Phaeoacremonium* species associated with decline of *prunus* species in Iran. *J. Plant Pathol.* **2017**, *99*, 571–581. [CrossRef]
39. Hashemi, H.; Mohammadi, H.; Abdollahzadeh, J. Symptoms and fungi associated with elm trees decline in Iran. *Eur. J. For. Res.* **2017**, *136*, 857–879. [CrossRef]
40. Panahandeh, S.; Mohammadi, H.; Gramaje, D. Trunk disease fungi associated with *Syzygium cumini* in Iran. *Plant Dis.* **2019**, *103*, 711–720. [CrossRef]
41. Mostert, L.; Groenewald, J.Z.; Summerbell, R.C.; Gams, W.; Crous, P.W. Taxonomy and pathology of *Togninia* (*Diaporthales*) and its *Phaeoacremonium* anamorphs. *Stud. Mycol.* **2006**, *54*, 1–115. [CrossRef]
42. Van Niekerk, J.M.; Crous, P.W.; Groenewald, J.Z.; Fourie, P.H.; Halleen, F. DNA phylogeny, morphology and pathogenicity of Botryosphaeria species on grapevines. *Mycologia* **2004**, *96*, 781–798. [CrossRef]
43. Harrington, T.C.; McNew, D.L. Phylogenetic analysis places the Phialophora-like anamorph genus *Cadophora* in the Helotiales. *Mycotaxon* **2003**, *87*, e151.
44. Gramaje, D.; Mostert, L.; Armengol, J. Characterization of *Cadophora luteo-olivacea* and *C. melinii* isolates obtained from grapevines and environmental samples from grapevine nurseries in Spain. *Phytopathol. Mediterr.* **2011**, *50*, S112–S126.
45. Mohammadi, H.; Sarcheshmehpour, M.; Mafi, E. Fungal trunk pathogens associated with wood decay of pistachio trees in Iran. *Span. J. Agric. Res.* **2015**, *2*, 6. [CrossRef]
46. Mousavi, M.; Mohammadi, H.; Soltaninejad, N. First report of *Phaeoacremonium aleophilum* (Ascomycota, Calosphaeriales) on almond and cherry trees in Kerman province. In Proceedings of the 21th Iranian Plant Protection Congress, Urmia, Iran, 23–26 August 2014; p. 28.
47. Sohrabi, M.; Mohammadi, H. *Phaeoacremonium* spp. and *Pleurostoma richardsiae* associated with trunk diseases of almond trees in Iran. In Proceedings of the 3rd Iranian Mycological Congress, University of Kurdistan, Sanandaj, Iran, 26–28 August 2017; p. 51.
48. Sohrabi, M.; Mohammadi, H. First report of *Lasiodiplodia theobromae* from almond trees in Iran. In Proceedings of the International Conference on Agricultural Science Medical Plants Traditional Medicine, Payam Noor University, Mashhad, Iran, 14–15 February 2018; pp. 1–11.
49. Sohrabi, M.; Mohammadi, H. Report on two species of *Phaeoacremonium* from walnut trees (*Juglans regia* L.) in Iran. In Proceedings of the 3rd Iranian Mycological Congress, University of Kurdistan, Sanandaj, Iran, 26–28 August 2017; p. 50.
50. Sohrabi, M.; Mohammadi, H.; Leon, M.; Armengol, J.; Banihashemi, Z. Detection of *Phaeoacremonium* spp. associated with decline of walnut trees (*Juglanconis regia*) in Kerman and Fars provinces. In Proceedings of the 4th Iranian Mycological Congress, Sari Agricultural Sciences and Nutural Resources University, Sari, Iran, 26–28 August 2019; p. 61.
51. Kadkhodahemmatabadi, S. Study of Disease Symptoms and Fungi Associated with Decline of Pomegranate and Fig Trees in Fars and Kerman Provinces. Master's Thesis, Shahid Bahonar University of Kerman, Kerman, Iran, 2018.
52. Whitelaw-Weckert, M.A.; Rahman, L.; Appleby, L.M.; Hall, A.; Clark, A.C.; Waite, H.; Hardie, W.J. Co-infection by Botryosphaeriaceae and Ilyonectria spp. fungi during propagation causes decline of young grafted grapevines. *Plant Pathol.* **2013**, *62*, 1226–1237. [CrossRef]
53. Essakhi, S.; Mugnai, L.; Crous, P.W.; Groenewald, J.Z.; Surico, G. Molecular and phenotypic characterization of novel *Phaeoacremonium* species associated with Petri disease and esca of grapevine. *Persoonia* **2008**, *21*, 119–134. [CrossRef] [PubMed]
54. Raimondo, M.L.; Lops, F.; Carlucci, A. *Phaeoacremonium italicum* sp nov., a new species associated with esca of grapevine in southern Italy. *Mycologia* **2014**, *106*, 1119–1126. [CrossRef]
55. Úrbez-Torres, J.R.; Haag, P.; Bowen, P.; O'Gorman, D.T. Grapevine trunk diseases in British Columbia: Incidence and characterization of the fungal pathogens associated with esca and Petri diseases of grapevine. *Plant Dis.* **2014**, *98*, 469–482. [CrossRef] [PubMed]
56. Sadeghi, F.; Samsampour, D.; Seyahooei, M.A.; Bagheri, A.; Soltani, J. Diversity and spatiotemporal distribution of fungal endophytes associated with Citrus reticulata cv. Siyahoo. *Curr. Microbiol.* **2019**, *76*, 279–289. [CrossRef] [PubMed]

57. Di Marco, S.; Calzarano, F.; Osti, F.; Mazzullo, A. Pathogenicity of fungi associated with a decay of kiwifruit. *Australas. Plant Pathol.* **2004**, *33*, 337–342. [CrossRef]
58. Spies, C.F.J.; Moyo, P.; Halleen, F.; Mostert, L. *Phaeoacremonium* species diversity on woody hosts in the Western Cape Province of South Africa. *Persoonia* **2018**, *40*, 26–62. [CrossRef]
59. Nigro, F.; Boscia, D.; Antelmi, I.; Ippolito, A. Fungal species associated with a severe decline of olive in southern Italy. *J. Plant Pathol.* **2013**, *95*, 668.
60. Rumbos, I. *Phialophora parasitica*, causal agent of cherry dieback. *J. Phytopathol.* **1986**, *117*, 283–287. [CrossRef]
61. Crous, P.W.; Gams, W. *Phaeomoniella chlamydospora* gen. et comb. nov., a causal organism of Petri grapevine decline and esca. *Phytopathol. Mediterr.* **2000**, *39*, 112–188.
62. Prodi, A.; Sandalo, S.; Tonti, S.; Nipoti, P.; Pisi, A. Phialophora-like fungi associated with kiwifruit elephantiasis. *J. Plant Pathol.* **2008**, *90*, 487–494.
63. Arzanlou, M.; Narmani, A.; Khodaei, S.; Moshari, S. Pome and stone fruit trees as possible reservoir hosts for *Phaeoacremonium* spp., the causal agents of grapevine esca disease, in Iran. *Arch. Phytopathol. Plant Protect.* **2014**, *47*, 717–727. [CrossRef]
64. Úrbez-Torres, J.R.; Peduto, F.; Vossen, P.M.; Krueger, W.H.; Gubler, W.D. Olive twig and branch dieback: Etiology, incidence, and distribution in California. *Plant Dis.* **2013**, *97*, 231–244. [CrossRef]
65. Hausner, G.; Eyjólfsdóttir, G.; Reid, J. Two additional species of the genus *Togninia*. *Can. J. Bot.* **1992**, *70*, 724–734. [CrossRef]
66. Calavan, E.C.; Wallace, J.M. *Hendersonula toruloidea* Nattrass on citrus in California. *Phytopathology* **1954**, *44*, 635–639.
67. Punithalingham, E. Plant diseases attributed to *Botryodiplodia theobromae* Pat. *Bibl. Mycol.* **1980**, *71*, 21–26.
68. Whiteside, J.O. Dothiorella Gummosis. In *Compendium of Citrus Diseases*, 1st ed.; Whiteside, J.O., Garnsey, S.M., Timmer, L.W., Eds.; American Phytopathological Society: St. Paul, MN, USA, 1980; p. 14.
69. Adesemoye, A.O.; Eskalen, A. First report of *Spencermartinsia viticola*, *Neofusicoccum australe*, and *N. parvum* causing branch canker of citrus in California. *Plant Dis.* **2011**, *95*, 770. [CrossRef]
70. Bautista-Cruz, M.A.; Almaguer-Vargas, G.; Leyva-Mir, S.G.; Colinas-León, M.T.; Correia, C.K.; Camacho-Tapia, M.; Robles-Yerena, L.; Michereff, S.J.; Tovar-Pedraza, J.M. Phylogeny, distribution, and pathogenicity of *Lasiodiplodia* species associated withcankers and dieback symptoms of Persian lime in Mexico. *Plant Dis.* **2018**, *103*. [CrossRef]
71. Hamrouni, N.; Nouri, M.T.; Trouillas, F.P.; Said, A.; Sadfi-Zouaoui, N.; Hajlaoui, M.R. Dothiorella gummosis caused by *Dothiorella viticola*, first record from citrus in Tunisia. *New Dis. Rep.* **2018**, *38*, 10. [CrossRef]
72. Abdollahzadeh, J.; Javadi, A.; Zare, R.; Phillips, A.J.L. A phylogenetic study of *Dothiorella* and *Spencermatinsia* species associated with woody plants in Iran, New Zealand, Portugal and Spain. *Persoonia* **2014**, *32*, 1–12.
73. Abdollahzadeh, J.; Hosseini, F.; Javadi, A. New records from Botryosphaeriaceae (Ascomycota) for mycobiota of Iran. *Mycol. Iran.* **2014**, *1*, 43–51.
74. Polizzi, G.; Aiello, D.; Vitale, A.; Giuffrida, F.; Groenewald, J.Z.; Crous, P.W. First report of shoot blight, canker, and gummosis caused by *Neoscytalidium dimidiatum* on citrus in Italy. *Plant Dis.* **2009**, *93*, 1215. [CrossRef]
75. Overton, E.B.; Steward, E.L.; Wenner, N.G. Molecular phylogenetics of grapevine decline fungi from Pennsylvania and New York. *Phytopathol. Mediterr.* **2005**, *44*, 90–91.
76. Travadon, R.; Lawrence, D.P.; Rooney-Latham, S.; Gubler, W.D.; Wilcox, W.F.; Rolshausen, P.E.; Baumgartner, K. *Cadophora* species associated with wood-decay of grapevine in North America. *Fungal Biol.* **2015**, *119*, 53–66. [CrossRef] [PubMed]
77. Spadaro, D.; Pellegrino, C.; Garibaldi, A.; Gullino, M.L. Development of SCAR primers for the detection of *Cadophora luteo-olivacea* on kiwifruit and pome fruit and of *Cadophora malorum* on pome fruit. *Phytopathol. Mediterr.* **2011**, *50*, 430–441.
78. De Sousa Santos, M.N. Contribuicão para conhecimento das relacoes *Quercus suber–Biscogniauxia mediterranea* (syn. *Hypoxylon mediterraneum*). *Silva Lusit.* **2003**, *11*, 21–29.
79. Vannini, A.; Scarascia Mugnozza, G. Water stress: A predisposing factor in the pathogenesis of *Hypoxylon mediterraneum* on *Quercus cerris*. *Eur. J. For. Pathol.* **1991**, *21*, 193–201. [CrossRef]
80. Vannini, A.; Biocca, M.; Paparatti, B. Contributo alla conoscenza del ciclo biologico di *Hypoxylon mediterraneum* (De Not.) Mill. su *Quercus cerris*. *Inf. Fitopatol.* **1996**, *56*, 53–55.
81. Læssøe, T.; Scheuer, C.; Granmo, A. *Biscogniauxia granmoi* (Xylariaceae) in Europe. *Österr. Z. Pilzkd.* **1999**, *8*, 139–147.

82. Vasilyeva, L.N. The taxonomic position of *Camarops polysperma* (Mont.) J. H. Miller and *Biscogniauxia* Kuntze in the Far East. *Mikol. Fitopatol.* **1988**, *22*, 388–396.
83. Pouzar, Z. Notes on taxonomy and nomenclature of *Nummularia* (Pyrenomycetes). *Ceska Mykol.* **1979**, *33*, 207–219.
84. Raimondo, M.L.; Lops, F.; Carlucci, A. Charcoal canker of pear, plum, and quince trees caused by *Biscogniauxia rosacearum* sp. nov. in Southern Italy. *Plant Dis.* **2016**, *100*, 1813–1822. [CrossRef]
85. Sritharan, T.; Savitri, K.N.; Jayasinghe, L.; Araya, H.; Fujimoto, Y. Isocoumarins and Dihydroisocoumarins from the endophytic fungus *Biscogniauxia capnodes* isolated from the fruits of *Averrhoa carambola*. *Nat. Prod. Commun.* **2019**, *14*. [CrossRef]
86. Henriques, J.; Nóbrega, F.; Sousa, E.; Lima, A. Analysis of the genetic diversity and phylogenetic relationships of *Biscogniauxia mediterranea* isolates associated with cork oak. *Phytoparasitica* **2016**, *44*, 19–34. [CrossRef]
87. Juybari, H.Z.; Tajick Ghanbary, M.A.; Rahimian, H.; Karimi, K.; Arzanlou, M. Seasonal, tissue and age influences on frequency and biodiversity of endophytis fungi of *Citrus sinensis* in Iran. *For. Pathol.* **2019**, e12559. [CrossRef]
88. Mirabolfathy, M.; Groenewald, J.Z.; Crous, P.W. The occurrence of charcoal disease caused by *Biscogniauxia mediterranea* on chestnut-leaved oak (*Quercus castaneifolia*) in the Golestan Forests of Iran. *Plant Dis.* **2011**, *95*, 876. [CrossRef]
89. Mirabolfathy, M.; Ju, Y.-M.; Hsieh, H.-M.; Rogers, J.D. *Obolarina persica* sp. nov., associated with dying Quercus in Iran. *Mycoscience* **2013**, *54*, 315–320. [CrossRef]
90. Rostamian, M.; Kavosi, M.R.; Bazgir, E.; Babanezhad, M. First report of *Biscogniauxia mediterranea* causing canker on wild almond (*Amygdalus scoparia*). *Australas. Plant Dis. Notes* **2016**, *11*, 30. [CrossRef]
91. Voglmayr, H.; Jaklitsch, W.M. *Stilbocrea walteri* sp. nov., an unusual species of Bionectriaceae. *Mycol. Prog.* **2019**, *18*, 91–105. [CrossRef] [PubMed]
92. Kinsey, G.C. *Phoma pinodella*. [Descriptions of Fungi and Bacteria]. *IMI Descr. Fungi Bact.* **2002**, *151*, 1505.
93. Ramos, A.P.; Talhinhas, P.; Sreenivasaprasad, S.; Oliveira, H. Characterization of *Colletotrichum gloeosporioides*, as the main causal agent of citrus anthracnose, and *C. karstii* as species preferentially associated with lemon twig dieback in Portugal. *Phytoparasitica* **2016**, *44*, 549–561. [CrossRef]
94. Sutton, B.C. The genus *Glomerella* and its anamorph *Colletotrichum*. In *Colletotrichum: Biology, Pathology and Control*; Bailey, J.A., Jeger, M.J., Eds.; CAB Int.: Wallingford, UK, 1992; pp. 1–26.
95. Damm, U.; Cannon, P.F.; Woudenberg, J.H.C.; Johnston, P.R.; Weir, B.S.; Tan, Y.P.; Shivas, R.G.; Crous, P.W. The *Colletotrichum boninense* species complex. *Stud. Mycol.* **2012**, *73*, 1–36. [CrossRef] [PubMed]
96. Lijuan, P.; Youlian, Y.; Kevin, D.H.; Bahkali, A.H.; Zuoyi, L. *Colletotrichum* species on *Citrus* leaves in Guizhou and Yunnan provinces, China. *Cryptogam. Mycol.* **2012**, *33*, 267–283. [CrossRef]
97. Taheri, H.; Javan-Nikkhah, M.; Elahinia, S.A.; Khodaparast, S.A.; Golmohammadi, G. Species of *Colletotrichum* associated with citrus trees in Iran. *Mycol. Iran.* **2016**, *3*, 1–14.
98. Sutton, B.C. *The Coelomycetes. Fungi Imperfecti with Pycnidia, Acervuli and Stromata*; Commonwealth Mycological Institute: Kew, UK, 1980.
99. Petrini, O.; Fisher, P.J. A comparative study of fungal endophytes in xylem and whole stem of *Pinus sylvestris* and *Fagus sylvatica*. Trans. *Br. Mycol. Soc.* **1988**, *91*, 233–238. [CrossRef]
100. Hormazabal, E.; Piontelli, E. Endophytic fungi from Chilean native gymnosperms: Antimicrobial activity against human and phytopathogenic fungi. *World J. Microbiol. Biotechnol.* **2009**, *25*, 813–819. [CrossRef]
101. Carlucci, A.; Lops, F.; Cibelli, F.; Raimondo, M.L. *Phaeoacremonium* species associated with olive wilt and decline in southern Italy. *Eur. J. Plant Pathol.* **2015**, *141*, 717–729. [CrossRef]
102. Abdollahi Aghdam, S.; Fotouhifar, K.-B. New reports of endophytic fungi associated with cherry (*Prunus avium*) and sour cherry (*Prunus cerasus*) trees in Iran. *Mycol. Iran.* **2016**, *3*, 75–85.
103. Ghobad-Nejhad, M.; Meyn, R.; Langer, E. Endophytic fungi isolated from healthy and declining Persian oak (*Quercus brantii*) in western Iran. *Nova Hedwig.* **2018**, *107*, 273–290. [CrossRef]
104. Pavlic, D.; Slippers, B.; Coutinho, T.A.; Wingfield, M.J. Botryosphaeriaceae occurring on native *Syzygium cordatum* in South Africa and their potential threat to *Eucalyptus*. *Plant Pathol.* **2007**, *56*, 624–636. [CrossRef]
105. Mohali, S.; Slippers, B.; Wingfield, M.J. Pathogenicity of seven species of the Botryosphaeriaceae on Eucalyptus clones in Venezuela. *Australas. Plant Pathol.* **2009**, *38*, 135–140. [CrossRef]

106. Úrbez-Torres, J.R.; Leavitt, G.M.; Guerrero, J.C.; Guevara, J.; Gubler, W.D. Identification and pathogenicity of *Lasiodiplodia theobromae* and *Diplodia seriata*, the causal agents of bot canker disease of grapevines in Mexico. *Plant Dis.* **2008**, *92*, 519–529. [CrossRef]
107. Michailides, T.J.; Morgan, D.P.; Felts, D. First report of *Botryosphaeria rhodina* causing shoot blight of pistachio in California. *Plant Dis.* **2002**, *86*, 1273. [CrossRef] [PubMed]
108. Phillips, A.J.L. The Botryosphaeria Site. 2006. Available online: http://www.crem.fct.unl.pt/botryosphaeria_site (accessed on 29 November 2016).
109. Crous, P.W.; Gams, W.; Wingfield, M.J.; Van Wyk, P.S. *Phaeoacremonium* gen. nov. associated with wilt and decline diseases of woody hosts and human infections. *Mycologia* **1996**, *88*, 786–796. [CrossRef]
110. Boerema, G.H.; Gruyter, J.D.; Noordeloos, M.E.; Hamers, M.E.C. *Phoma Identification Manual*; CABI Publishing: Wallingford, UK, 2004.
111. Cai, L.; Hyde, K.D.; Taylor, P.W.J.; Weir, B.S.; Waller, J.; Abang, M.M.; Zhang, J.Z.; Yang, Y.L.; Phoulivong, S.; Liu, Z.Y.; et al. A polyphasic approach for studying Colletotrichum. *Fungal Divers.* **2009**, *39*, 183–204.
112. Aveskamp, M.M.; Verkley, G.J.M.; Gruyter, J.D.; Murace, M.A.; Perelló, A.; Woudenberg, H.C.; Groenewald, J.Z.; Crous, P.W. DNA phylogeny reveals polyphyly of *Phoma* section *Peyronellaea* and multiple taxonomic novelties. *Mycologia* **2009**, *101*, 363–382. [CrossRef]
113. Aveskamp, M.M.; de Gruyter, J.; Woudenberg, J.H.C.; Verkley, G.J.M.; Crous, P.W. Highlights of the Didymellaceae: A polyphasic approach to characterise *Phoma* and related pleosporalean genera. *Stud. Mycol.* **2010**, *65*, 1–60. [CrossRef]
114. Weir, B.S.; Johnston, P.R.; Damm, U. The *Colletotrichum gloeosporioides* species complex. *Stud. Mycol.* **2012**, *73*, 115–180. [CrossRef]
115. White, T.J.; Bruns, T.; Lee, S.; Taylor, J. Amplification and direct sequencing of fungal ribosomal RNA genes for phylogenetics. In *PCR Protocols: A Guide to Methods and Applications*; Innis, M.A., Gelfand, D.H., Snisky, J.J., White, T.J., Eds.; Academic: San Diego, CA, USA, 1990; pp. 315–322.
116. Carbone, I.; Kohn, L.M. A method for designing primer sets for speciation studies in filamentous ascomycetes. *Mycologia* **1999**, *91*, 553–556. [CrossRef]
117. Glass, N.L.; Donaldson, G.C. Development of primer sets designed for use with the PCR to amplify conserved genes from filamentous ascomycetes. *Appl. Environ. Microbiol.* **1995**, *61*, 1323–1330. [CrossRef] [PubMed]
118. O'Donnell, K.; Cigelnik, E. Two divergent intragenomic rDNA ITS2 types within a monophyletic lineage of the fungus *Fusarium* are nonorthologous. *Mol. Phylogenet. Evol.* **1997**, *7*, 103–116. [CrossRef]
119. Katoh, K.; Toh, H. Parallelization of the MAFFT multiple sequence alignment program. *Bioinformatics* **2010**, *26*, 1899–1900. [CrossRef] [PubMed]
120. Rambaut, A. *Sequence Alignment Editor. Version 2.0*; University of Oxford: Oxford, UK, 2002.
121. Swofford, D.L. *PAUP* 4.0: Phylogenetic Analysis Using Parsimony (* and Other Methods)*; Sinauer Associates: Sunderland, MA, USA, 2003.
122. Kumar, S.; Stecher, G.; Tamura, K. MEGA7: Molecular Evolutionary Genetics Analysis version 7.0 for bigger datasets. *Mol. Biol. Evol.* **2016**, *33*, 1870–1874. [CrossRef] [PubMed]
123. Darriba, D.; Taboada, G.L.; Doallo, R.; Posada, D. jModelTest 2: More models, new heuristics and parallel computing. *Nat. Methods* **2012**, *9*, 772. [CrossRef]

plants

MDPI

Article

Water Stress Enhances the Progression of Branch Dieback and Almond Decline under Field Conditions

Carlos Agustí-Brisach [1,*], David Moldero [2], María del Carmen Raya [1], Ignacio J. Lorite [3], Francisco Orgaz [2] and Antonio Trapero [1]

1 Departamento de Agronomía, ETSIAM, Universidad de Córdoba, Campus de Rabanales, Edif. C4, 14071 Córdoba, Spain; ag2raorm@uco.es (M.d.C.R.); trapero@uco.es (A.T.)
2 Instituto de Agricultura Sostenible (IAS), Consejo Superior de Investigaciones Científicas (CSIC), Campus Alameda del Obispo, Avda Menéndez Pidal s/n, 14004 Córdoba, Spain; dmoldero@ias.csic.es (D.M.); orgaz@ias.csic.es (F.O.)
3 IFAPA-Centro Alameda del Obispo, Junta de Andalucía, Avda Menéndez Pidal s/n, 14004 Córdoba, Spain; ignacioj.lorite@juntadeandalucia.es
* Correspondence: cagusti@uco.es; Tel.: +34-957-218530

Received: 23 July 2020; Accepted: 13 September 2020; Published: 16 September 2020

Abstract: Branch dieback and tree decline have been described as a common complex disease worldwide in woody crops, with Botryosphaeriaceae and Diaporthaceae being considered the most frequent fungi associated with the disease symptoms. Their behaviour is still uncertain, since they are considered endophytes becoming pathogenic in weakened hosts when stress conditions, such as water deficiency occur. Therefore, the main goal of this study was to determine if water stress enhances general decline on weakened almond trees subjected to different irrigation treatments under natural field conditions. In parallel, the occurrence of fungal species associated with almond decline was also determined in relation to disease progression by fungal isolation, and morphological and molecular based-methods. The symptoms of branch dieback and general decline were observed over time, mainly in the experimental plots subjected to high water deficiency. Botryosphaeriaceae were the most consistently isolated fungi, and *Botryosphaeria dothidea* was the most frequent. *Collophorina hispanica* was the second most frequent species and *Diaporthe* and *Cytospora* species were isolated in a low frequency. Most of them were recovered from both asymptomatic and symptomatic trees, with their consistency of isolation increasing with the disease severity. This work reveals the need to elucidate the role of biotic and abiotic factors which increase the rate of infection of fungal trunk pathogens, in order to generate important knowledge on their life cycle.

Keywords: fungal trunk pathogens; occurrence; *Prunus dulcis*; water deficiency; weakened hosts

1. Introduction

Almond [*Prunus dulcis* (Mill.) D.A. Webb] represents the second highest woody crop by acreage in Spain after cultivated olive (*Olea europaea* subsp. *europaea* L.). To date, Spain currently leads the world in almond cultivation, with 657,768 hectares of cultivated almond (34% of the global surface) and 339.033 tonnes of production [1,2]. In this country, Andalusia region (southern Iberian Peninsula) represents the 30% of the Spanish cultivated almond surface [2].

Until recently, almond crop has been associated with traditional dry farming systems in marginal areas of southern Spain with unfavorable conditions to produce high yields. However, due to the global economic impact of almond kernels, as well as the necessity to find extensive alternative crops in Andalusia, almond plantings are increasingly being established in regions with better favorable conditions (i.e., moderate-warm temperatures, high humidity, irrigation-water resources, etc.). This change also involves different cultural practices than those used in traditional systems which are more likely to increase yields. They include dense planting, high levels of irrigation-water and fertilization, pruning- and harvest-mechanization, and a high number of pesticide treatments preventing pest and plant diseases [3]. As a consequence of this new scenario, the occurrence of secondary almond diseases, as well as emerging ones, have been reported recently in the new almond growing regions across the Guadalquivir Valley in Andalusia region [3–5].

Among the emerging diseases already described in this geographic area [3–5], a new tree decline syndrome stands out in the new intensive almond plantings. It includes a broad diversity of symptoms, such as gummosis, shoot blight, defoliation, branch dieback, canker formation, internal wood discoloration and tree death. The first studies determining its etiology suggest that it is a complex disease probably associated with Botryosphaeriaceae Theiss. and Syd. fungi, among other secondary pathogens [5]. On the other hand, other syndromes, such as a branch dieback and cankers associated with *Diaporthe amygdali* (Delacr.) Udayanga, Crous and K.D. Hyde, or the foamy canker, have also been observed in the new almond plantings in Andalusia. In particular, foamy canker always occurs when the vigorous hybrid Garnem is used as rootstock [5], but the causal agent of these syndrome has not yet been described anywhere, due to the impossibility of reproducing the symptoms with the microorganisms isolated from the diseased trees [5,6]. However, the etiology of all these syndromes in the environmental conditions of Andalusia is still uncertain, since little attention has been given to their low occurrence, until recently. The only previous studies describing symptoms of almond decline in Spain were conducted in Mallorca (Balearic Islands, western Mediterranean Sea) [7–9]. These authors indicated that fungi belonging to Botryosphaeriaceae were the main causal agents associated with the disease, among other secondary fungal species belonging to the genera *Collophorina* (=*Collophora*) Damm and Crous, *Diaporthe* Nitschke, *Eutypa* Tul. and C. Tul. or *Phaeoacremonium* W. Gams, Crous and M.J. Wingf. Branch dieback and tree decline has been described as a common syndrome worldwide in a broad diversity of woody crops including grapevine [10–12], olive [13–15] and tree nuts [6,15–17]. In any cases, the main fungal species associated with tree decline belongs to Botryosphaeriaceae and/or Diaporthaceae Höhn. ex Wehm., with the first ones being the most aggressive [6,9,15–18]. However, the role of this wide diversity of fungi that has been causing tree decline is still uncertain. Most of the fungi are characterized by remaining latent in the infected tissues for a long period of time (endophytic phase), but they become pathogenic in weakened hosts when stress conditions occur [19,20]. In fact, it is rare to find weakened or stressed trees that are not infected by dieback and canker fungi, while their occurrence and aggressiveness is low in healthy plants [19].

In this sense, previous studies evaluating the effect of water stress on development of canker diseases have been conducted under semi-controlled conditions. Crist and Schoeneweiss [21] demonstrated that canker formation and colonization of bark and wood on birch tree (*Betula alba* L.) occurred when seedlings inoculated with *Botryosphaeria dothidea* (Moug.) Ces. and De Not. were subjected to defoliation stress, increasing in severity with time of exposure to stress. Later studies with this same pathogen also demonstrated that the lesions developed on inoculated plants of peach [*Prunus persica* (L.) Batsch] were larger on water-stressed plants in comparison with those on non-stressed ones [22]. Similar studies have also been conducted to determine the effect of water stress on the aggressiveness of other pathogens, which are different to those described previously. For example, Maxwell et al. [23] evaluated the influence of water stress on Septoria canker, caused by *Septoria musiva* Peck in Populus stems. This study showed that cankers on inoculated water-stressed trees were significantly larger than those on non-stressed ones.

However, to date, there is no scientific evidence on the question of whether water stress could enhance the progression of branch dieback and general decline on weakened almond trees under field conditions. Since the occurrence of decline syndromes is growing in the new almond plantings in southern Spain along the last few years [5], determining whether water stress enhances the incidence and severity of almond decline is essential. The current scenario that we face to in the new almond plantings is subjected to two-limiting conditions, which could favor the disease development, including; (1) the typical environmental conditions in southern Spain are characterized by scarce rains and warm temperatures during summer (from May to September), which predispose plants to water deficiency for a long time; and (2) the need to optimize water-irrigation treatments within the frame of eco-friendly agriculture towards a sustainable use of water resources. Therefore, the main goal of this study was to determine whether water stress enhances the general decline of weakened almond trees, subjected to different irrigation treatments, under natural field conditions. In parallel, the occurrence of fungal species, associated with branch dieback and almond decline, was also determined in relation to the disease progress by fungal isolation, and morphological and molecular based-methods for their identification.

2. Results

2.1. *Effect of Water Stress on Branch Dieback of Almond under Natural Field Conditions*

At the beginning of the evaluation period (June 2018), the number of almond trees of each category among the 80 evaluated trees was as follow: Category 0 = 34 trees 0 (asymptomatic trees with 0% of affected surface by branch dieback), Category 1 = 30 trees (<25% of affected surface by branch dieback), Category 2 = 8 trees (25–50% of affected surface by branch dieback); Category 3 = 1 tree (51–75% of affected surface by branch dieback); Category 4 = 0 trees (76–90% of affected surface by branch dieback); and Category 5 = 7 trees (>90% affected surface by branch dieback or dead trees). In general, Disease Severity (DS) progresses in significantly higher values of relative area under the disease progress curve (RAUDPC; $P = 0.0173$) and final disease severity ($P = 0.0012$) when almond trees were subjected to T3 (Severe Regulated Deficit Irrigation; RAUDPC = 47.2 ± 8.7%; Final disease severity = 80.4 ± 3.6%), followed by T2 (Moderate Sustained Deficit Irrigation; RAUDPC = 31.3 ± 5.1%; Final disease severity = 71.3 ± 9.1%) and T1 (Moderate Regulated Deficit Irrigation; RAUDPC = 28.5 ± 4.5%; Final disease severity = 59.1 ± 6.5%) (Figures 1 and 2). Almond trees used as control (T0) showed the lowest values of RAUDPC (22.2 ± 1.5%), as well as the lowest values of final disease severity (42.2 ± 4.1%) (Figures 1 and 3). Control trees did not show internal wood discoloration.

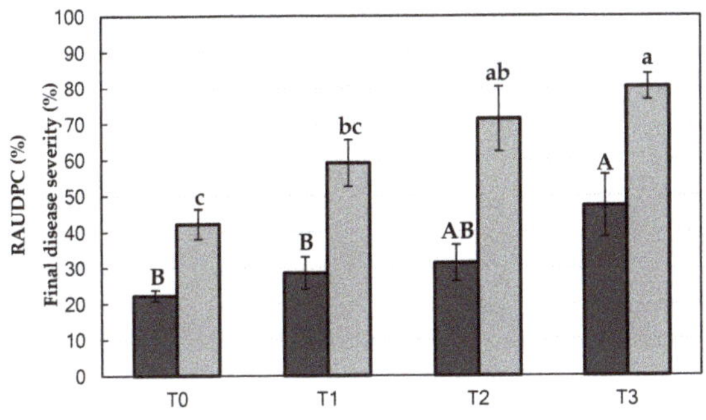

Figure 1. Disease severity [RAUDPC (%; dark grey columns) and Final disease severity (%; light grey columns)] of branch dieback of almond trees under natural conditions in an experimental field (Córdoba, Andalusia region, southern Spain) subjected to four irrigation treatments from April 2013 to October 2019 (T0: Control; T1: Moderate Regulated Deficit Irrigation; T2: Moderate Sustained Deficit Irrigation; T3: Severe Regulated Deficit Irrigation). The disease severity assessments were conducted from June 2018 (next spring after first symptoms of branch dieback occur) to September 2019 (end of the experiment). For each disease parameter, columns represent the means of sixteen trees and vertical bars are the standard errors of the means. Columns with different capital or lowercase letters differ significantly for RAUDPC, or Final disease severity, respectively, according to Fisher's LSD test at P = 0.05.

Figure 2. Disease progress and symptoms of branch dieback on one almond tree subjected to Severe Regulated Deficit Irrigation (T3) monitored from June 2018 to September 2019. Assessment times and rating-scales values were: (**a**) June 2018-2.0; (**b**) September 2018-3.0; (**c**) June 2019-3.0; (**d**) September 2019-4.0; (**e**–**g**) symptoms of branch dieback and internal wood discoloration in affected branches.

Figure 3. Disease progress on one almond tree from Control (T0) monitored from June 2018 to September 2019. Assessments times and rating-scales values were: (**a**) June 2018-0.0; (**b**) September 2018-1.0; (**c**) June 2019-1.0; (**d**) September 2019-2.0.

2.2. *Occurrence, Consistency and Frequency of Isolated Fungi*

Fungal species, associated with branch dieback and almond decline, were isolated from all the categories of severity evaluated, with the exception from the trees belonging to the category 5 from which only saprophytes (i.e., *Alternaria* spp. Nees, *Penicillium* spp. Link, *Sordaria* spp. Ces. and De

Not., etc.) were recovered. The consistency of isolation of each isolated fungi is shown in Table 1. In general, the occurrence of fungal species and the consistency of their isolation increased with the DS, with trees belonging to the categories 3 and 4 showing the highest number of fungal species, as well as the highest consistency of isolation. However, three fungal species were isolated from trees belonging to category 0, while only one fungal species was recovered from trees belonging to category 1. The consistency of isolation in these two categories was somewhat lower (≤5.4%) than those obtained from the remaining ones (up to 16.6%) (Figure 4). The total fungal biomass in the sampled trees per category was three fungal species in the trees belonging to Category 0, one fungal species in the trees belonging to Category 1, three fungal species in the trees belonging to the Category 2, and four fungal species in the trees belonging to the Category 3 and 4. Therefore, there was no linear correlation between the in planta abundance (biomass) of the studied fungal species and the severity of the category of severity ($r = -0.6455$; $P = 0.2394$).

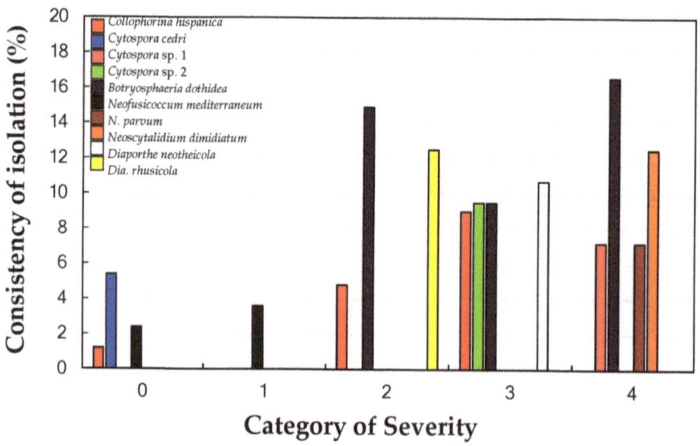

Figure 4. Consistency of isolation (Y-axis; Av. %) of the fungal species identified in this study associated with branch dieback of almond in each category of severity (0 = 0%, 1 = < 25%, 2 = 25–50%, 3 = 51–75%, 4 = 75–90% of affected surface by branch dieback) in the whole of the experiment. For each category and fungal species, columns represent the total consistency of isolation along the two years (2018–2019), in which the disease severity was evaluated [Consistency of isolation = (N°. of positive wood pieces/168) × 100; where 168 is the total attempts of isolation (wood pieces) per category of disease severity in the whole of the experiment obtained as follow: 21 wood pieces per tree × 2 trees of each disease category × 2 years of evaluation × 2 sampling times per year].

Fungal species, belonging to Botryosphaeriaceae, were the most frequent and they also showed the highest consistency of isolation of the whole of the experiment (Table 1). *Botryosphaeria dothidea* was the most frequent species, since it was isolated from the 50.0% of the sampled trees, from trees belonging to the categories 2, 3 and 4. This was followed by *Collophorina hispanica* (Gramaje, Armengol and Damm) Damm and Crous, which was isolated from the 41.6% of the sampled trees, from trees belonging to the categories 0, 2 and 3. The species belonging to *Cytospora* Ehrenb. were also isolated from trees belonging to the categories 0 (*Cytospora cedri* Syd., P. Syd. & E.J. Butler), 3 (*Cytospora* sp. 2) and 4 (*Cytospora* sp. 1) with a low frequency (8.3%). The frequency of the remaining species was also low [*Dia. neotheicola* A.J.L. Phillips and J.M. Santos (8.3%), *Dia. rhusicola* Crous (8.3%), *N. mediterraneum* Crous, M.J. Wingf. and A.J.L. Phillips (16.7%), *N. parvum* (Pennycook and Samuels) Crous, Slippers & A.J.L. Phillips (8.3%) and *Neoscytalidium dimidiatum* (Penz.) Crous and Slippers (8.3%)]. Co-infections in the same tree and sampling moment occurred only one time for the following combinations: *C. hispanica* and *Cytospora* sp.; and *Cytospora* sp. and *N. parvum*.

Table 1. Fungal isolates used in the phylogenetic analysis and their corresponding GenBank accession numbers.

Species	Isolate [1]	Consistency of Isolation (%) [2]	Host/Cultivar	Collector	Date [3]	GenBank Accession no. [4]			
						ITS	TUB	EF	GAPDH
Botryosphaeriaceae analyses									
Botryosphaeria dothidea	ColPat-607	3.6	*Prunus dulcis* cv. Lauranne	C. Agustí-Brisach & A. Trapero	06/12/2018	MT303980	MT309728	-	-
	ColPat-610	22.6	*Prunus dulcis* cv. Lauranne	C. Agustí-Brisach & A. Trapero	06/12/2018	MT303982	MT309730	-	-
	ColPat-652	7.2	*Prunus dulcis* cv. Lauranne	C. Agustí-Brisach & A. Trapero	09/13/2018	MT303983	MT309731	-	-
	ColPat-653	6.4	*Prunus dulcis* cv. Lauranne	C. Agustí-Brisach & A. Trapero	09/13/2018	MT303984	MT309732	-	-
	ColPat-654	8.2	*Prunus dulcis* cv. Lauranne	C. Agustí-Brisach & A. Trapero	09/13/2018	MT303985	MT309733	-	-
	ColPat-657	4.8	*Prunus dulcis* cv. Lauranne	C. Agustí-Brisach & A. Trapero	09/13/2018	MT303986	MT309734	-	-
	ColPat-658	6.4	*Prunus dulcis* cv. Lauranne	C. Agustí-Brisach & A. Trapero	09/13/2018	MT303987	MT309735	-	-
	ColPat-764	11.9	*Prunus dulcis* cv. Lauranne	C. Agustí-Brisach & A. Trapero	06/13/2019	MT303988	MT309736	-	-
	ColPat-765	9.1	*Prunus dulcis* cv. Lauranne	C. Agustí-Brisach & A. Trapero	06/13/2019	MT303989	MT309737	-	-
	ColPat-794	16.0	*Prunus dulcis* cv. Lauranne	C. Agustí-Brisach & A. Trapero	09/11/2019	MT303991	MT309739	-	-
	ColPat-795	23.8	*Prunus dulcis* cv. Lauranne	C. Agustí-Brisach & A. Trapero	09/11/2019	MT303992	MT309740	-	-
	ColPat-796	19.1	*Prunus dulcis* cv. Lauranne	C. Agustí-Brisach & A. Trapero	09/11/2019	MT303993	MT309741	-	-
	ColPat-797	17.9	*Prunus dulcis* cv. Lauranne	C. Agustí-Brisach & A. Trapero	09/11/2019	MT303994	MT309742	-	-
	CBS 100564, PD 97,14304	-	*Paeonia* sp	P. Vink	nd	KX464085	KX464781	-	-
	PD4, 3626	-	*Prunus dulcis*	T.J. Michailides	8/2005	GU251091	GU251751	-	-
	PD33,3657	-	*Prunus dulcis*	T.J. Michailides	8/2005	GU251093	GU251753	-	-
	PD68,3623	-	*Prunus dulcis*	T.J. Michailides	8/2005	GU251095	GU251755	-	-
	PD107, 809	-	*Prunus dulcis*	T.J. Michailides	8/2005	GU251097	GU251757	-	-
	PD122, A2.1	-	*Prunus dulcis*	T.J. Michailides	5/2007	GU251098	GU251758	-	-
	PD146, A27	-	*Prunus dulcis*	T.J. Michailides	5/2007	GU251099	GU251759	-	-
Diplodia juglandis	**CBS 188.87**	-	*Juglans regia*	nd	nd	EU673316	EU673119	-	-
Diplodia mutila	**6B99**	-	*Juglans regia*	nd	5/31/2011	KF778791	KF778886	-	-
	CBS 112553[T]	-	*Vitis vinifera*	A.J.L. Phillips	nd	AY259093	DQ458850	-	-
Diplodia seriata	**CBS 112555[T]**	-	*Vitis vinifera*	A. J. L. Phillips	nd	AY259094	DQ458856	-	-
	3H18	-	*Juglans regia*	nd	nd	KF778796	KF778891	-	-
	PD34, 3381	-	*Prunus dulcis*	T.J. Michailides	7/2004	GU251111	GU251771	-	-
	PD50, 3348	-	*Prunus dulcis*	T.J. Michailides	8/2004	GU251113	GU251773	-	-

Table 1. *Cont.*

Species	Isolate [1]	Consistency of Isolation (%) [2]	Host/Cultivar	Collector	Date [3]	GenBank Accession no. [4]			
						ITS	TUB	EF	GAPDH
	Mz-F1	-	*Malus domestica*	nd	nd	KU942427	KU976444	-	-
Dothiorella iberica	**CBS 115041^T**	-	*Quercus ilex*	J. Luque	12/2009	AY573202	EU673096	-	-
	5G97	-	*Juglans regia*		12/13/2010	KF778808	KF778903	-	-
	UCD1448SLO	-	nd	nd	nd	EF202009	EF202016	-	-
	UCRDI3	-	*Prunus dulcis*	nd	nd	KP012591	KP067201	-	-
Dothiorella sarmentorum	**CBS 164.33**	-	nd	nd	nd	KX464127	KX464881	-	-
	PD78, 3797	-	*Prunus dulcis*	T.J. Michailides	8/2006	GU251169	GU251829	-	-
	PD79, 3795	-	*Prunus dulcis*	T.J. Michailides	8/2006	GU251170	GU251830	-	-
Lasiodiplodia citricola	**6I34**	-	*Juglans regia*	nd	10/6/2011	KF778809	KF778904	-	-
Lasiodiplodia theobromae	**CBS 124.13**	-	nd	J.J. Taubenhaus	nd	DQ458890	DQ458858	-	-
Macrophomina phaseolina	**PD112, A28.1**	-	*Prunus dulcis*	T.J. Michailides	5/2007	GU251105	GU251765	-	-
Neofusicoccum australe	**Y264-21-1**	-	*Vitis vinifera*	nd	nd	JF437920	JF437922	-	-
Neofusicoccum mediterraneum	ColPat-605	7.2	*Prunus dulcis* cv. Lauranne	C. Agustí-Brisach & A. Trapero	06/12/2018	MT303979	MT309727	-	-
	ColPat-799	4.8	*Prunus dulcis* cv. Lauranne	C. Agustí-Brisach & A. Trapero	09/11/2019	MT303995	MT309743	-	-
	CBS 121718^T; PD312	-	*Eucalyptus* sp.	nd	6/2006	GU251176	GU251836	-	-
	PD48, 3483	-	*Prunus dulcis*	T.J. Michailides	9/2004	GU251186	GU251846	-	-
	PD49, 3227	-	*Prunus dulcis*	T.J. Michailides	6/2004	GU251187	GU251847	-	-
	PD55, 2953	-	*Prunus dulcis*	T.J. Michailides	1/2004	GU251189	GU251849	-	-
	Bot-04	-	*Vitis vinifera* cv. Pedro Ximénez	C. Agustí- Brisach and A. Trapero	2016	MG745841	MG745803	-	-
	1H96	-	*Juglansregia*	nd	9/15/2006	KF778811	KF778906	-	-
	CAA 002	-	*Pistacia vera* cv. Kerman	T.J. Michailides	nd	EU017537	KX505925	-	-
	Y546-2-1	-	*Vitis vinifera*	nd	nd	JF437919	JF437921	-	-
Neofusicoccum nonquaesitum	**PD86, A9**	-	*Prunus dulcis*	T.J. Michailides	5/2007	GU251156	GU251816	-	-
	PD90, A42	-	*Prunus dulcis*	T.J. Michailides	5/2007	GU251157	GU251817	-	-
Neofusicoccum parvum	ColPat-608	14.3	*Prunus dulcis* cv. Lauranne	C. Agustí-Brisach & A. Trapero	06/12/2018	MT303981	MT309729	-	-
	CMW9081^T	-	*Pinus nigra*	G. J. Samuels	nd	AY236943	AY236917	-	-
	1L83	-	*Juglansregia*	nd	11/4/2005	KF778854	KF778949	-	-
	PD17, 3621	-	*Prunus dulcis*	T.J. Michailides	8/2005	GU251143	GU251803	-	-
	PD39, 3656	-	*Prunus dulcis*	T.J. Michailides	8/2005	GU251144	GU251804	-	-
Neoscytalidium dimidiatum	ColPat-792	25.0	*Prunus dulcis* cv. Lauranne	C. Agustí-Brisach & A. Trapero	09/11/2019	MT303990	MT309738	-	-
	Kale4-C	-	*Prunus armeniaca*	E. Oksal	2018	MK788362	MK803352	-	-
	Arp2-D	-	*Vitis vinifera*	E. Oksal	2018	MK813852	MK816354	-	-
	2-D61	-	*Ficus carica*	M. Nouri	2016	MG021572	MG021515	-	-
	CBS 145.78	-	nd	nd	nd	MH861121	KF531796	-	-
	COUFAL0146	-	*Nopalea rochenillifera*	nd	nd	MH251955	MH251971	-	-
N. novaehollandiae	**CBS122071**	-	*Crotalaria medicaginea*	nd	nd	EF585540	-	EF585580	-
	CBS122610	-	*Acacia synchronicia*	nd	nd	EF585536	-	EF585578	-
Diaporthe asheicola	**CBS 136967**	-	*Vaccinium ashei*	nd	nd	KJ160562	KJ160518	-	-

Table 1. *Cont.*

Species	Isolate [1]	Consistency of Isolation (%) [2]	Host/Cultivar	Collector	Date [3]	GenBank Accession no. [4]			
						ITS	TUB	EF	GAPDH
Diaporthaceae analyses									
Diaporthe acaciigena	**CBS 129521^{T}**	-	*Acacia retinodes*	P.W. Crous, I.G. Pascoe & J. Edwards	nd	KC343005	KC343973	KC343731	-
Diaporthe alleghaniensis	**CBS 495.72^{T}**	-	*Betuta alleghaniensis*	nd	nd	FJ889444	KC343975	GQ250298	-
Diaporthe alnea	**CBS 146.46^{T}**	-	*Alnus* sp.	S. Truter	nd	KC343008	KC343976	KC343734	-
Diaporthe ambigua	**CBS 114015**	-	nd	nd	nd	MH862953	KC343978	KC343736	
Diaporthe arctii	**DP0482**	-	*Arctium lappa*	W. Jaklitsch	nd	KJ590736	KJ610891	KJ590776	-
Diaporthe australafricana	**CBS 111886^{T}**	-	*Vitis vinifera*	L. Mostert	nd	KC343038	KC344006	KC343764	-
Diaporthe chamaeropis	**CBS 753.70**	-	*Spartium junceum*	J.A. von Arx	nd	KC343049	KC344017	KC343775	-
Diaporthe cinerascens	**CBS 719.96**	-	*Ficus carica*	E. Ilieva	nd	KC343050	KC344018	KC343776	-
Diaporthe cuppatea	**CBS 117499**	-	*Aspalathus linearis*	J.C. Janse van Rensburg	nd	MH863021	KC344025	KC343783	-
Diaporthe cynaroidis	**CBS 122676^{T}**	-	*Protea cynaroides*	S. Marincowitz	nd	NR111846	KC344026	KC343784	-
Diaporthe eres	**CBS 287.74**	-	*Sorbus aucuparia*	W.M. Loerakker	nd	KC343084	KC344052	KC343810	-
Diaporthe hickoriae	**CBS 145.26^{T}**	-	*Carya glabra*	L.E. Wehmeyer	nd	NR103699	KC344086	GQ250309	-
Diaporthe inconspicua	**CBS 133813^{T}**	-	*Maytenus ilicifolia*	R.R. Gomes	nd	NR111849	KC344091	KC343849	-
Diaporthe infecunda	**CBS 133812^{T}**	-	*Schinus terebinthifolius*	J. Lima	nd	NR111850	KC344094	KC343852	-
Diaporthe lusitanicae	**CBS 123212**	-	nd	nd	nd	MH863279	KC344104	KC343862	-
Diaporthe neotheicola	ColPat-762	21.4	*Prunus dulcis* cv. Lauranne	C. Agustí-Brisach & A. Trapero	06/13/2019	MT304007	MT309745	MT309762	-
	ColPat-763	17.1	*Prunus dulcis* cv. Lauranne	C. Agustí-Brisach & A. Trapero	06/13/2019	MT304008	MT309746	MT309763	-
	CBS 111553^{T}	-	*F. vulgare*	Alan Phillips	nd	NR145303	KC344069	KC343827	-
	6I30	-	*Juglansregia*	T.J. Michailides	10/6/2011	KF778871	KF778966	KF779061	-
	CAA816	-	*Vaccinium corymbosum*	nd	nd	MK792314	MK837934	MK828083	
	ColPat-445	-	*Juglans regia* cv. Tulare	C. Agustí-Brisach & A. Trapero	07/14/2017	MK522106	MK447993	MK490932	
Diaporthe novem	**CBS 127270^{T}**	-	*Glycine max*	T. Duvnjak	nd	NR111855	KC344124	KC343882	
Diaporthe phaseolorum	**AR4203**	-	*Phaseolus vulgaris*			KJ590738	KJ610893	KJ590739	-
Diaporthe rhusicola	ColPat-606	25.0	*Prunus dulcis* cv. Lauranne	C. Agustí-Brisach & A. Trapero	06/12/2018	MT304006	MT309744	MT309761	-
	6I14	-	*Prunus dulcis*	T.J. Michailides	9/12/2011	KF778872	KF778967	KF779062	
	6I15	-	*Prunus dulcis*	T.J. Michailides	9/12/2011	KF778873	KF778968	KF779063	-
	6I31	-	*Juglans regia*	T.J. Michailides	10/06/2011	KF778874	KF778969	KF779064	-
	6I43	-	*Juglans regia*	T.J. Michailides	10/06/2011	KF778875	KF778970	KF779065	-
Diaporthe vaccinii	**CBS 160.32^{T}**	-	*Oxycoccus macrocarpos*	C.L. Shear	nd	NR103701	KC344196	KC343954	
Diaporthella corylina	**CBS 121124**	-	*Corylus* sp.	nd	nd	KC343004	KC343972	KC343730	-
Phomopsis amygdali	**CBS 126679^{T}**	-	*Prunus dulcis*	nd	nd	KC343022	KC343990	KC343748	-
	CBS 115620	-	*Prunus persica*	nd	nd	KC343020	KC343988	KC343746	-

Table 1. *Cont.*

Species	Isolate [1]	Consistency of Isolation (%) [2]	Host/Cultivar	Collector	Date [3]	GenBank Accession no.[4]			
						ITS	TUB	EF	GAPDH
Tympanidaceae analyses									
Collophorina africana	**CBS 120872**^{T}		*Prunus salicina*	U. Damm	nd	GQ154570	-	GQ154643	GQ154648
	GLMC 1736		*Prunus domestica*	nd	nd	MK314542	-	MK314507	MK314474
Collophorina badensis	**GLMC 1684**^{T}		*Prunus domesica*	nd	nd	MK314546	-	MK314503	MK314482
Collophorina capensis	**CBS 120879**		*Prunus salicina*	U. Damm	nd	GQ154571	-	GQ154644	GQ154649
Collophorina germanica	**GLMC 1445**^{T}		*Prunus avium*	nd	nd	MK314550	-	MK314515	MK314477
Collophorina hispanica	ColPat-651	2.4	*Prunus dulcis* cv. Lauranne	C. Agustí-Brisach & A. Trapero	09/13/2018	MT303996	-	MT309747	MT309754
	ColPat-655	9.5	*Prunus dulcis* cv. Lauranne	C. Agustí-Brisach & A. Trapero	09/13/2018	MT303997	-	MT309748	MT309755
	ColPat-759	9.5	*Prunus dulcis* cv. Lauranne	C. Agustí-Brisach & A. Trapero	06/13/2019	MT303998	-	MT309749	MT309756
	ColPat-760	9.5	*Prunus dulcis* cv. Lauranne	C. Agustí-Brisach & A. Trapero	06/13/2019	MT303999	-	MT309750	MT309757
	ColPat-761	8.3	*Prunus dulcis* cv. Lauranne	C. Agustí-Brisach & A. Trapero	06/13/2019	MT304000	-	MT309751	MT309758
	ColPat-800	2.4	*Prunus dulcis* cv. Lauranne	C. Agustí-Brisach & A. Trapero	09/11/2019	MT304001	-	MT309752	MT309759
	ColPat-801	3.6	*Prunus dulcis* cv. Lauranne	C. Agustí-Brisach & A. Trapero	09/11/2019	MT304002	-	MT309753	MT309760
	CBS 128566	-	*Prunus dulcis*	J. Armengol	2010	JN808839	-	JN808850	JN808843
	CBS 128568^T	-	*Prunus dulcis*	J. Armengol	2010	JN808841	-	JN808852	JN808845
	CBS 128569	-	*Prunus dulcis*	J. Armengol	2010	MH864962	-	JN808853	JN808846
Collophorina neorubra	**GLMC 929**^{T}	-	*Prunus avium*	nd	nd	MK314533	-	MK314511	MK314485
Collophorina paarla	**CBS 120877**	-	*Prunus salicina*	U. Damm	nd	GQ154586	-	GQ154645	GQ154651
Collophorina rubra	**CBS 120873**^{T}**, STE-U 6109**	-	*Prunus persica*	U. Damm	nd	NR119747		JN808855	JN808848
Cadophora luteo-olivacea	**CBS 141.41**^{T}	-	*Prunus dulcis*	nd	nd	AY249066	-	KM497089	JN808849
Valsaceae analyses									
Cytospora acaciae	**CBS 468.69**	-	*Ceratonia siliqua*	nd	nd	MH859354	-	KX965181	-
Cytospora amygdali	**LH356**	-	*Prunus dulcis*	nd	nd	MG971852	-	MG971658	-
	LH357^T**, CBS 144233**	-	*Prunus dulcis*	nd	nd	MG971853	-	MG971659	-
Cytospora austromontana	**CBS 116821**	-	*Eucalyptus pauciflora*	nd	nd	KY051796	-	KX965068	-
Cytospora cabornacea	**CBS 219.54**	-	*Ulmus* sp.	nd	nd	DQ243805	-	KX965164	-
Cytospora californica	**KARE1105**	-	*Prunus dulcis*	nd	nd	MG971947	-	MG971663	-
Cytospora cedri	ColPat-604	10.8	*Prunus dulcis* cv. Lauranne	C. Agustí-Brisach & A. Trapero	06/12/2018	MT304003	-	MT311983	-
	CBS 196.97	-	nd	nd	nd	KY051906	-	KX965154	-
Cytospora ceratosperma	**CBS 512.76**	-	*Fagus sylvatica*	nd	nd	KY051941	-	KX965184	-
Cytospora chrysosperma	**9E-33, CBS 144242**	-	*Camellia* sp.	nd	nd	MG971892	-	MG971602	-
Cytospora eucalypti	**KARE1585, CBS 144241**	-	*Prunus dulcis*	nd	nd	MG971907	-	MG971617	-

Table 1. *Cont.*

Species	Isolate [1]	Consistency of Isolation (%) [2]	Host/Cultivar	Collector	Date [3]	GenBank Accession no.[4]			
						ITS	TUB	EF	GAPDH
Cytospora granati	**6F-45^T, CBS 144237**	-	*Punica granatum*	nd	nd	MG971799	-	MG971514	-
Cytospora joaquinensis	**KARE975^T, CBS 144235**	-	*Populus deltoides*	nd	nd	MG971895	-	MG971605	-
Cytospora longispora	**10F-57^T, CBS 144236**	-	*Prunus domestica*	nd	nd	MG971905	-	MG971615	-
Cytospora mali	**CBS 109499**	-	*Malus* sp.	nd	nd	KY051769	-	KX965048	-
Cytosppora oleicola	**KARE1021^T, CBS 144248**	-	*Olea europaea*	nd	nd	MG971944	-	MG971660	-
Cytospora parakantschavelii	**KARE974, CBS 144243**	-	*Populus deltoides*	nd	nd	MG971898	-	MG971608	-
Cytospora parapistaciae	**KARE232**	-	*Pistacia vera*	nd	nd	MG971807	-	MG971522	-
	KARE270^T, CBS 144506	-	*Pistacia vera*	nd	nd	MG971804	-	MG971519	-
Cytospora parasitica	**CFCC 53173**	-	*Berberis* sp.	nd	nd	MK673070	-	MK672957	-
Cytospora pavettae	**CBS 145562**	-	*Pavetta revoluta*	M.J. Wingfield	nd	MK876386	-	MK876497	-
Cytospora pistaciae	**KARE443^T, CBS 144238**	-	*Pistacia vera*	nd	nd	MG971802	-	MG971517	-
Cytospora plurivora	**KARE1452^T, CBS 144239**	-	*Olea europaea*	nd	nd	MG971861	-	MG971572	-
Cytospora populicola	**KARE973^T, CBS 144240**	-	*Populus deltoides*	nd	nd	MG971891	-	MG971601	-
Cytospora punicae	**7C-09**	-	*Punica granatum*	nd	nd	MG971939	-	MG971650	-
Cytospora rhodophila	**CBS 190.42**	-	*Syringa* sp.	nd	nd	KY051901	-	KX965147	-
Cytospora ribis	**CBS 187.36**	-	*Ribes rubrum*	nd	nd	DQ243810	-	KX965144	-
	CFCC 50039	-	*Platycladus orientalis*	Xinlei Fan	nd	KR045642	-	KU710931	-
Cytospora sacculus	**CFCC 89626**	-	*Juglans regia*	Xinlei Fan	nd	KR045647	-	KU710934	-
	CBS 116855	-	*Quercus alba*	nd	nd	KY051824	-	KX965091	-
Cytospora sorbicola	**4L-58**	-	*Prunus domestica*	nd	nd	MG971839	-	MG971553	-
Cytospora sp. 1	ColPat-609	14.3	*Prunus dulcis* cv. Lauranne	C. Agustí-Brisach & A. Trapero	06/12/2018	MT304004	-	MT311984	-
Cytospora sp. 2	ColPat-656	2.0	*Prunus dulcis* cv. Lauranne	C. Agustí-Brisach & A. Trapero	09/13/2018	MT304005	-	MT311985	-
Diaporthella corylina	**CBS 121124**	-	*Corylus* sp.	L.N. Vassiljeva	nd	KC343004	-	KC343730	-
Valsa germanica	**CBS 195.42**	-	nd	nd	nd	KY051902	-	KX965151	-

[1] Sequences from GenBank used in the phylogenetic analysis indicated in bold type. T = Ex-type isolates; AR, DP: Isolates in culture collection of Systematic Mycology and Microbiology Laboratory, USDA-ARS, Beltsville, Maryland, USA; CAA = A. Alves, Universidade de Aveiro, Portugal; CBS: Culture collection of the Centraalbureau voor Schimmelcultures, Fungal Biodiversity Centre, Utrecht, The Netherlands; CFCC = China Forestry Culture Collection Center; CMW = Culture collection of the Forestry and Agricultural Biotechnology Institute (FABI), University of Pretoria, Pretoria, South Africa; ColPat = 'Colección Patología', Department of Agronomy, University of Cordoba, Spain; KARE= Collections of the Department of Plant Pathology at the Kearney Agricultural Research and Extension Centre of the University of California, Parlier, CA.; PD: Plant Protection Service, Wageningen, The Netherlands; STE-U = University of Stellenbosch, South Africa. [2] The consistency of isolation (%) of each isolate was calculated as the number of positive attempts of isolation (wood pieces) of a given fungus divided by the total attempts of isolation in the whole of the experiment [Consistency of isolation = (N°. of positive wood pieces /168) × 100; where 168 is the total attempts of isolation (wood pieces) per category of disease severity in the whole of the experiment obtained as follow: 21 wood pieces per tree × 2 trees of each disease category × 2 years of evaluation × 2 sampling times per year]. [3] Collection date: month/day/year; n/d: non-determined. [4] ITS = internal transcribed spacer, TUB = β-tubulin-2 gene regions, EF = translation elongation 1-α, GAPDH = 200-bp intron of the glyceraldehyde-3-phosphate dehydrogenase.

2.3. Molecular Identification of Isolated Fungi

For all Datasets, the topology obtained by Maximum Parsimony (MP) was confirmed with those obtained by BI analysis. The model used in BI analysis, and the gene boundaries, the number of total characters (T), parsimony-informative characters (PI), parsimony-uninformative characters (PNI) and conserved sites (C) processed in each maximum parsimony analysis, as well as TL, consistency index (CI), retention index (RI), homoplasy index (HI) and rescaled consistency index (RC) values obtained from the one most parsimonies trees in each Dataset are shown in Table 2.

Botryosphaeriaceae analyses (*Dataset I-A*). Most of the Botryosphaeriaceae isolates (13 out of 17 isolates) were grouped in a well-supported clade with GenBank reference sequences of *B. dothidea* [bootstrap support (BS; %)/Bayesian posterior probability (PP):100/1.00]. The remaining isolates were identified as *N. mediterraneum* (ColPat-605 and ColPat-799; BS/PP:77–83/0.99–0.88), *N. parvum* (ColPat-608; BS/PP:99/1.00), and *Neoscytalidium dimidiatum* (Penz.) Crous and Slippers (ColPat-792) (BS/PP:100/1.00) (Figure 5a). To confirm the identification of this last isolate, an additional phylogeny was conducted by means the combined alignment of ITS and EF loci, including reference isolates of *Neoscytalidium novaehollandiae* Pavlic, T.I. Burgess and M.J. Wingf. (*Dataset I-B*). The MP analyses showed nine most parsimonious, and one of those is shown in Figure 5b.

Diaporthaceae analyses (*Dataset II*). Our isolates clustered in two well-supported clades with reference sequences of *Dia. neotheicola* (ColPat-762 and ColPat-763; BS/PP:98/1.00) and *Dia. rhusicola* (ColPat-606; BS/PP:99/1.00) (Figure 6).

Tympanidaceae analysis (*Dataset III*). All the isolates belonging to Tympanidaceae clustered together in a well-supported clade with GenBank reference sequence of *Collophorina hispanica* (=*Collophora hispanica*; BS/PP:100/1.00) (Figure 7).

Valsaceae analysis (*Dataset IV*). Among the three isolates belonging to Valsaceae included in this study, only one (ColPat-604) was grouped in a well-supported clade with a GenBank reference sequence of *Cytospora cedri* Syd., P. Syd. and E.J. Butler (BS/PP:100/1.00). However, it was not possible to distinguish the remaining two isolates (ColPat-609 and ColPat-656) at the species level into the genus *Cytospora*, and they were identified as *Cytospora* sp. 1 (ColPat-609) and *Cytospora* sp. 2 (ColPat-656) (Figure 8).

Table 2. Number of taxa, genes and statistical information of the different analyses performed in this study: Bayesian (IB) and Maximum Parsimony Analyses (MP).

Dataset/Phylogenetic Analyses	Number of Taxa		Gene		Bayesian Analyses	Maximum Parsimony Analyses									
						Characters [2]				MP Tree [3]					
	In Study	GenBank (incl.outgroups)	Combination	Boundaries	Best Fit Model [1]	T	PI	PNI	C	N°.Trees	TL	CI	RI	HI	RC
I-A/Botryosphaeriaceae	17	43	ITS/ TUB	1-503/ 504-921	K2+G/ T92+G	733	119	133	481	4	375	0.677	0.940	0.323	0.636
I_B/*Neoscytalidium dimidiatum*	1	8	ITS/ EF	1-490/ 491-711	T92/ K2	684	9	244	431	9	258	0.857	0.833	0.143	0.714
II/Diaporthaceae	3	30	EF/TUB/ ITS	1-254/ 255-739/ 740-1227	K2+G/ T92+I/ K2+G+I	993	301	124	568	1	1046	0.551	0.769	0.449	0.424
III/Tympanidaceae	7	12	ITS/ EF/ GAPDH	1-482/ 483-667/ 668-805	K2+G/ K2+I/ K2+I	719	100	143	476	10	334	0.850	0.919	0.150	0.782
IV/Valsaceae	3	32	ITS/ EF/	1-557/ 558-821	K2+G+I/ HK4+G	613	171	73	369	1	743	0.469	0.742	0.531	0.348

[1] Best fit nucleotide substitution models determined by MEGA v. 7.0, used for each gene partition to perform Bayesian Inference analyses using MrBayes v.3.2.6. [2] Numbers of total characters (positions) in the final dataset (T), parsimony-informative characters (PI), parsimony-uninformative characters (PNI) and conserved sites (C), processed in each analysis; all positions containing gaps and missing data were eliminated. [3] Total of equally most parsimonious trees obtained for each MP analyses: N° of Tree, tree length (TL), consistency index (CI), retention index (RI), homoplasy index (HI) and rescaled consistency index (RC).

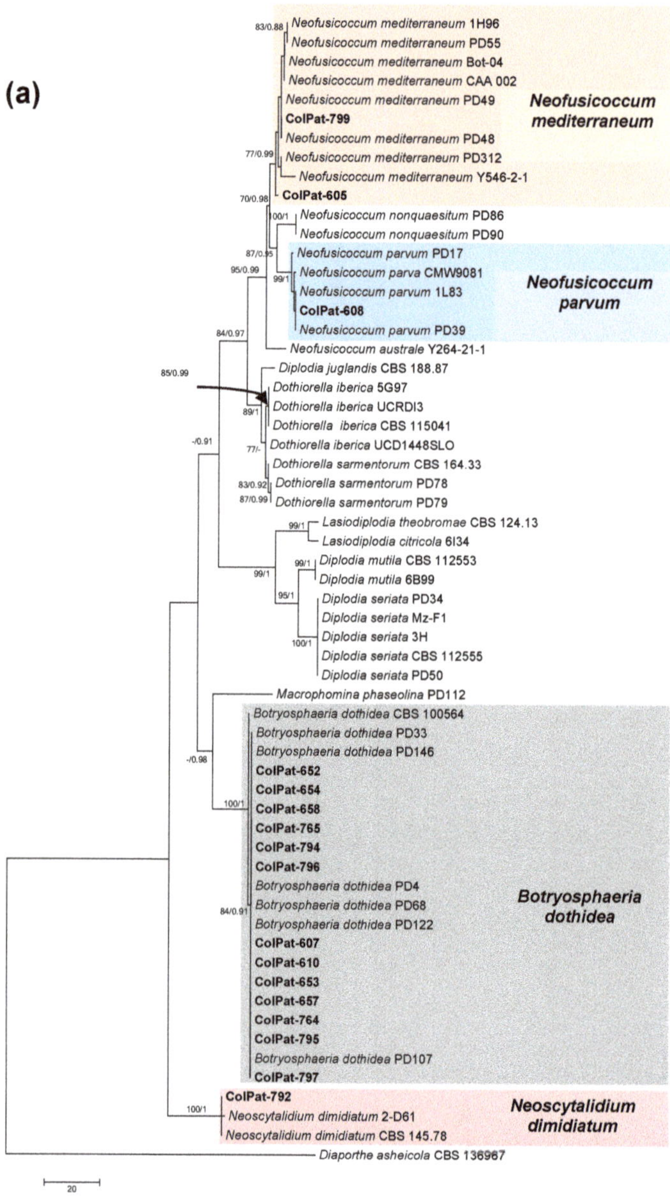

Figure 5. *Cont.*

(b)

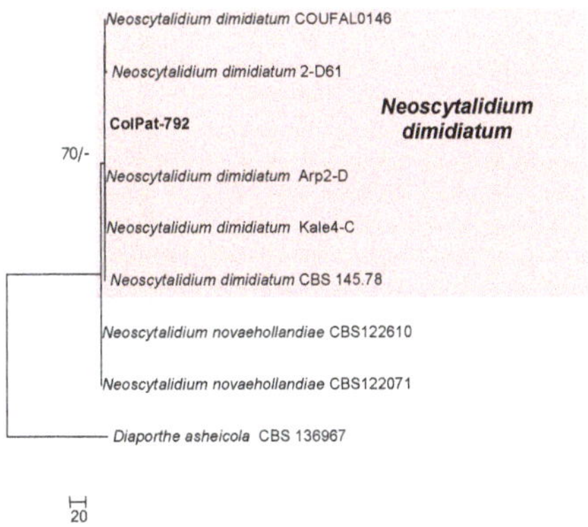

Figure 5. (**a**) The first of the four most parsimonious trees (TL = 375; CI = 0.677; RI = 0.940; HI = 0.323; RC = 0.636) obtained by Maximum Parsimony (MP) analyses of combined ITS+TUB sequence alignment of species belonging to Botryosphaeriaceae; (**b**) One of the nine MP trees (TL = 258; CI = 0.854; RI = 0.833; HI = 0.143; RC = 0.714) obtained using the combined ITS+TUB+EF sequence alignment of species belonging to *Neoscytalidium*. Bootstrap support values [MP, >70%] and Bayesian posterior probabilities [PP, >0.8] are shown at the nodes. *Diaporthe asheicola* L. Lombard & Crous CBS 136967 was used as the outgroup. Studied isolates in bolt.

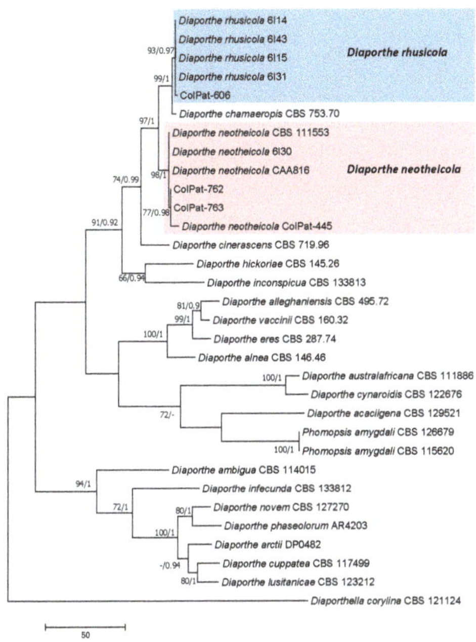

Figure 6. The most parsimonious tree (TL = 1046; CI = 0.551; RI = 0.769; HI = 0.449; RC = 0.424) obtained by Maximum Parsimony analyses of the combined EF+TUB+ITS sequence alignment of species belonging to Diaporthaceae. Bootstrap support values [MP, >70%] and Bayesian posterior probabilities [PP, >0.8] are shown at the nodes. *Diaporthella corylina* Lar.N. Vassiljeva CBS 121124 was used as the outgroup. Studied isolates in bolt.

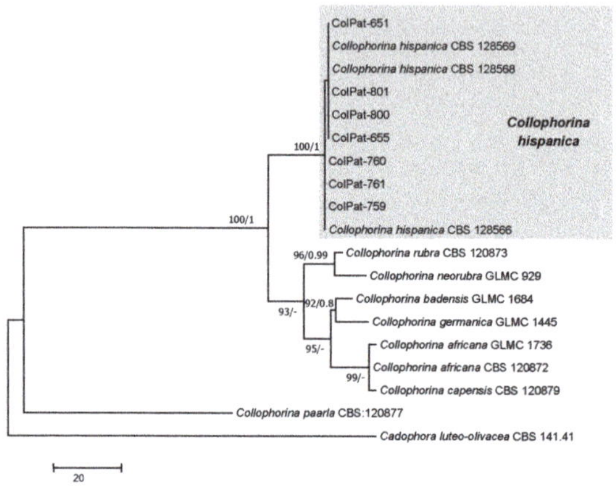

Figure 7. The first of the 10 most parsimonious tree (TL = 334; CI = 0.850; RI = 0.919; HI = 0.150; RC = 0.782) obtained by Maximum Parsimony analyses of the combined ITS + EF + GADPH sequence alignment of species belonging to Tympanidaceae. Bootstrap support values [MP, >70%] and Bayesian posterior probabilities [PP, >0.8] are shown at the nodes. *Cadophora luteo-olivacea* (J.F.H. Beyma) T.C. Harr. and McNew CBS 141.41 was used as the outgroup. Studied isolates in bolt.

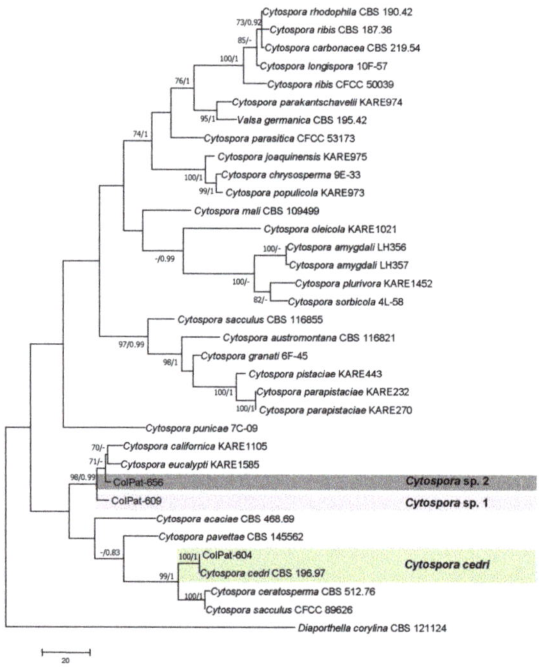

Figure 8. The most parsimonious tree (TL = 743; CI = 0.469; RI = 0.742; HI = 0.531; RC = 0.348) obtained by Maximum Parsimony analyses of the combined ITS+EF sequence alignment of species belonging to Valsaceae. Bootstrap support values [MP, >70%] and Bayesian posterior probabilities [PP, >0.8] are shown at the nodes. *Diaporthella corylina* CBS 121124 was used as the outgroup. Studied isolates in bolt.

3. Discussion

Studying whether the effect of abiotic factors, such as water stress enhances the incidence and development of branch dieback and decline syndromes on weakened trees, is essential in improving our understanding of the endophytic behaviour of fungi associated with this complex disease. In fact, to date, the role of the fungal trunk pathogens, causing tree decline, is still uncertain, since their aggressiveness could vary markedly depending on abiotic (i.e., ecological, environmental and agronomical aspects) and/or biotic (i.e., plant-pathogen interactions) factors. Consequently, several authors consider that most of the fungal trunk pathogens are secondary or opportunistic, causing damage when biotic or abiotic circumstances occur [11,19,24,25].

The environmental and agronomic conditions regarding the availability of irrigation-water resources in southern Spain could be a limiting factor, enhancing the development of branch dieback and decline syndromes on fruit and nut crops. Therefore, we have evaluated the effect of water stress enhancing the disease development on weakened almond trees under natural field conditions. The first symptoms of branch dieback occurred in late-summer autumn 2017 in an eight-year old experimental field, subjected to four different irrigation treatments, since 2013. The symptoms included branch dieback, canker formation, internal wood discoloration and general decline (Figure 2d–g) were observed mainly in the experimental plots subjected to high water-stressed conditions (T2, T3). In fact, the DS progress was significantly higher in almond trees subjected to T3 than in those subjected to T0 after two consecutive years of periodic evaluations. Our results are in accordance with those previously obtained by several authors under the control conditions, which showed that stem cankers, developed by *B. dothidea* or *S. musiva* on water-stressed plants of peach, or *Populus*, respectively, were higher than those developed on non-water-stressed plants [22,23]. On the other hand, almond trees under full irrigation (T0) also showed minimum levels of dieback symptoms. Although, no internal wood discoloration was observed, several fungal species, such as *C. hispanica*, *Cy. cedri* and *N. mediterraneum* were isolated from those trees. These results reinforce the hypothesis that these fungi could cause latent infections in asymptomatic or lesser-symptomatic trees. In parallel, the conclusions obtained in this study should be considered to discard the high levels of irrigation water as potential abiotic factor associated with the prevalence of the disease in the newly established almond growing regions in southern Spain, as we initially hypothesized in the introduction. To the best of our knowledge, this is the first approach, which has demonstrated the endophytic behaviour of fungal trunk pathogens on weakened trees, subjected to water stress under natural field conditions.

Concerning the occurrence of fungal species associated with branch dieback and almond decline, the following seven species belonging to four different families were identified: Botryosphaeriaceae: *B. dothidea*, *N. mediterraneum*, *N. parvum* and *Neoscytalidum* sp.; Diaporthaceae: *Dia. neotheicola* and *Dia. rhusicola*; Tympanidaceae: *C. hispanica*; and, Valsaceae: *Cytospora cedri* and *Cytospora* spp. Among them, *B. dothidea*, *C. hispanica*, *Dia. neotheicola*, *N. mediterraneum* and *N. parvum* have been previously described associated with branch dieback and decline on weakened almond trees in Spain [7,8,26]. Moreover, the pathogenicity of most of these species has been previously demonstrated in almond trees in Spain [5,8,9,26]. On the other hand, *Dia. rhusicola* and species belonging to *Cytospora* and *Neoscytalidium* genera, are associated with branch dieback and tree decline in other nut crops, such as English walnut or pistachio [6,15–17], but to our knowledge, these species have not been previously reported in association with branch dieback and almond decline in Spain. However, their pathogenicity to almond should be demonstrated in the future to confirm they are canker pathogens of almond.

Botryosphaeriaceae were the most frequent isolated fungi and they also showed the highest consistency of isolation in the whole of the experiment, with *B. dothidea* being the species most frequently isolated. The differences in consistency of isolation of Botryosphaeriaceae fungi from weakened almonds can occur, depending on the scenario where the surveys are conducted, but in general, *B. dothidea* and *Neofusicoccum* species are usually the most frequent [9,25]. Likewise, according to the literature, our results also suggest that Botryosphaeriaceae spp. found on weakened almond trees are able to endanger the productivity and longevity of orchards in Spain, as well as in other countries [9].

Among Botryosphaeriaceae fungi, notice that *B. dothidea* has been reported worldwide causing canker diseases in a broad range of woody crops, including different *Prunus* spp. [6,7,27]. However, the role of this fungus as a trunk pathogen is still uncertain, given it has been reported as a latent pathogen of global importance for its endophytic behaviour in woody plant health [20]. In fact, studies conducted recently in southern Spain, which compared the pathogenicity of *B. dothidea* on inoculated detached and attached shoots of almond, the English walnut and pistachio, demonstrated that, in every case, the fungus is significantly higher aggressive on detached shoots than on attached [5,16,17]. It was confirmed that *B. dothidea* could remain latent in woody plants until trees become weakened as a consequence of different biotic and/or abiotic factors.

With respect to Diaporthaceae, *Dia. neotheicola* and *Dia. rhusicola* showed a low consistency of isolation and their occurrence was also low. These two species have been previously reported, associated with branch dieback and shoot blight of English walnut in California [28] and southern Spain [16], and the first one was also isolated from pistachio in southern Spain [17] and recently reported associated with twig cankers and shoot blight of almond in Spain [26]. Usually, Diaporthaceae species occur simultaneously with Botryosphaeriaceae in the same orchards, with Botryosphaeriaceae being always the most frequent [16,17,28]. In addition, studies conducted in California by Agustí-Brisach et al. [29] suggest that coinfections between Botryosphaeriaceae and Diaporthaceae species result in antagonistic interactions on infection and disease development on English walnut. But, *Dia. amygdali*, which is a common species associated with branch dieback and cankers of almond [5,7], was not found in this experiment.

It is interesting to note that, in this study, *C. hispanica* was the second most frequent species isolated from weakened almond trees after *B. dothidea*. Our results are in concordance with those found by Olmo et al. [8], who indicated that this slow-growing species is common in declined almond trees. However, it is usually excluded in the diagnosis process probably because its presence goes unnoticed, due to its slow growth [8].

Finally, *Cytospora* species were also isolated in low consistency from the trees of three different categories of severity. According to our results, *Cytospora* spp. have also been reported associated with canker diseases in weakened tree nuts (English walnut and pistachio) in southern Spain showing less frequency and aggressiveness, and often simultaneously with Botryosphaeriaceae and/or Diaporthaceae fungi [16,17]. However, several *Cytospora* spp. have been already reported in California as canker pathogens of several fruit and nut crops including *Prunus* spp. such as almond, apricot or peach [30].

The isolations made during this study suggest that most of these fungi can occur in both asymptomatic and symptomatic trees, but their frequency of isolation increases with increasing DS. However, the consistency of isolation was low for all the species in the whole of the experiment, showing the highest values in weakened almond trees, belonging to categories 3 and 4 (51 to 90% of final disease severity). This information reinforces the hypothesis that the fungal species associated with tree decline could have a major endophytic behaviour, and its aggressiveness is probably enhanced by abiotic factors, such as water stress on previously infected and weakened trees. In general, studies on the etiology of fungal trunk diseases describe a broad list of fungi associated with the disease, but do not usually considering the strict pathogenic behaviour of each. Nevertheless, this work reveals the need to go on elucidating the role of biotic and abiotic factors, enhancing the infection of fungal trunk pathogens and disease development on woody crops towards generating important knowledge on their life cycle. Therefore, focus the research on such relevant challenge will provide a better understanding of the biology of fungi associated with tree decline syndrome. It will build a strong foundation for developing effective management approaches against the disease, by taking into consideration the optimum water management.

4. Materials and Methods

4.1. Experimental Field, Irrigation Treatments and Experimental Design

The present study was conducted in a nine- to 10-years-old experimental field of almond cv. Guara grafted onto GF-677 rootstock (5.5 ha; 7 × 6 tree spacing; 238 trees/ha) belonging to the Andalusian Institute of Agricultural and Fisheries Research and Training (IFAPA in Spanish) Centre 'Alameda del Obispo' located in Córdoba (Andalusia region, Spain; 37.8°N, 4.8°W), whose soil was classified as a Typic Xerofluvent of sandy loam texture and exceeds 1.5 m depth. The climate of this region is the typical Mediterranean climate, characterized by hot and dry summers (T^a Av. 27.0 °C; T^a min Av. = 19.3 °C; T^a max Av. = 36.7 °C), mild winters, with 600 mm of annual rainfall average, concentrated from October to April (T^a Av. 13.0 °C; T^a min Av. = 7.9 °C; T^a max Av. = 21.1 °C).

The experimental field used in this study was established in February 2009. Pruning for tree formation was done along the two first years, and then there never were pruning interventions. Control management strategies to prevent pest (Acetamiprid 20%; Deltametrine 2.5%) and diseases (Boscalid 26.7% + Pyraclostrobin 6.7%; Thiram 50%; Tebuconazole 50% + Tryfloxistrobin 25%) were done according to a treatment-calendar based on the weather conditions which could favour the typical almond pest and diseases of this area. Weeds were controlled by mowing and herbicide applications (Glyphosate 36%; Oxifluorphen 24%). Mineral fertilization was calculated and applied following the recommendations of the California Fertilization Guidelines for Almonds (https://apps1.cdfa.ca.gov/FertilizerResearch/docs/Almonds.html). Therefore, this experimental field is representative of the edapho-climatic characteristics and the standard crop management of the new almond plantings in Andalusia [31].

The irrigation system was formed by two pressure compensating drip irrigation laterals, spaced 1 m from the tree rows, and all the trees were fully irrigated until the irrigation treatments began, as described below. From April 2013 to October 2019 (before and along this present study), the experimental field was subjected to four irrigation treatments: (*i*) *Control* (T0): The trees were irrigated to cover their full water requirements (ET), which was calculated using the relationship between ground cover (GC) and a transpiration coefficient, proposed by Espadafor et al. [32]. An additional 15% of that quantity was supplied to account for the evaporation from emitter wet surfaces under the trees using Bonachela et al. [33] model; (*ii*) *Moderate Regulated Deficit Irrigation* (T1): seasonal irrigation was 65% of T0, but the deficit was mainly concentrated during kernel filling stage, where almonds are less affected by water stress. Specifically, the irrigation supplied was: 70% of T0 in spring; 40% in kernel filling stage (from middle-July to harvest period in middle-August) and 100% in the postharvest period; (*iii*) *Moderate Sustained Deficit Irrigation* (T2): This treatment consisted of 65% of T0 steadily throughout the irrigation season. In total, a similar amount of irrigation water to that of the T1 was supplied; and (*iv*) *Severe Regulated Deficit Irrigation* (T3): This treatment received 30% of seasonal irrigation in relation to T0, following a similar water allocation strategy as in T1. Irrigation was 40% of T0 in spring and after harvest, and only 15% during the kernel-filling stage.

A randomized complete block design, with four replicated blocks, each consisting of four irrigation treatments, was used in this experiment. There were 16 trees per treatment plot, from which the four central trees were used for experimental measurements and the remaining 12 ones served as guard line. Therefore, the experiment included a total of 256 trees (4 blocks × 4 irrigation treatments × 16 trees per elementary plot), from which 80 trees were evaluated. Weather data were collected from an automated weather station located at 300 m apart from the orchard.

4.2. Disease Severity Assessment and Data Analysis

In the summer-autumn of 2017, the experimental trees subjected to the different irrigation treatments started to weaken, showing the first symptoms of branch dieback and general decline. Since then, the progress of this syndrome was monitored over time by periodic assessments from June 2018 to September 2019. Two assessments per year were conducted, with a total of four assessments.

DS was assessed based on the estimation of the percentage of the affected surface of the tree canopy using a 0–5 rating scale. Each scale value was referred as 'category' of severity (six categories in total) for further purposes of this study (*see* Section 4.3. *Sampling and fungal isolation*). The values of this scale have a linear relationship with the percentage of affected tissues (leaves and shoots) in order to satisfy the homogeneity of variances and normality for suitable statistical analysis [34]. The equivalences between the values of the scale and the percentage of affected surface of the tree canopy are: 0 = 0%, 1 = < 25%, 2 = 25–50%, 3 = 51–75%, 4 = 75–90%, 5 = ≥ 90%. The DS was assessed in June and September of each year (four evaluations in total), before, and after, harvest, respectively, and all the blocks of the whole of the experiment were evaluated each time. The relative area under the disease progress curve (RAUDPC) was calculated by the trapezoidal integration method from the disease severity values over time [35].

The dependent variables 'final disease severity (%) and RAUDPC (%) were subjected to ANOVA to determine the differences in DS between irrigation treatments. Data were tested for normality and homogeneity of variances, and logarithmically transformed where necessary. Treatment means for the global analyses were compared using Fisher's protected LSD test [36]. All the data were analysed using Statistix 10 [37].

4.3. *Sampling and Fungal Isolation*

Two almond trees per each category of disease severity (12 trees in total) were selected to temporarily monitor the fungi isolated from affected tissues. From each tree, branches and shoots showing dieback and cankers were collected in each disease assessment time. Samples were kept at 4 °C until being processed in the laboratory.

For fungal isolation, the outer bark of affected wood samples was removed, and were subsequently washed under running tap water. Little wood pieces were collected from the margin of the affected area of symptomatic samples or randomly selected across the wood section, in the case of asymptomatic samples (Category 0). All the wood pieces were surface disinfected by dipping into a 10% (vol/vol) solution of commercial bleach (Cl at 50 g l^{-1}) for 2 min. Subsequently, they were air dried on sterile filter paper and plated onto malt extract agar (MEA) [20 g of MEA (Merck KGaA, Darrmstadt, Germany), 20 g of agar (Rokoagar AF LAB, ROKO Industries, Llanera, Asturias, Spain; 1 l of sterile distilled water (SDW)] supplemented with 0.5 g l^{-1} of streptomycin sulphate (Sigma-Aldrich, St. Louis, MO, USA) (MEAS). From each category of severity, a total of 168 wood pieces, obtained from the margin of the affected tissues, were plated on Petri dishes for fungal isolation [attempts of isolation (wood pieces) onto Petri dishes: 2 trees per disease category × 3 Petri dishes per tree × 7 wood pieces per Petri dish × 4 sampling moments]. Petri dishes were incubated for 5 to 14 days at 25 °C in darkness and they were examined daily removing the isolation attempts (wood pieces) contaminated by saprophytes (i.e., *Alternaria* spp. *Penicillium* spp., etc.) by cutting the agar up to 1-cm-radio beyond the colony margin using a sterile scalpel. It was necessary to prevent the contamination of the whole agar surface of the Petri dishes before our target fungi developed onto MEAS, due to the low mycelial growth rate of some expected fungi for isolation.

When the colonies were large enough to be examined, hyphal tips from the margin of the fungal colonies were transferred to the potato dextrose agar (PDA; Difco Laboratories®, Detroit) in order to obtain pure cultures. They were all incubated as previously described, and were grouped into four fungal groups (families) according to colony colour and mycelial growth development of each: Botryosphaeriaceae (light to dark grey, fast growing mycelium), *Collophorina* spp. (reddish to beige, very slow growing mycelia), *Cytospora* (beige to olive grey, middle-slow growing mycelium), and *Diaporthe* (beige-white, middle-fast growing mycelium). These preliminary morphological observations were helpful in selecting 30 representative isolates that were subsequently identified by molecular tools (Table 1). All the isolates were single-spored by a serial dilution method and they were registered and maintained at 4 °C in darkness (Fungal collection of the Department of Agronomy, University of Cordoba, Spain).

4.4. Assessment of Consistency and Frequency of Isolated Fungi

The consistency of isolation (%) of each isolate was calculated as the number of positive attempts of isolation (wood pieces) of a given fungus divided by the total attempts of isolation in the whole of the experiment [Consistency of isolation = (N°. of positive wood pieces /168) × 100; where 168 is the total attempts of isolation (wood pieces) per category of disease severity in the whole of the experiment obtained as follow; 21 wood pieces per tree × 2 trees of each disease category × 2 years of evaluation × 2 sampling times per year]. The frequency of isolation (%) of each fungal species was estimated as the ratio between the number of trees from which each species was isolated and the total of sampled trees (12 trees). Additionally, the *in planta* abundance (biomass) of the studied fungal species and the severity of the dieback symptoms (category) was also compared by Pearson's linear correlation ($n = 5$; data from Category 5 was excluded since only saprophytes were isolates) using Statistix 10 [37].

4.5. Molecular Identification of Isolated Fungi

4.5.1. DNA Extraction

Mycelial tissues of the 30 isolated fungi (Table 1) previously grown on PDA were ground by means the FastPrep®-24 grinder machine (MP Biomedicals, Santa Ana, CA, USA). Subsequently, genomic DNA was extracted using the E.Z.N.A® Fungal DNA Kit (OMEGA BioTek, Norcross, GA, USA). A MaestroNano® spectrophotometer (MaestroGen, Taiwan) was used to determine the concentration and purity of the extracted DNA.

4.5.2. PCR Analysis and Sequencing

The 5.8S nuclear ribosomal gene with two flanking internal transcribed spacers (ITS) was amplified for all the 30 isolated fungi. Subsequently, part of the beta-tubulin (TUB) gene, part of the translation elongation factor 1-alpha (EF) and/or a 200-bp intron of the glyceraldehyde-3-phosphate dehydrogenase (GAPDH) were amplified for the different isolates according to the necessities to complete the further phylogenetic analysis. To this end, the protocols described in the literature for each family and genus were followed to identify our fugal isolates (Tables 1 and 3). The PCRs were performed in a total volume of 25 µl [20 ng of genomic DNA, 5 µl of 5× My Taq Reaction Buffer and 0.13 µl of My Taq DNA Polymerase (Bioline)]. Additionally, 0.4 or 0.2 µM (each) primer was added for the ITS; or for the TUB, EF, and GAPDH PCRs, respectively. A negative control was included in all PCRs using ultrapure water instead of DNA. Primer pairs and PCR cycling programs used to amplify each locus are shown in Table 3. Ultrapure water was used instead of DNA as negative control. A MyCycler™ Thermal Cycler (BIO-RAD) was used to conduct the PCRs.

Electrophoresis of the amplification products from PCR was conducted on a 1.5% (w/v) agarose gel stained with RedSafe™ (Intron Biotechnology). A 100-bp DNA molecular weight marker (Ladder-GTP, gTPbio) was used, and the agarose gel was visualized under UV. Finally, the PCR products were purified by means the MEGAquick-spin™ Total Fragment DNA Purification kit (INTRON Biotechnology). The resulting amplicons were sequenced in both directions [Central Service Support Research (SCAI) of the UCO (Spain)].

4.5.3. Phylogenetic Analysis

Consensus sequences from DNA sequences generated with forward and reverse primers were obtained with the SeqMan software (DNASTART Lasergen SeqMan® v. 7.0.0, Madison, WI, USA). They were compiled into a single FASTA file format. Subsequently, they were BLAST searched in GenBank (http://www.ncbi.nlm.nih.gov/genbank/) to determine the close related species for each sequence.

Firstly, a neighbor-joining (NJ) analysis was performed individually for each locus. It was useful to determine whether the sequence datasets were congruent and combinable (*data not shown*). To this end, the maximum composite likelihood method with 2000 bootstrap replications was used. Genetic distances were calculated using the Kimura 2-parameter mode and tree topologies of 70% reciprocal bootstrap generated individually for each locus were compared visually. The data of different loci were combined into single concatenated datasets when no supported nodes were in conflict.

Independent phylogenetic analyses were conducted for the isolates of each fungal group (family), previously established according to their main morphological characteristics and Blast analysis. The combined alignment of the ITS and TUB loci was analysed in order to infer the phylogeny of isolates belonging to Botryosphaeriaceae (*Dataset I-A*). Additionally, a little phylogeny combining ITS and EF loci was also conducted into Botryosphaeriaceae group to confirm the identification of *Neoscytalidium* sp. Isolate (*Dataset I-B*). In the case of Diaporthaceae, the combined alignment of the EF, TUB and ITS loci was conducted (*Dataset II*). Isolates belonging to Tympanidaceae were identified by means the combined alignment of the ITS, EF and GAPDH loci (*Dataset III*). Finally, the combined alignment of the ITS and EF loci was performed to infer the phylogeny of the isolates belonging to Valsaceae (*Dataset IV*). For each multilocus alignment, data of the reference taxa (including outgroup) downloaded from GenBank and the number of the taxa included in this study are shown in Table 1, and Table 2, respectively.

The reference Genbank taxa were selected based on their high similarity with our query sequences using MegaBLAST [38] and they were added and aligned with our sequences by Clustal W. Maximum parsimony (MP) analyses were conducted using MEGA version 7.0 software [38], and they were performed by means the Tree-Bisection-Regrafting (TBR) algorithm with search level one. The initial trees were obtained by the random addition of sequences (10 replicates). The gaps and missing data were treated as complete deletions. A total of 1,000 bootstrap replications were done to ensure the robustness of the topology [39]. Tree length (TL), consistency index (CI), retention index (RI), homoplasy index (HI) and rescaled consistency index (RC) were calculated for each resulting MP tree.

Additionally, Markov chain Monte Carlo (MCMC) methods were used to perform Bayesian inference (BI) analyses by means the software MrBayes v.3.2.6 [40]. They were useful for estimating the posterior probability of trees. The best fit models of the evolution used for each gene partition were also determined by MEGA v. 7.0 [38]. Two analyses with four MCMC chains each were run simultaneously for 1×10^7 generations, starting from a random tree topology. The trees were sampled every 100 generations, and the "temperature" parameter was set to 0.2. The first 25% of the saved trees was discarded as the burn-in phase of the analysis. The sequences derived in this study were uploaded at GenBank (Table 1).

Table 3. Primer pairs and PCR conditions used for the amplification of the genes included in this study.

Gene [1]	Primer Pairs	PCR Cycling Program (T^a-Time)						References
		Initial Denaturation	Amplification				Final extension	
			N° of Cycles	Denaturation	Annealing	Extension		
ITS	ITS4/ITS5	95 °C-3 min	35	95 °C-30 s	48 °C-30 s	72 °C-45 s	72 °C-10 min	[41]
TUB	Bt2a/Bt2b	95 °C-3 min	35	95 °C-15 s	55 °C-15 s	72 °C-45 s	72 °C-7 min	[42]
EF	EF1-728F/EF1-986R	95 °C-3 min	35	95 °C-30 s	50 °C-30 s	72 °C-45 s	72 °C-10 min	[43]
GAPDH	GDF1/GDR1	94 °C-5 min	40	95 °C-15 s	52 °C-15 s	72 °C-10 s	72 °C-7 min	[44]

[1] ITS = internal transcribed spacer; TUB = β-tubulin; EF = translation elongation factor 1-α; GAPDH = a 200-bp intron of the glyceraldehyde-3-phosphate dehydrogenase.

Author Contributions: Field evaluations and sampling, fungal isolation and morphological identification, fungal collection, writing-original draft preparation, editing, C.A.-B.; field evaluations, monitoring irrigation treatments over time, writing-review and editing, D.M.; phylogenetic investigation, formal analyses, and phylogenetic visualization/data presentation, GenBank sequences submission, writing review and editing M.d.C.R.; experimental field management, funding acquisition, project administration, I.J.L.; conceived the idea of the research, designed the scientific experiments and methodology, writing-review and editing, supervision, project administration, F.O. and A.T. All authors have read and agreed to the published version of the manuscript.

Funding: This research was partially funded by Spanish Ministry of Science, Innovation and Universities (MICINN), project AGL2015-66141-R, co-financed by the European Union FEDER Funds, and by European Regional Development Fund, project Innova-Clima (PR.AVA.AVA2019.051). C.A.-B. and D.M. are the holders of a 'Juan de la Cierva-Incorporación' postdoctoral fellowship and a 'Formación de Personal Investigador' pre-doctoral fellowship, respectively, both funded by MICINN.

Conflicts of Interest: The authors declare no conflict of interest.

References

1. Food and Agricultural Organisation of the United Nations (FAO) Website. 2017. Available online: http://www.fao.org/faostat/en/#data/QC (accessed on 21 April 2020).
2. Ministry of Agriculture, Fisheries and Food (MAPA; Spain). 2018. Available online: https://www.mapa.gob.es/es/estadistica/temas/estadisticas-agrarias/agricultura/superficies-producciones-anuales-cultivos/ (accessed on 21 April 2020).
3. Ollero-Lara, A.; Agustí-Brisach, C.; Lovera, M.; Roca, L.F.; Arquero, O.; Trapero, A. Field susceptibility of almond cultivars to the four most common aerial fungal diseases in southern Spain. *Crop Prot.* **2019**, *121*, 18–27. [CrossRef]
4. López-Moral, A.; Raya-Ortega, M.C.; Agustí-Brisach, C.; Roca, L.F.; Lovera, M.; Luque, F.; Arquero, O.; Trapero, A. Morphological, pathogenic and molecular characterization of *Colletotrichum acutatum* isolates causing almond anthracnose in Spain. *Plant Dis.* **2017**, *101*, 2034–2045. [CrossRef]
5. Agustí-Brisach, C.; Roca, L.F.; Antón-Domínguez, B.I.; López-Moral, A.; Raya, M.C.; Lovera, M.; Arquero, O.; Trapero, A. Decaimiento del almendro en plantaciones jóvenes en Andalucía. *Vida Rural* **2020**, *476*, 36–43.
6. Teviotdale, B.L.; Michailides, T.J.; Pscheidt, J.W. *Compendium of Nut Crop Diseases in Temperate Zones*; APS Press: Saint Paul, MN, USA, 2002.
7. Gramaje, D.; Agustí-Brisach, C.; Pérez-Sierra, A.; Moralejo, E.; Olmo, D.; Mostert, L.; Damm, U.; Armengol, J. Fungal trunk pathogens associated with wood decay of almond trees on Mallorca (Spain). *Persoonia* **2012**, *28*, 1–13. [CrossRef]
8. Olmo, D.; Armengol, J.; León, M.; Gramaje, D. Pathogenicity testing of lesser-known fungal trunk pathogens associated with wood decay of almond trees. *Eur. J. Plant Pathol.* **2015**, *143*, 607–611. [CrossRef]
9. Olmo, D.; Armengol, J.; León, M.; Gramaje, D. Characterization and pathogenicity of Botryosphaeriaceae species isolated from almond trees on the island of Mallorca (Spain). *Plant Dis.* **2016**, *100*, 2483–2491. [CrossRef]
10. Úrbez-Torres, J.R. The status of Botryosphaeriaceae species infecting grapevines. *Phytopathol. Mediterr.* **2011**, *50*, S5–S45.
11. Bertsch, C.; Ramírez-Suero, M.; Magnin-Robert, M.; Larignon, P.; Chong, J.; Abou-Mansour, E.; Spagnolo, A.; Clément, C.; Fontaine, F. Grapevine trunk disease: Complex and still poorly understood. *Plant Pathol.* **2013**, *62*, 243–265. [CrossRef]
12. Gramaje, D.; Úrbez-Torres, J.R.; Sosnowski, M.R. Managing grapevine trunk diseases with respect to etiology and epidemiology: Current strategy and future prospects. *Plant Dis.* **2018**, *102*, 12–39. [CrossRef]
13. Úrbez-Torres, J.R.; Peduto, F.; Vossen, P.M.; Krueger, W.H.; Gubler, W.D. Olive twig and branch dieback: Etiology, incidence, and distribution in California. *Plant Dis.* **2013**, *97*, 231–244. [CrossRef]
14. Moral, J.; Agustí-Brisach, C.; Pérez Rodríguez, M.; Xavíer, C.; Raya, M.C.; Rhouma, A.; Trapero, A. Identification of fungal species associated with branch dieback of olive and resistance of table cultivars to *Neofusicoccum mediterraneum* and *Botryosphaeria dothidea*. *Plant Dis.* **2017**, *101*, 306–316. [CrossRef]
15. Moral, J.; Morgan, D.; Trapero, A.; Michailides, T.J. Ecology and epidemiology of nut crops and olive diseases caused by Botryosphaeriaceae fungi. *Plant Dis.* **2019**, *103*, 1809–1827. [CrossRef]

16. López-Moral, A.; Lovera, M.; Raya, M.C.; Cortés-Cosano, N.; Arquero, O.; Trapero, A.; Agustí-Brisach, C. Etiology of branch dieback and shoot blight of English walnut caused by Botryosphaeriaceae and *Diaporthe* fungi in southern Spain. *Plant Dis.* **2020**, *104*, 533–550. [CrossRef]
17. López-Moral, A.; Raya, M.C.; Ruíz-Blancas, C.; Medialdea, I.; Lovera, M.; Arquero, O.; Trapero, A.; Agustí-Brisach, C. Aetiology of branch dieback, panicle and shoot blight of pistachio associated with fungal trunk pathogens in southern Spain. *Plant Pathol.* **2020**. [CrossRef]
18. Phillips, A.J.L.; Alves, A.; Abdollahzadeh, J.; Slippers, B.; Wingfield, M.J.; Groenewald, J.Z.; Crous, P.W. The Botryosphaeriaceae: Genera and species known from culture. *Stud. Mycol.* **2013**, *76*, 51–167. [CrossRef] [PubMed]
19. Schoeneweiss, D.F. The role of environmental stress in diseases of woody plants. *Plant Dis.* **1981**, *65*, 308–314. [CrossRef]
20. Marsberg, A.; Kemler, M.; Jami, F.; Nagel, J.H.; Postma-Smidt, A.; Naidoo, S.; Wingfield, M.J.; Crous, P.W.; Spataforas, J.W.; Hesse, C.N.; et al. *Botryosphaeria dothidea*: A latent pathogen of global importance to woody plant health. *Mol. Plant Pathol.* **2017**, *18*, 477–488. [CrossRef]
21. Crist, C.R.; Schoeneweiss, D.F. The influence of controlled stresses on susceptibility of European white birch stems to attack by *Botryosphaeria dothidea*. *Phytopathology* **1975**, *65*, 369–373.
22. Pusey, P.L. Influence of water stress on susceptibility of nonwounded peach bark to *Botryosphaeria dothidea*. *Plant Dis.* **1989**, *73*, 1000–1003. [CrossRef]
23. Maxwell, D.L.; Kruger, E.L.; Stanosz, G.R. Effects of water stress on colonization of poplar stems and excised leaf disks by *Septoria musiva*. *Phytopathology* **1997**, *87*, 381–388. [CrossRef]
24. Hofstetter, V.; Buyck, B.; Croll, D.; Viret, O.; Couloux, A.; Gindro, K. What if Esca disease of grapevine were not a fungal disease? *Fungal Divers.* **2012**, *54*, 51–67. [CrossRef]
25. Gramaje, D.; Baumgartner, K.; Halleen, F.; Mostert, L.; Sosnowski, M.R.; Úrbez-Torres, J.R.; Armengol, J. Fungal trunk diseases: A problem beyond grapevines? *Plant Pathol.* **2016**, *65*, 355–356. [CrossRef]
26. León, M.; Berbegal, M.; Rodríguez-Reina, J.M.; Elena, G.; Abad-Campos, P.; Ramón-Albalat, A.; Olmo, D.; Vicent, A.; Luque, J.; Miarnau, X.; et al. Identification and characterization of *Diaporthe* spp. associated with twig cankers and shoot blight of almonds in Spain. *Agronomy* **2020**, *10*, 1062.
27. Sutton, B.C. *The Coelomycetes. Fungi Imperfecti with Pycnidia, Acervuli and Stromata*; Commonwealth Mycological Institute: Kew, UK, 1980.
28. Chen, S.F.; Morgan, D.P.; Hasey, J.K.; Anderson, K.; Michailides, T.J. Phylogeny, morphology, distribution, and pathogenicity of Botryosphaeriaceae and Diaporthaceae from English walnut in California. *Plant Dis.* **2014**, *98*, 636–652. [CrossRef]
29. Agustí-Brisach, C.; Moral, J.; Felts, D.; Trapero, A.; Michailides, T.J. Interaction between *Diaporthe rhusicola* and *Neofusicoccum mediterraneum* causing branch dieback and fruit blight of English walnut in California, and effect of pruning wounds to the infection. *Plant Dis.* **2019**, *103*, 1196–1205. [CrossRef]
30. Lawrence, D.P.; Holland, L.A.; Nouri, M.T.; Travadon, R.; Abramians, A.; Michailides, T.J.; Trouillas, F.P. Molecular phylogeny of *Cytospora* species associated with canker diseases of fruit and nut crops in California, with the descriptions of ten new species and one new combination. *IMA Fungus* **2018**, *9*, 333–369. [CrossRef]
31. Arquero, O.; Belmonte, A.; Casado, B.; Cruz-Blanco, M.; Espadafor, M.; Fernández, J.L.; Gallego, J.C.; García, A.; Lorite, I.J.; Lovera, M.; et al. *Manual del almendro. Consejería de Agricultura, Pesca y Desarrollo Rural*; Ediciones Junta de Andalucía: Sevilla, Spain, 2013.
32. Espadafor, M.; Orgaz, F.; Testi, L.; Lorite, I.J.; Villalobos, F.J. Transpiration of young almond trees in relation to intercepted radiation. *Irrig. Sci.* **2001**, *33*, 265–275. [CrossRef]
33. Bonachela, S.; Orgaz, F.; Villalobos, F.J.; Fereres, E. Soil evaporation from drip-irrigated olive orchards. *Irrig. Sci.* **2001**, *20*, 65–71. [CrossRef]
34. Moral, J.; Trapero, A. Assessing the susceptibility of olive cultivars to anthracnose caused by *Colletotrichum acutatum*. *Plant Dis.* **2009**, *93*, 1028–1036. [CrossRef]
35. Campbell, C.L.; Madden, L.V. *Introduction to Plant Disease Epidemiology*; John Wiley & Sons, Ltd.: New York, NY, USA, 1990.
36. Steel, R.G.D.; Torrie, J.H. *Bioestadística*, 2nd ed.; McGraw-Hill: Bogotá, Colombia, 1985.
37. *Statistix10. User's Manual*; Analytical Software: Tallahassee, FL, USA, 2013.
38. Kumar, S.; Stecher, G.; Tamura, K. MEGA7: Molecular Evolutionary Genetics Analysis version 7.0 for bigger datasets. *Mol. Biol. Evol.* **2016**, *33*, 1870–1874. [CrossRef]

39. Felsenstein, J. Confidence limits on phylogenies: An approach using the bootstrap. *Evolution* **1985**, *39*, 783–791. [CrossRef] [PubMed]
40. Ronquist, F.; Teslenko, M.; Van Der Mark, P.; Ayres, D.L.; Darling, A.; Höhna, S.; Larget, B.; Liu, L.; Suchard, M.A.; Huelsenbeck, J.P. MrBayes 3.2: Efficient Bayesian phylogenetic inference and model choice across a large model space. *Syst. Biol.* **2012**, *61*, 539–542. [CrossRef]
41. White, T.J.; Bruns, T.; Lee, S.; Taylor, J. Amplification and direct sequencing of fungal ribosomal RNA genes for phylogenetics. In *PCR Protocols: A Guide to Methods and Applications*; Innis, M.A., Gelfand, D.H., Snisky, J.J., White, T.J., Eds.; Academic: San Diego, CS, USA, 1990; pp. 315–322.
42. Glass, N.L.; Donaldson, G. Development of primer sets designed for use with PCR to amplify conserved genes from filamentous ascomycetes. *Appl. Environ. Microb.* **1995**, *61*, 1323–1330. [CrossRef]
43. Carbone, I.; Kohn, L.M. A method for designing primer sets for speciation studies in filamentous ascomycetes. *Mycologia* **1999**, *91*, 553–556. [CrossRef]
44. Guerber, J.C.; Liu, B.; Correll, J.C.; Johnston, P.R. Characterization of diversity in Colletotrichum acutatum sensu lato by sequence analysis of two gene introns, mtDNA and intron RFLPs, and mating compatibility. *Mycologia* **2003**, *95*, 872–895. [CrossRef]

plants

MDPI

Article

Early Season Symptoms on Stem, Inflorescences and Flowers of Grapevine Associated with *Botryosphaeriaceae* Species

Pedro Reis [1,*], Ana Gaspar [1], Artur Alves [2], Florence Fontaine [3], Inês Lourenço [4], José Saramago [4], Mariana Mota [1] and Cecília Rego [1]

1 LEAF—Linking Landscape, Environment, Agriculture and Food, School of Agriculture, University of Lisbon, 1349-017 Lisbon, Portugal; anapatriciagaspar@hotmail.com (A.G.); mariana@isa.utl.pt (M.M.); crego@isa.ulisboa.pt (C.R.)

2 CESAM—Centre for Environmental and Marine Studies, Department of Biology, University of Aveiro, 3810-193 Aveiro, Portugal; artur.alves@ua.pt

3 SFR Condorcet FR CNRS 3417, Université de Reims Champagne-Ardenne, Résistance Induite et Bioprotection des Plantes EA 4707, BP 1039, 51687 Reims Cedex 2, France; florence.fontaine@univ-reims.fr

4 BASF Portuguesa, S.A. Rua 25 de Abril, 1, 2689-538 Prior Velho, Portugal; ines.lourenco@basf.com (I.L.); jose.saramago@basf.com (J.S.)

* Correspondence: pedroreis@isa.ulisboa.pt

Received: 21 September 2020; Accepted: 22 October 2020; Published: 24 October 2020

Abstract: Botryosphaeria dieback caused by several *Botryosphaeriaceae* species is one of the most important grapevine trunk diseases affecting vineyards worldwide. These fungi cause wedge-shaped perennial cankers and black streaking of the wood and have also been associated with intervein leaf chlorosis, dried or mummified berries, and eventually, the death of the plant. Early season symptoms may sometimes be disregarded by growers, being mistaken with symptoms from other diseases such as downy mildew or botrytis rot. Currently, few studies are available to determine what species may be causing these early season symptoms in grapevines. During the 2018 season, during the flowering period, grapevine samples showing necrosis on green shoots, dried inflorescences, and flowers, were collected in vineyards throughout the central regions of Portugal. Isolations were performed from symptomatic organs, and twenty-three isolates of *Botryosphaeriaceae* were selected. An analysis of the ITS and part of the translation elongation factor 1-α sequences was performed, revealing that the two main species apparently responsible for these symptoms were *Diplodia seriata* and *Neofusicoccum parvum*. In pathogenicity tests conducted on 1-year-old plants grown under controlled conditions in a greenhouse and on field-grown clusters, symptoms were reproduced, confirming the pathogenic behavior of the selection of isolates.

Keywords: grapevine; *Diplodia seriata*; *Neofusicoccum parvum*; pathogenicity; diagnostic

1. Introduction

Grapevine trunk diseases (GTDs) are one of the most critical problems affecting viticulture worldwide, causing yield reduction and increased production costs adding to the existing cost of the annual vineyard operations. In fact, these diseases pose a significant threat to sustainable viticulture worldwide since the cost associated with GTD losses has been increasing during recent years. For example, in Spain, there was an increase of 1.8% of incidence in vineyards in 2003 up to 10.5% in 2007 [1]. In France, it is estimated that the equivalent of 1 billion euros is lost every year due to these diseases [2], while in Australia, the economic impact can reach up to 8.3 billion AUD [3]. GTDs include three main fungal diseases—eutypa dieback, esca disease and botryosphaeria dieback—that involve one or several species of fungi [4–6]. In Portugal, botryosphaeria dieback and Esca are considered the major diseases affecting adult grapevines, causing considerable economic losses, in all the country's vine growing regions [2].

Botryosphaeria dieback is, currently, one of the most important GTDs worldwide [4], caused by fungi in the *Botryosphaeriaceae* family. Fungi belonging to this family have been found worldwide as endophytes, saprophytes or pathogenic in many perennial and ornamental plants [7–10].

Twenty-six species in different *Botryosphaeriaceae genera* have been associated with botryosphaeria dieback in grapevines [11–17]. The most common species isolated from grapevines are *Diplodia seriata* [18–24], *Diplodia mutila* [25,26], *Neofusicoccum parvum* [27], and *Lasiodiplodia theobromae* [16,28–30].

In different countries, species occurring on grapevines have been shown to differ in pathogenicity, in their epidemiology, and in the symptoms produced [31]. Therefore, the symptomatology of botryosphaeria dieback is somewhat complex. However, the most common symptoms include no or limited burst, perennial cankers, trunk dieback, wood necrosis, vascular streaking, and plant death [16,28,31–34]. Infection of grapevines by *Botryosphaeriaceae* can also lead to leaf and berry symptoms, where yellowish-orange or wine-red spots on leaf margins and blades can appear, depending on the cultivar (white or red) [4,20,35–37].

During the 2018 season, growers and field technicians in the central region of Portugal, started to observe symptoms characterized by necrosis on the base of green shoots, which could lead to the complete detachment of the shoot later in the season, wilting of the apex of the shoot, wilting of leaves, and more important, necrosis on the peduncle and rachis of the developing clusters, drying of inflorescences and flowers. These symptoms appeared even after the planned fungicide applications against other grapevine diseases, which was puzzling for the growers. Samples were sent for analysis in our laboratory and revealed the presence of *Botryosphaeriaceae* on the tissues analyzed. Although the epidemiology and distribution of *Botryosphaeria* canker in the wood of grapevines as well as other hosts have been widely studied, the relationship between *Botryosphaeriaceae* and these symptoms have not yet been investigated and no studies were carried out to understand which species may be causing these early season symptoms. The objective of this study was thus to identify the pathogens of these early season symptoms in vineyards of Portuguese central regions by means of morphological features, DNA analysis, and pathogenicity tests.

2. Results

2.1. Sampling and Fungal Isolation

A total of 42 *Botryosphaeriaceae* were isolated from the samples received from the 17 vineyards, and from several different symptomatic organs, including apex and base of the shoots, clusters, and leaves. Other fungi isolated from these samples were *Alternaria alternata*, *Phomopsis* spp. and *Penicillium* spp. The majority of *Botryosphaeriaceae* isolates were obtained from the symptomatic tissue collected at the apex of the shoot (36%), followed by the base of the shoots (33%). Isolates obtained from the rachis of the clusters represented 17% while 9% were from the clusters (flowers and inflorescences) and only 5% from the leaves. From this set of isolates, a total of twenty-three isolates (Table 1) were selected based on their cultural characteristics, with the attention of keeping at least one isolated obtained from each vineyard. These isolates were characterized and identified based on morpho-cultural characters, DNA sequencing, and pathogenicity tests.

Table 1. *Botryosphaeriaceae* isolates obtained during the study, respective GenBank accession numbers, and percent identity when compared to reference sequences.

Host			Origin		GenBank Accession Number		Percent Identity *	
Isolate Number	(*V. vinifera* cvs.)	Isolation Region		Species	ITS	Tef1-α	ITS	*Tef1-α*
Bt201	Seara Nova	Apex of the shoot	Vineyard 1	*Diplodia seriata*	MT786219	MW018672	100%	99%
Bt202	Alicante	Base of the shoot	Vineyard 2	*Diplodia seriata*	MT786220	MW018673	99%	99%
Bt203	Castelão	Rachis	Vineyard 3	*Diplodia seriata*	MT786221	MW018674	99%	100%
Bt204	Syrah	Base of the shoot	Vineyard 4	*Diplodia seriata*	MT786222	MW018675	100%	99%
Bt205	Castelão	Cluster	Vineyard 5	*Neofusicoccum parvum*	MT786223	MW018676	100%	99%
Bt206	Castelão	Leaf stem	Vineyard 5	*Diplodia seriata*	MT786224	MW018677	98%	99%
Bt207	Castelão	Base of the shoot	Vineyard 6	*Diplodia seriata*	MT786225	MW018678	100%	100%
Bt208	Aragonez	Apex of the shoot	Vineyard 7	*Diplodia seriata*	MT786226	MW018679	100%	99%
Bt209	Aragonez	Rachis	Vineyard 8	*Diplodia seriata*	MT786227	MW018680	100%	100%
Bt210	Castelão	Apex of the shoot	Vineyard 5	*Diplodia seriata*	MT786228	MW018681	99%	99%
Bt211	Arinto	Base of the shoot	Vineyard 9	*Neofusicoccum parvum*	MT786229	MW018682	99%	99%
Bt212	Castelão	Base of the shoot	Vineyard 10	*Diplodia seriata*	MT786230	MW018683	100%	100%
Bt213	Seara Nova	Apex of the shoot	Vineyard 1	*Diplodia seriata*	MT786231	MW018684	100%	99%
Bt214	Castelão	Apex of the shoot	Vineyard 11	*Diplodia seriata*	MT786232	MW018685	100%	100%
Bt215	Seara Nova	Rachis	Vineyard 12	*Diplodia seriata*	MT786233	MW018686	98%	100%
Bt216	Alicante	Apex of the shoot	Vineyard 13	*Neofusicoccum parvum*	MT786234	MW018687	100%	98%
Bt217	Aragonez	Apex of the shoot	Vineyard 14	*Neofusicoccum parvum*	MT786235	MW018688	100%	98%
Bt218	Aragonez	Base of the shoot	Vineyard 15	*Diplodia mutila*	MT786236	MW018689	100%	100%
Bt219	Alicante	Cluster	Vineyard 2	*Neofusicoccum parvum*	MT786237	MW018690	99%	100%
Bt220	Seara Nova	Base of the shoot	Vineyard 1	*Diplodia seriata*	MT786238	MW018691	99%	99%
Bt221	Seara Nova	Apex of the shoot	Vineyard 1	*Diplodia seriata*	MT786239	MW018692	100%	99%
Bt222	Castelão	Apex of the shoot	Vineyard 16	*Diplodia seriata*	MT786240	MW018693	100%	100%
Bt223	Alicante	Rachis	Vineyard 17	*Diplodia seriata*	MT786241	MW018694	99%	99%

* Reference sequences used—*D. seriata* ITS—AY259094, tef1-α—AY573220 [38]; *N. parvum* ITS—AY259098, tef1-α—AY573221 [38]; *D. mutila* ITS—KJ361837, tef1-α—KJ361829 [39].

2.2. Morphological Characterization and DNA Analysis for Fungal Identification

Isolates of *Botryosphaeriaceae* could be split into six groups based on colony morphology, after 8 days of growth on PDA. After promoting sporulation of the isolates under study on 2% water agar medium with autoclaved pine needles, three groups were distinguished based on the morphology of conidia. One group containing 12 isolates produced conidia initially hyaline, becoming dark brown, aseptate and ovoid, with average dimensions of (20.95–) 23.60 ± 1.15 (–26.46) × (7.79–) 9.18 ± 0.5 (10.61), which were determined to be *D. seriata* based on conidia morphology and DNA analysis (Table 2). A second group with 3 isolates produced ellipsoidal conidia, with flat base, unicellular and hyaline, which could develop 1 to 2 septa over time, with average dimensions of (14.18–) 16.39 ± 1.95 (–19.46) × (4.50–) 5.19 ± 0.13 (–6.02) which were determined to be *N. parvum*, based on conidial morphology and DNA analysis (Table 2). Finally, there was one isolate that formed hyaline and aseptate conidia, oblong to ovoid with both ends broadly rounded and with dimensions of (20.80–) 23.50 ± 1.41 (–25.05), which was determined to be *D. mutila* based on above-mentioned parameters (Table 2). There were seven isolates that did not sporulate until the end of this work, but their species identification was determined through DNA analysis.

Table 2. Conidial dimension of the *Botryosphaeriaceae* species under study.

Species/Isolate	Conidial dimensions	
	Length (μm)	Width (μm)
D. seriata		
Bt201	(21.27–) 25.22 ± 1.89 (–28.69)	(8.33–) 9.91 ± 0.88 (–11.75)
Bt202	*	
Bt203	(20.75–) 22.50 ± 1.12 (–24.23)	(6.71–) 8.17 ± 0.57 (–9.16)
Bt204	(20.43–) 22.65 ± 1.56 (–26.20)	(7.83–) 9.06 ± 0.76 (–10.93)
Bt206	(20.99–) 23.84 ± 1.52 (–26.36)	(7.71–) 9.23 ± 0.95 (–10.77)
Bt207	(19.61–) 23.58 ± 1.95 (–27.30)	(7.49–) 9.29 ± 0.81 (–11.26)
Bt208	(21.72–) 25.26 ± 1.41 (–28.73)	(9.53–) 11.13 ± 1.01 (–13.37)
Bt209	(19.61–) 22.17 ± 1.52 (24.53)	(7.20–) 9.07 ± 0.78 (–10.55)
Bt210		
Bt212	(21.27–) 23.91 ± 1.36 (–26.22)	(7.34–) 8.34 ± 0.60 (–9.36)
Bt213	(23.28–) 25.99 ± 1.58 (–29.18)	(7.73–) 9.03 ± 0.56 (–9.84)
Bt214		
Bt215	(21.02–) 24.58 ± 1.47 (–26.99)	(10.14–) 11.21 ± 0.66 (–12.83)
Bt220		
Bt221	(20.80–) 23.28 ± 1.35 (–25,53)	(8.59–) 9.99 ± 0.60 (–11.10)
Bt222		
Bt223	(21.42–) 22.94 ± 1.34 (–26.37)	(8.97–) 9.72 ± 0.66 (–11.40)
N. parvum		
Bt205	(15.51–) 17.18 ± 1.02 (–19.92)	(4.56–) 5.22 ± 0.43 (–6.11)
Bt211	(15,.2–) 18.28 ± 1.75 (–22,.4)	(4.73–) 5.33 ± 0.31 (–5.94)
Bt216	(11.39–) 13.71 ± 1.28 (–16.13)	(4.20–) 5.02 ± 0.45 (–6.02)
Bt217		
Bt219		
D. mutila	(20.80–) 23.50 ± 1.41 (–25.05)	(9.47–) 11.15 ± 1.57 (–16.25)
Bt218	(21.27–) 25.22 ± 1.89 (–28.69)	(8.33–) 9.91 ± 0.88 (–11.75)

* Isolate without sporulation.

2.3. Pathogenicity Tests

2.3.1. Pathogenicity Tests on Tendrils and Leaves

All the isolates tested were pathogenic towards the Aragonez 1-year-old grafted cuttings used in the greenhouse experiment, and they were able to, in some extent, reproduce symptoms closely resembling the ones observed on the field survey, namely necrotic tendrils, apex of the shoots and leaves (Figure 1A–C). Isolate Bt204 identified as *D. seriata*, was the isolate where the highest percentage (50%) of plants showed the above-described symptoms, followed by isolate Bt216 identified as *N. parvum* (40%) (Figure 2). The isolate with the lowest percentage of infected plants (20%) was Bt218 which was identified as *D. mutila*. No control plants showed any symptoms of disease, growing normally during the experimental period.

Figure 1. Symptoms observed on tendril and green shoots: (**A**) necrotic tendrils obtained after inoculation with *D. mutila* (Bt218); (**B**) necrotic apex of the shoot after inoculation with *D. seriata* (Bt204); (**C**) necrotic leaf after inoculation with *N. parvum* (Bt216); (**D**) lesion obtained after inoculation with *D. seriata* (Bt204); (**E**) lesion obtained after inoculation with *N. parvum* (Bt216); (**F**) control inoculation.

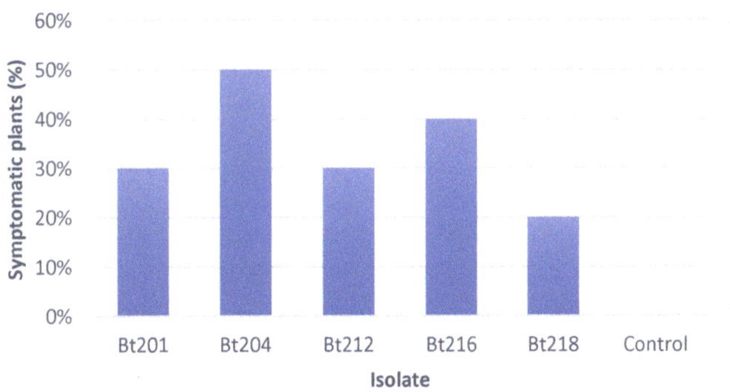

Figure 2. Percentage of plants showing symptoms on tendrils and leaves, after inoculation with the *Botryosphaeriaceae* isolates under study. Ten plants were used per isolate while control plants were sprayed with sterile distilled water.

2.3.2. Pathogenicity Tests on Green Stems

All isolates under study proved to be pathogenic towards 1-year-old grafted Aragonez grapevines, by being able to produce lesions in the inoculated tissues (Figure 1D–F). All the isolates showed significant statistical differences when compared to the control plants (Figure 3), being the largest average lesions recorded for isolate Bt216 (*N. parvum*) which showed significant differences towards the remaining isolates under study. The smallest average lesions were recorded for three isolates identified as *D. seriata* (Bt201, Bt204 and Bt212), while isolate Bt218 (*D. mutila*) showed lesions on average between those produced by *N. parvum* and *D. seriata* (Figure 3)

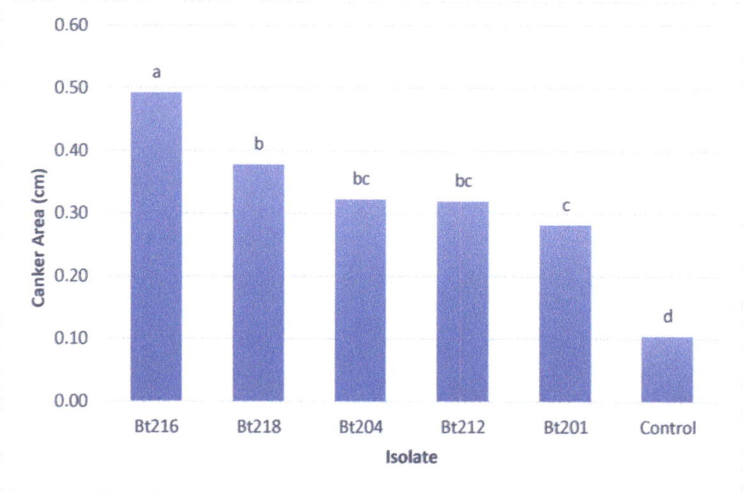

Figure 3. Mean canker areas (cm) in green stem caused by artificial inoculation with *Botryosphaeriaceae* isolates under study. Different letters in column correspond to significant differences ($p < 0.05$) based on ranks assessed by Kruskal-Wallis analysis. Ten plants were used per isolate while control plants were sprayed with sterile distilled water.

2.3.3. Field Pathogenicity Tests on Clusters

All isolates used in this study were able to cause symptoms of dried berries and inflorescences on inoculated clusters (phenological stage EL 25 to 27) of Cabernet Sauvignon (Figure 4A,B). Although only an exploratory statistical analysis was possible based on the infection scale established, due to the low number of repetitions (n), all isolates showed a significant different proportion of symptomless (1-ranked) samples ($p < 0.05$) towards the control, therefore confirming the pathogenicity of the isolates under study when artificially infecting field-grown clusters. Lower p-values ($p = 0.0049$) for isolates Bt201 and Bt218 may suggest a stronger ability of these two isolates in causing symptoms on clusters when comparing to the remaining isolates which recorded a higher p value ($p = 0.0182$). In order to address more in detail this topic, a boxplot analysis was performed (Figure 5) and results showed that among the different isolates, Bt216 showed the lowest average infected area, as nearly all the samples were ranked as a 2-class, and no sample ranked more than 3-class, corresponding to less than 25% of affected area. For the isolates Bt212, Bt218 and Bt204, half of the samples, showed more that 25% of the cluster area with infection symptoms. Within these three isolates, Bt218 seems to be less aggressive, as no samples were quoted as 3-class or more severe. Bt201 suggested once more to showed tendentially higher infected area values.

Figure 4. (**A**) best cluster from each inoculation/isolate; (**B**) worst cluster from each inoculation/isolate.

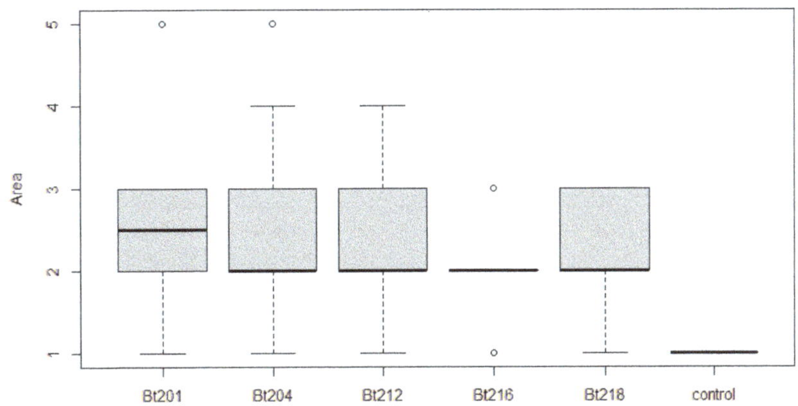

Figure 5. Boxplot of the percentage of infected area on clusters artificially inoculated with *Botryosphaeriaceae* isolates under study. Scale of evaluation of area infected—1 = no symptoms; 2 = 1–5%; 3 = 5–25%; 4 = 25–50%; 5 = >50%. The median is represented by the solid line. Top and bottom lines of the box correspond to the 25th and 75th percentiles of the data, respectively. Circles represent outliers.

For all the pathogenicity experiments, pathogens were recovered from symptomatic tissues of all infected plants, while no *Botryosphaeriaceae* isolates were re-isolated from control plants.

3. Discussion

This is the first study aimed at describing grapevine early season symptoms associated with Botryosphaeriaceous fungi in Portugal, including necrosis and wilting of the apex of the green shoots, wilting of leaves, necrosis on the peduncle and rachis of the developing clusters and, drying of inflorescences and flowers. Morphological studies and DNA sequence analysis allowed to identify the presence of three different species of *Botryosphaeriaceae*: *D. seriata*, *N. parvum* and *D. mutila*. The most common species found in our study causing the described symptoms was *D. seriata*, since 74% of the isolates belonged to this species, which agrees with previous studies such as Auger et al. [40] in Chile, Luque et al. [29] in Spain and Carlucci et al. [41] in Italy, this being one of the most common species associated with botryosphaeria dieback in vineyards. All the species identified in the present study have been previously identified in other grape-growing regions worldwide, and they have been linked with a broad range of symptoms, including leaf spots, fruit rot, shoot dieback, bud necrosis, vascular discoloration of the wood and perennial cankers [16]. All isolates tested showed the ability to cause symptoms to some extent, similar to those observed in the field. Regarding the ability for causing necrosis on tendrils and leaves, all the isolates showed few differences on the percentage of infected plants, being the only exception, *D. mutila* which was able to only cause symptoms on 20% of the plants. Considering the potential of causing lesions on green stems, the isolate belonging to the species *N. parvum* (Bt216) was able to induce the largest average lesions, whereas the lowest average lesions were recorded for the isolates belonging to *D. seriata*. These results are in accordance with previously described results regarding aggressiveness of *Botryosphaeriaceae* species [16], in which *N. parvum* is considered to be highly aggressive towards grapevine, while *D. seriata* is considered to be only mildly aggressive. In this case, *D. mutila* showed intermediate average lesions values contrary to what was observed on the pathogenicity tests on both tendrils and leaves, where this isolate appeared to be the least aggressive. More epidemiology studies should be performed on the behavior of these species to understand if there is a difference of aggressiveness towards different types of grapevine tissues, or if the differences observed were due to experimental design. Considering the ability to cause symptoms on clusters, such as dried berries and inflorescences, the highest average infected area was recorded for *D. seriata* and *D. mutila*. These results are contrary to the ones obtained for the other pathogenicity tests. This may be not due to the aggressiveness of the isolates/species, but to the differences on experimental conditions. These pathogenicity tests were conducted on field-grown grapevines while the other tests were conducted on grapevines growing in a greenhouse-controlled environment, not to mention the differences in plant age.

Although the main economic impact fungi of the *Botryosphaeriaceae* family are associated with the trunk and cane symptoms, damage by these fungi showing up so early in the growing season should not be overlooked, since they could perform an important role in their epidemiology and become a source of inoculum for wound infections leading to trunk diseases. *Botryosphaeriaceae* fungi occur in most parts of the world and are found as endophytes or parasites and saprophytes on a vast number of both annual and perennial plants [16]. Infection by *Botryosphaeriaceae* is considered to occur mainly through pruning wounds [16,42,43], since cankers start to develop from wounds on leaves, branches or stems. However, several studies have shown that these fungi can infect through lenticels, stomata, or other openings on healthy plants [44–48]. Nowadays, pruning wounds are considered as the main door of infection for *Botryosphaeriaceae* on grapevine. Nonetheless, Shafi et al. [48] recently showed by fluorescence microscopy that these pathogens can remain latent on the grapevine bark, and even without any type of wounding, germinating conidia and mycelium could be observed near lenticels, as well as mycelia in the underlying wood, demonstrating that the pathogens had entered through the lenticels. Therefore, our observation suggests that *Botryosphaeriaceae* fungi may thus have the ability to colonize and infect healthy grapevine tissues, even without wounding, which may be one of the reasons for the symptoms observed on these vineyards during the spring of 2018.

Pycnidia of *Botryosphaeriaceae* associated with dieback disease can be detected in old pruning wounds, infected spurs, embedded in the bark of the cordons or trunk of infected grapevines, and also on pruning debris left in the vineyard [16,49–51]. In France, Kuntzmann et al. [52] reported that conidia of *Botryosphaeriaceae* were released during the whole vegetative period, but *D. mutila* released its spores later that *D. seriata*, indicating that either these fungi differ in their ability to grow and sporulate or that they merely differ in their response to meteorological conditions. These authors also reported that in their study 50% of the spores of *D. mutila* were captured during late summer and autumn while most of the conidia of *D. seriata* were captured during the spring months. This is in accordance to our findings, especially when taking into consideration that our sampling took place between the months of March and July, since most of the species found during our study, belonged to *D. seriata*, while only one isolate of *D. mutila* was recorded. As stated before, our study was conducted when grapevines were at the E-L 23–25 phenological stage which corresponds exactly to the flowering period which has been reported previously by Spagnolo et al. [53] to be the most sensitive period to infection by botryosphaeria dieback agents, as a consequence of the high metabolic activity leaning towards the development of flowers. Therefore, we believe that the conjunction of all these factors, presence of inoculum due to precipitation during the spring months, associated with the phenological stage of the grapevines and the ability of these fungi in infecting through other pathways other that wounds, may have a strong influence on the manifestation of early season symptoms caused by *Botryosphaeriaceae*. Thus, planning further research is strongly advised on this subject with special attention for epidemiology and pathogenicity studies to determine the infection pathway and infection moments regarding environmental conditions.

To the best of our knowledge, this is the first study aimed at describing these *Botryosphaeriaceae* symptoms in grapevines. In conclusion, our work has demonstrated that Botryosphaeria dieback fungi, mostly associated with wood cankers, appear to have the potential to cause serious early season symptoms, since the pathogenicity tests conducted with all the species found were able to reproduce these symptoms on several different grapevine organs. Nevertheless, we strongly believe that further research is needed on this subject by collecting more samples from different regions of Portugal and by testing a wider range of isolates, to try to understand what are the most common species involved in the expression of these symptoms, and what is or if there is an influence of geographical location and climate conditions [54,55]. The knowledge and clarification of the symptoms caused by these fungi and the development of proper diagnostic may help growers not to confuse them with symptoms of other diseases, and to set up a proper management plan.

4. Materials and Methods

4.1. Sampling and Fungal Isolation

During the early spring of the 2018 season, at the phenological stages E-L 23–25 [56], samples from 17 vineyards spread throughout the central regions of Portugal, namely Lisboa, Tejo, and Alentejo, were received at Instituto Superior de Agronomia (ISA). These samples showed necrosis on shoots and dried inflorescences and flowers, and isolations were made by cutting several pieces from symptomatic organs. Pieces collected were surface disinfected with a 7% sodium hypochlorite solution, rinsed in sterile distilled water (SDW) and plated onto 9 mm Petri dishes containing Potato Dextrose Agar (PDA, BD Difco, Sparks, MD, USA) amended with chloramphenicol (PanReac AppliChem, Darmstadt, Germany) at 250 mg/L. After incubation at 25 °C for one week, Petri dishes were assessed for the presence of *Botryosphaeriaceae* colonies which were sub-cultured onto fresh PDA dishes and, again incubated at 25 °C for one week, in darkness. All isolates obtained were stored in the collection of the ISA, Lisbon, Portugal, and were afterwards characterized morphologically, as well as properly identified by DNA sequence analyses.

4.2. Morphological Characterization and DNA Analysis for Fungal Identification

4.2.1. Morphological Characterization

Isolates under study were, plated onto 6 mm Petri dishes containing 2% water agar with autoclaved pine needles (*Pinus pinea*) and incubated at 25 °C under fluorescent light, in order to promote sporulation [27,28,34]. Pycnidia were mounted on microscope slides in a solution of lactophenol blue, and digital images were recorded with a Leica DFC295 camera on a Leica DM 2500 microscope at a 400× amplification. Twenty conidia were measured with the Leica Suite v3.16 program, for each isolate under study. Dimensions of the conidia are given as the range of dimensions with minimum and maximum dimensions in parentheses followed by mean and standard deviation.

4.2.2. DNA Analysis

DNeasyTM Plant Mini Kit by Qiagen® (Venlo, The Netherlands) was used to extract genomic DNA from 8-day-old cultures grown in PDA and incubated at 25 °C, in the darkness. The ITS region was amplified using primers ITS5 and ITS4 [57], while the primers EF1-688F and EF1–1251R [58] were used to amplify part of the elongation factor 1α gene. The PCR mixtures contained 1 × PCR buffer REDTaq Ready Mix (Sigma-Aldrich, Saint Louis, MO, USA), 3 mM $MgCl_2$, 0.4 mM dNTP mix, 12.5 pmol of each primer, 0.06 unit/µL of Taq Polymerase and 25–50 ng of template DNA. Each reaction volume was made up to 25 µL with sterile ultrapure water. Negative controls with sterile ultrapure water instead of the template DNA were used in every reaction. The amplification conditions for ITS were as follows: initial denaturation of 5 min at 95 °C, followed by 40 cycles of 30 s at 94 °C, 30 s at 58 °C, 1 min and 40 s at 72 °C and a final extension period of 10 min at 72 °C. For the amplification of part of the tef1-α gene, the conditions were, an initial denaturation of 2 min at 94 °C, followed by 35 cycles of 30 s at 94 °C, 45 s at 55 °C, 1 min at 72 °C and a final extension period of 10 min at 72 °C. Each amplicon was separated by electrophoresis at 120 V for 30 min in a 1% agarose gel in 0.5 × TBE buffer. Gels were stained with 3 µL of GreenSafe Premium (Nzytech, Lisbon, Portugal), and were visualized using with a UV transilluminator to assess PCR amplification.

The amplified PCR fragments were purified using an Illustra ExoProStar Enzymatic PCR and Sequencing Clean-up Kit (GE Life Sciences, Buckinghamshire, UK) and both strands of the PCR products were sent for sequencing at STABVIDA (Lisbon, Portugal). Sequences obtained were edited and aligned using MEGA7 [59] to find a consensus sequence. These sequences were then compared with sequences from GenBank in BLAST searches, and species identification was obtained when at least 98% of similarity was found.

4.3. *Pathogenicity Tests*

Five representative isolates were selected from the fungal collection under study for pathogenicity tests, with the attention of selecting at least one isolate from the three different species of *Botryosphaeriaceae* identified three isolates of *Diplodia seriata* (Bt201, Bt204 and Bt212), one isolate of *Neofusicoccum parvum* (Bt216) and one isolate of *Diplodia mutila* (Bt218). To reproduce all the symptoms found in the field, three different pathogenicity tests were designed, being two performed on grapevines kept in a greenhouse and one on grapevines established in the field. Greenhouse pathogenicity tests were conducted on 1-year-old grafted Aragonez (=Tempranillo) plants, since is the most planted cultivar in Portugal and field tests were conducted on clusters of Cabernet Sauvignon which is currently the most planted cultivar worldwide.

4.3.1. Pathogenicity tests on Tendrils and Leaves

Isolates were plated in Petri dishes containing 2% water agar with autoclaved pine needles (*Pinus pinea*) and incubated at 25 °C under fluorescent light, to promote sporulation [27,28,34]. Conidia were harvested from these plates by collecting the pycnidia formed on the pine needles into a 1.5 mL Eppendorf tube containing sterile distilled water and crushing them with the help of a pestle. These spore suspensions were filtered through cheesecloth and the concentrations were adjusted to 10^5 spores/mL. To ensure full coverage of the tissues to be inoculated, 2 mL of each spore suspension was sprayed on the green tissues (leaves and tendrils) of 1-year-old grafted cuttings of cultivar Aragonez (=Tempranillo), individually potted in 1 L free draining bags containing a sandy soil mixture kept in a ventilated greenhouse at 24 °C under natural light. The aerial part of the plants was covered with a plastic bag for 3 days to promote infection, and the plants were assessed for the development of symptoms, namely necrosis on any of the inoculated organs or drying of leaves and tendrils, one week after inoculation. Ten plants were used for each isolate, while control plants were sprayed with sterile distilled water. The percentage of plants showing symptoms for each isolate was recorded.

4.3.2. Pathogenicity Tests on Green Stems

One-year old grafted cuttings of cultivar Aragonez (=Tempranillo), individually potted in 1 L free draining bags containing a sandy soil mixture, kept in a ventilated greenhouse at 24 °C under natural light, were inoculated following the method described by Reis et al. [35]. Ten plants were used for each isolate, and the assessment for symptoms development, including external lesions or cankers, was performed one month after inoculation, by measuring the width and length of the lesions and calculating the elliptical area of the lesion. All statistical analysis was performed using the R program (www.r-project.org). Assumptions for variance analysis was assessed and when all the assumptions were not accomplished, the influence of distinct levels of one factor was assessed using the non-parametrical test of Kruskal-Wallis. In this case, when the significant differences were found ($p < 0.05$), the comparison between the distinct level was made using the ranks.

4.3.3. Field Pathogenicity Tests on Clusters

Clusters of cultivar Cabernet Sauvignon on EL –25 to 27 stages, were selected from several field-grown grapevines in a vineyard located at ISA, Lisbon, Portugal. Artificial inoculations were performed with spore suspensions obtained as referred earlier for the pathogenicity tests on green tissues. Again, to ensure full coverage of the tissues to be inoculated, 1 ml of each spore suspension was sprayed on each cluster, which were covered individually with plastic bags for 3 days. Ten replicates were used for each isolate, and control cluster were sprayed with sterile distilled water. After one week, clusters were assessed for the development of symptoms such as dried or necrotic berries and inflorescences. These symptoms were quantified by using an adaptation of the EPPO protocol for evaluation of fungicides against *Botryotinia fuckeliana* on grapevine [60], where the percentage of the cluster area infected was assessed according to the following scale: 1 = no symptoms; 2 = 1–5%; 3 = 5–25%; 4 = 25–50%; 5 = >50%. The total percentage of clusters showing symptoms for each isolate was also recorded. The virulence of the different isolates was assessed through a multiple proportion test, comparing the proportion of symptomless samples of each modality towards the control. Boxplots for the infection rank of each isolate were defined to compare aggressiveness of the different isolates. All statistical analysis was performed using the R program (www.r-project.org).

In order to fulfil Koch's postulates, samples were collected from all the different types of symptoms/lesions observed, and placed on PDA (Difco, USA, BD) amended with chloramphenicol (PanReac AppliChem, Darmstadt, Germany) at 250 mg/L to recovered the inoculated fungi.

Author Contributions: P.R, C.R., A.A and F.F. designed the experiments; P.R. and A.G. implemented the methodology; I.L. and J.S. collected the samples on the field; P.R., M.M. and A.G. did the formal analysis; P.R. did the original draft preparation; A.A., F.F. and C.R. did the review and editing; A.A., F.F. and C.R. did the supervision; A.A. and C.R. did the funding acquisition. All authors have read and agreed to the published version of the manuscript.

Funding: This research was funded by FEDER funding through COMPETE program (POCI-01-0145-FEDER-016788) and Programa Operacional Regional de Lisboa-POR Lisboa (LISBOA-01-0145-FEDER-016788) and by national funding through FCT within the research project ALIEN (PTDC/AGR-PRO/2183/2014). The authors are thankful to FCT/MCTES for financing CESAM (UIDB/50017/2020 + UIDP/50017/2020), LEAF—Linking Landscape, Environment, Agriculture and Food (UID/AGR/04129/2013), and the PhD grant of Pedro Reis (SFRH/BD/131766/2017).

Conflicts of Interest: The authors declare no conflict of interest. The funders had no role in the design of the study; in the collection, analyses, or interpretation of data; in the writing of the manuscript, or in the decision to publish the results.

References

1. Rubio, J.J.; Garzón, E. Las enfermedades de madera de vid como amenaza del sector vitícola. *Rev. Winetech* **2011**, *11*, 18–21.
2. De la Fuente, M.; Fontaine, F.; Gramaje, D.; Armengol, J.; Smart, R.; Nagy, Z.A.; Rego, C.; Corio-Costet, M. *Grapevine Trunk Diseases. A Review*, 1st ed.; OIV Publications: Paris, France, 2016.

3. Ridgway, H.J.; Baskarathevan, J.; Amponsah, N.; Jaspers, M.V.; Jones, E.E. The identity, distribution and diversity of botryosphaeriaceous species in New Zealand vineyards—A national perspective. *Phytopathol. Mediterr.* **2014**, *53*, 565.
4. Bertsch, C.; Ramirez-Suero, M.; Magnin-Robert, M.; Larignon, P.; Chong, J.; Abou-Mansour, E.; Spagnolo, A.; Clément, C.; Fontaine, F. Trunk diseases of grapevine: Complex and still poorly understood syndromes described as eutypa dieback, esca and black dead arm. *Plant Pathol.* **2012**, *62*, 243–265. [CrossRef]
5. Bruez, E.; Lecomte, P.; Grosman, J.; Doublet, B.; Bertsch, C.; Fontaine, F.; Da Costa, J.P.; Ugaglia, A.; Teissedre, P.L.; Guerin-Dubrana, L.; et al. Overview of grapevine trunk diseases in France in the early 2000s. *Phytopathol. Mediterr.* **2013**, *52*, 262–275.
6. Mondello, V.; Songy, A.; Battiston, E.; Pinto, C.; Coppin, C.; Trotel-Aziz, P.; Clément, C.; Mugnai, L.; Fontaine, F. Grapevine trunk diseases: A review of fifteen years of trials for their control with chemicals and biocontrol agents. *Plant Dis.* **2018**, *102*, 1189–1217. [CrossRef]
7. Smith, H.; Wingfield, M.J.; Crous, P.W.; Coutinho, T.A. *Sphaeropsis sapinea* and *Botryosphaeria dothidea* endophytic in *Pinus* spp. in South Africa. *S. Afr. J. Bot.* **1996**, *62*, 86–88. [CrossRef]
8. Smith, H.; Wingfield, M.J.; Petrini, O. *Botryosphaeria dothidea* endophytic in *Eucalyptus grandis* and *Eucalyptus nitens* in South Africa. *Forest. Ecol. Manag.* **1996**, *89*, 189–195. [CrossRef]
9. Van Niekerk, J.M.; Fourie, P.H.; Halleen, F.; Crous, P.W. *Botryosphaeria* spp. as grapevine trunk disease pathogens. *Phytopathol. Mediterr.* **2006**, *45*, S43–S54.
10. Bellé, A.; Comont, G.; Nivault, A.; Abou-Mansour, E.; Coppin, C.; Dufour, M.C.; Corio-Costet, M.F. Life traits of four *Botryosphaeriaceae* species and molecular responses of different grapevine cultivars or hybrids. *Plant Pathol.* **2017**, *66*, 763–776. [CrossRef]
11. Gramaje, D.; Úrbez-Torres, J.R.; Sosnowsky, M.R. Managing grapevine trunk diseases with respect to etiology and epidemiology: Current strategies and future prospects. *Plant Dis.* **2018**, *102*, 12–39. [CrossRef]
12. Pitt, W.M.; Huang, R.; Steel, C.C.; Savocchia, S. Pathogenicity and epidemiology of *Botryosphaeriaceae* species isolated from grapevines in Australia. *Australas. Plant Pathol.* **2013**, *42*, 573–582. [CrossRef]
13. Pitt, D.G.; Lanteigne, L.; Hoepting, M.K.; Plamondon, J. Effects of precommercial thinning on the forest value chain in north-western New Brunswick: Part 1—Roundwood production and stumpage value. *For. Chron.* **2013**, *89*, 446–457. [CrossRef]
14. Pitt, W.M.; Úrbez-Torres, J.R.; Trouillas, F.P. *Dothiorella vidmadera*, a novel species from grapevines in Australia and notes on *Spencermartinsia*. *Fungal Divers.* **2015**, *61*, 209–219. [CrossRef]
15. Rolshausen, P.E.; Akgül, D.S.; Perez, R.; Eskalen, A.; Gispert, C. First report of wood canker caused by *Neoscytalidium dimidiatum* on grapevine in California. *Plant Dis.* **2013**, *97*, 1511. [CrossRef]
16. Úrbez-Torres, J.R. The status of *Botryosphaeriaceae* species infecting grapevines. *Phytopathol. Mediterr.* **2011**, *50*, S5–S45.
17. Yang, T.; Groenewald, J.Z.; Cheewangkoon, R.; Jami, F.; Abdollahzadeh, J.; Lombard, L.; Crous, P.W. Families, genera and species of Botryosphaeriales. *Fungal Biol.* **2017**, *121*, 322–346. [CrossRef]
18. Cristinzo, G. Gravil attachi di Botryosphaeria obtusa su vite provincial di Insernia. *Inf. Fitopatol.* **1978**, *6*, 21–23.
19. Castillo-Pando, M.; Sommers, A.; Green, C.D.; Priest, M.; Sriskanthades, M. Fungi associated with dieback of Semillon grapevines in the Hunter Valley of New South Wales. *Australas. Plant Pathol.* **2001**, *30*, 59–63. [CrossRef]
20. Larignon, P.; Fulchic, R.; Laurent, C.; Dubos, B. Observation of black dead arm in French vineyards. *Phytopathol. Mediterr.* **2001**, *40*, S336–S342.
21. Phillips, A.J.L.; Crous, P.W.; Alves, A. *Diplodia seriata*, the anamorph of *"Botryosphaeria" obtusa*. *Fungal Divers.* **2007**, *25*, 141–155.
22. Rovesti, L.; Montermini, A. Un deprimento della vitte causato de *Sphaeropsis malorum* in provincia di Reggio Emilia. *Inf. Fitopatol.* **1987**, *1*, 59–61.
23. Savocchia, S.; Steel, C.C.; Stodart, B.J.; Somers, A. Pathogenicity of *Botryosphaeria* species from declining grapevines in sub-tropical of Eastern Australia. *Vitis* **2007**, *46*, 27–32.
24. Úrbez-Torres, J.R.; Leavitt, G.M.; Guerrero, J.C.; Guevara, J.; Gubler, W.D. Identification and pathogenicity of *Lasiodiplodia theobromae* and *Diplodia seriata*, the causal agents of bot canker disease of grapevines in Mexico. *Plant Dis.* **2008**, *92*, 519–529. [CrossRef] [PubMed]

25. Lehoczky, J. Black dead arm disease of grapevine caused by *Botryosphaeria stevensii* infection. *Acta Phytopathol. Acad. Sci. Hung.* **1974**, *9*, 319–327.
26. Taylor, A.; Hardy, G.E.; St, J.; Wood, P.; Burgess, T. Identification and pathogenicity of Botryosphaeria species associated with grapevine decline in Western Australia. *Australas. Plant Pathol.* **2005**, *34*, 187–195. [CrossRef]
27. Crous, P.W.; Slippers, M.J.; Rheeder, J.; Marasas, W.F.O.; Phillips, A.J.L.; Alves, A.; Burguess, T.; Barber, P.; Groenewald, J.Z. Phylogenetic lineages in the *Botryosphaeriaceae*. *Stud. Mycol.* **2006**, *55*, 235–253. [CrossRef] [PubMed]
28. Phillips, A.J.L. *Botryosphaeria* species associated with diseases of grapevines in Portugal. *Phytopathol. Mediterr.* **2002**, *41*, 3–18.
29. Luque, J.; Martos, S.; Aroca, A.; Raposo, R.; Garcia-Figueres, F. Symptoms and fungi associated with mature grapevine plants in Northeast Spain. *J. Plant Pathol.* **2009**, *91*, 381–390.
30. Rodríguez-Gálvez, E.; Maldonado, E.; Alves, A. Identification and pathogenicity of *Lasiodiplodia theobromae* causing dieback of table grapes in Peru. *Eur. J. Plant Pathol.* **2015**, *141*, 477–489. [CrossRef]
31. Van Niekerk, J.M.; Crous, P.W.; Groenewald, J.Z.; Fourie, P.H.; Halleen, F. DNA phylogeny, morphology and pathogenicity of *Botryosphaeria* species on grapevines. *Mycologia* **2004**, *96*, 781–798. [CrossRef]
32. Larignon, P.; Dubos, B. The villainy of Black Dead Arm. *Wines Vines* **2001**, *82*, 86–89.
33. Linaldeddu, B.T.; Deidda, A.; Scanu, B.; Francheschini, A.; Serra, S.; Berraf-Tebbal, A.; Zouaoui Boutiti, M.; Ben Jamâa, M.L.; Phillips, A.J.L. Diversity of *Botryosphaeriaceae* species associated with grapevine and other woody hosts in Italy, Algeria and Tunisia, with descriptions of *Lasiodiplodia exigua* and *Lasiodiplodia mediterranea* sp. Nov. *Fungal Divers.* **2015**, *71*, 201–214. [CrossRef]
34. Úrbez-Torres, J.R.; Leavitt, G.M.; Voegel, T.; Gubler, W.D. Identification and distribution of Botryosphaeria species associated with grapevine cankers in California. *Plant Dis.* **2006**, *90*, 1490–1503. [CrossRef] [PubMed]
35. Reis, P.; Magnin-Robert, M.; Nascimento, T.; Spagnolo, A.; Abou-Mansour, E.; Fioretti, C.; Clément, C.; Rego, C.; Fontaine, F. Reproducing Botryosphaeria Dieback Foliar Symptoms in a Simple Model System. *Plant Dis.* **2016**, *100*, 1071–1079. [CrossRef]
36. Reis, P.; Pierron, P.; Larignon, L.; Lecomte, P.; Abou-Mansour, E.; Farine, S.; Bertsch, C.; Jaques, A.; Trotel-Aziz, P.; Fontaine, F. *Vitis* methods to understand and develop strategies for diagnosis and sustainable control of grapevine trunk diseases. *Phytopathology* **2019**, *109*, 916–931. [CrossRef]
37. Spagnolo, A.; Larignon, P.; Magnin-Robert, M.; Hovasse, A.; Cilindre, C.; Van Dorsselaer, A.; Clément, C.; Schaffer-Reiss, C.; Fontaine, F. Flowering as the most highly sensitive period of grapevine (*Vitis vinifera* cv. *Mourvèdre*) to the botryosphaeria dieback agents of *Neofusicoccum parvum* and *Diplodia seriata* infection. *Int. J. Mol. Sci.* **2014**, *15*, 9644–9669. [CrossRef]
38. Phillips, A.J.L.; Alves, A.; Abdollahzadeh, J.; Slippers, B.; Wingfield, M.J.; Groenwald, J.Z.; Crous, P.W. The *Botryosphaeriaceae*: Genera and species known from culture. *Stud. Mycol.* **2013**, *76*, 51–167. [CrossRef]
39. Alves, A.; Linaldeddu, B.T.; Deidda, A.; Scanu, B.; Phillips, A.J.L. The complex of *Diplodia* species associated with Fraxinus and some other woody hosts in Italy and Portugal. *Fungal Divers.* **2014**, *67*, 143–156. [CrossRef]
40. Auger, J.; Esterio, M.; Ricke, G.; Pérez, I. Black dead arm and basal canker of *Vitis vinifera* cv. Red Globe caused by *Botryosphaeria obtusa* in Chile. *Plant Dis.* **2004**, *88*, 1286. [CrossRef]
41. Carlucci, A.; Cibelli, F.; Lops, F.; Raimondo, M.L. Characterization of *Botryosphaeriaceae* species as casual agents of trunk diseases on grapevines. *Plant Dis.* **2015**, *99*, 1678–1688. [CrossRef]
42. Von Arx, J.A.; Muller, E. Die Gattungen der amerosporen Pyrenomyceten. *Beriage Kryptogamenflora Schweiz* **1954**, *11*, 1–434.
43. Smith, H.; Kemp, G.H.J.; Wingfield, M.J. Canker and die-back of Eucalyptus in South Africa caused by *Botryosphaeria dothidea*. *Plant Pathol.* **1994**, *43*, 1031–1034. [CrossRef]
44. Kim, K.W.; Park, E.W.; Ahn, K.-K. Pre-penetration behaviour of *Botryosphaeria dothidea* on apple fruits. *Plant Pathol. J.* **1999**, *15*, 223–227.
45. Michailides, K. Pathogenicity, distribution, sources of inoculum, and infection courts of *Botryosphaeria dothidea* on pistachio. *Phytopathology* **1991**, *81*, 566–573. [CrossRef]
46. Smith, R.J.; Verdegaal, P.; Gubler, W.D. *Botryosphaeriaceae* species spore-trapping studies in California vineyards. *Plant Dis.* **2010**, *94*, 717–724.
47. Slippers, B.; Wingfield, M.J. *Botryosphaeriaceae* as endophytes and latent pathogens of woody plants: Diversity, ecology and impact. *Fungal Biol.* **2007**, *21*, 90–106. [CrossRef]

48. Shafi, A.; Ridgway, H.; Jaspers, M.; Jones, E. Saprophytic colonization of the bark by *Neofusicoccum* species mediates subsequent infection of grapevines through wounds. *Phytopathol. Mediterr.* **2019**, *58*, 395–449.
49. Elena, G.; Luque, J. Pruning debris of grapevine as a potential inoculum source of *Diplodia seriata*, causal agent of *Botryospaheria* dieback. *Eur. J. Plant Pathol.* **2016**, *144*, 803–810. [CrossRef]
50. Hewitt, W.B. Diplodia cane dieback and bunch rot. In *Compendium of Grape Diseases*; Pearson, R.C., Goheen, A.C., Eds.; APS Press: St. Paul, MN, USA, 1988; pp. 25–26.
51. Van Niekerk, J.M.; Calitz, F.J.; Halleen, F.; Fourie, P.H. Temporal spore dispersal patterns of grapevine trunk pathogens in South Africa. *Eur. J. Plant Pathol.* **2010**, *127*, 375–390. [CrossRef]
52. Kuntzmann, P.; Villaume, S.; Bertsch, C. Conidial dispersal on *Diplodia* species in a French vineyard. *Phytopathol. Mediterr.* **2009**, *48*, 150–154.
53. Spagnolo, A.; Mondello, V.; Larignon, P.; Villaume, S.; Rabenoelina, F.; Clément, C.; Fontaine, F. Defense responses in grapevine (cv. Mourvèdre) after inoculation with the Botryosphaeria dieback pathogens *Neofusicoccum parvum* and *Diplodia seriata* and their relationship with flowering. *Int. J. Mol. Sci.* **2017**, *18*, 1–12. [CrossRef] [PubMed]
54. Fischer, M.; Peighami Ashnaei, S. Grapevine, esca complex, and environment: The disease triangle. *Phytopathol. Mediterr.* **2019**, *58*, 17–37.
55. Songy, A.; Fernandez, O.; Clément, C.; Larignon, P.; Fontaine, F. Grapevine trunk diseases under thermal and water stresses. *Planta* **2019**, *249*, 1655–1679. [CrossRef]
56. Dry, P.; Coombe, B. Grapevine growth stages—The modified E-L system. In *Viticulture 1—Resources*, 2nd ed.; Winetitles: Broadview, Australia, 2004.
57. White, T.J.; Bruns, T.; Lee, S.; Taylor, J. Amplified and direct sequencing of fungal ribosomal RNA genes for phylogenetics. In *PCR Protocols: A Guide to Methods and Applications*; Innis, M.A., Gelfand, D.H., Sninsky, J.J., White, T.J., Eds.; Academic: San Diego, CA, USA, 1990; pp. 315–322.
58. Alves, A.; Crous, P.W.; Correia, A.; Phillips, A.J.L. Morphological and molecular data reveal cryptic speciation in *Lasiodiplodia theobromae*. *Fungal Divers.* **2008**, *28*, 1–13.
59. Kumar, S.; Stecher, G.; Tamura, K. MEGA7: Molecular Evolutionary Genetics Analysis version 7.0 for bigger datasets. *Mol. Biol. Evol.* **2016**, *33*, 1870–1874. [CrossRef]
60. OEPP/EPPO. *EPPO Standards PP1, Efficacy Evaluation of Plant Protection Products, Vol. 2, Fungicides & Bactericides*; OEPP/EPPO: Paris, France, 2004; pp. 22–24.

plants MDPI

Article

Production of Phytotoxic Metabolites by Botryosphaeriaceae in Naturally Infected and Artificially Inoculated Grapevines

Pierluigi Reveglia [1,2,3,*], Regina Billones-Baaijens [1], Jennifer Millera Niem [1,4], Marco Masi [2], Alessio Cimmino [2], Antonio Evidente [2] and Sandra Savocchia [1]

1 National Wine and Grape Industry Centre, School of Agricultural and Wine Sciences, Charles Sturt University, Locked Bag 588, Wagga Wagga, NSW 2678, Australia; rbaaijens@csu.edu.au (R.B.-B.); jmniem@up.edu.ph (J.M.N.); ssavocchia@csu.edu.au (S.S.)
2 Dipartimento di Scienze Chimiche, Universita' di Napoli Federico II, Complesso Universitario Monte Sant'Angelo, Via Cintia 4, 80126 Napoli, Italy; marco.masi@unina.it (M.M.); alessio.cimmino@unina.it (A.C.); evidente@unina.it (A.E.)
3 Department of Clinical and Experimental Medicine, University of Foggia, Viale Pinto 1, 71121 Foggia, Italy
4 UPLB Museum of Natural History, University of the Philippines—Los Baños, College, Laguna 4031, Philippines
* Correspondence: pierluigi.reveglia@unifg.it

Citation: Reveglia, P.; Billones-Baaijens, R.; Millera Niem, J.; Masi, M.; Cimmino, A.; Evidente, A.; Savocchia, S. Production of Phytotoxic Metabolites by Botryosphaeriaceae in Naturally Infected and Artificially Inoculated Grapevines. *Plants* **2021**, *10*, 802. https://doi.org/10.3390/plants10040802

Academic Editor: Carlos Agustí-Brisach

Received: 20 March 2021
Accepted: 12 April 2021
Published: 19 April 2021

Abstract: Grapevine trunk diseases (GTDs) are considered a serious problem to viticulture worldwide. Several GTD fungal pathogens produce phytotoxic metabolites (PMs) that were hypothesized to migrate to the foliage where they cause distinct symptoms. The role of PMs in the expression of Botryosphaeria dieback (BD) symptoms in naturally infected and artificially inoculated wood using molecular and analytical chemistry techniques was investigated. Wood samples from field vines naturally infected with BD and one-year-old vines inoculated with *Diplodia seriata*, *Spencermartinsia viticola* and *Dothiorella vidmadera* were analysed by cultural isolations, quantitative PCR (qPCR) and targeted LC-MS/MS to detect three PMs: (*R*)-mellein, protocatechuic acid and spencertoxin. (*R*)-mellein was detected in symptomatic naturally infected wood and vines artificially inoculated with *D. seriata* but was absent in all non-symptomatic wood. The amount of (*R*)-mellein detected was correlated with the amount of pathogen DNA detected by qPCR. Protocatechuic acid and spencertoxin were absent in all inoculated wood samples. (*R*)-mellein may be produced by the pathogen during infection to break down the wood, however it was not translocated into other parts of the vine. The foliar symptoms previously reported in vineyards may be due to a combination of PMs produced and climatic and physiological factors that require further investigation.

Keywords: *Vitis vinifera*; Botryosphaeria dieback; foliar symptoms; (*R*)-mellein; qPCR; LC-MS/MS

1. Introduction

Grapevine is one of the most economically important crops worldwide, with approximately 71% of the world grape production being used for wine production [1]. A variety of fungal diseases threaten viticultural regions all over the world, compromising the yield and quality of the wine [2–5]. Among them, grapevine trunk diseases (GTDs), caused by one or several xylem-inhabiting fungi result in a progressive decline of vines, loss in productivity and eventually death of the vines [6].

Over the past few decades considerable knowledge has been gained on identifying GTDs, and therefore the frequency of symptoms reported due to these diseases has increased [6,7]. Recently, significant findings have been obtained on grapevine leaf stripe disease, a widely spread wood disease of Esca complex, regarding phytotoxic metabolites, symptom expression and their management [8,9]. However, the relationship of GTDs with biotic and abiotic stresses [10–13], the expression of symptoms and their effective management requires further investigation [6,14,15].

The main GTDs that threaten vineyards worldwide are Eutypa dieback (ED), Esca complex diseases, and Botryosphaeria dieback (BD), all of them caused by pathogenic fungi belonging to different families. Vines infected with ED, Esca complex, and BD usually present external symptoms such as necrotic buds, cane and trunk dieback, chlorotic leaves and leaf necrosis, and internal wood symptoms such as wedge-shaped necrotic lesions, arch-shaped necrotic lesions, brown streaking and a blackened cortex [6].

The interest in BD has increased substantially over the past decades due to increased incidence in vineyards worldwide [4,6,7,15,16]. Internal symptoms of BD include brown streaks and wedge-shaped discolorations in the wood, while external symptoms include death of the canes, shoots, and buds, stunting of shoots, delayed budburst, and foliar symptoms [4,17,18]. The appearance of foliar symptoms in grapevines infected with Esca complex and ED is frequently reported in the northern hemisphere [6,19–22]. Foliar symptoms in vines infected with BD have also been reported in Europe [17,23,24]. Nevertheless, and to our knowledge, BD foliar symptoms have not been observed in Australian vineyards to date [13].

Many foliar symptoms are usually associated with phytotoxic metabolites (PMs) produced by the causal fungi [25]. A general hypothesis suggests that PMs are translocated far from the inoculum without being detoxified, resulting in the development of foliar symptoms [26]. However, no conclusive data has been reported to support this hypothesis to date [25,27–30]. Therefore, the role of PMs in the expression of foliar symptoms requires further investigation. European isolates of Botryosphaeriaceae involved in BD have been reported to produce phytotoxins in vitro [31–33] and PMs produced by Botryosphaeriaceae pathogens have also been detected in infected wood with BD and Esca symptoms [34].

Considering this background and the absence of foliar symptoms in Australian vineyards, the following questions arose: (i) are the Botryosphaeriaceae spp. that infect grapevines in Australia capable of producing PMs in vitro?; (ii) are the PMs produced in vitro further produced and translocated in planta when the pathogens attack the host?; and (iii) are the PMs involved in symptom expression?

In 2016, studies were undertaken to investigate the ability of the most prevalent Australian Botryosphaeriaceae to produce PMs in vitro. These studies revealed all BD pathogens produced PMs in vitro. Some of the purified PMs were shown to cause phytotoxicity in detached leaves [35–40].

Based on these previous results, the objective of this study was to investigate the production of PMs by Botryosphaeriaceae pathogens in naturally and artificially infected grapevines, and their likelihood of translocation in woody tissue. A multifaceted approach using a combination of molecular (quantitative PCR) and analytical chemistry (LC/MS-MS) techniques were used for these investigations.

2. Results

2.1. Artificially Inoculated Vines

2.1.1. Wood Symptoms

All inoculated vines did not exhibit foliar symptoms or external necrosis and showed healthy development during the entire duration of the experiment. At 6 months post-inoculation, representative vines inoculated with the pathogens, exhibited vascular staining and discoloration of the wood close to the inoculation point (Figure S1) when the bark was removed, and the trunk was cut into cross sections.

However, the lesion lengths were not significantly different between varieties or any of the inoculation treatments, including the non-inoculated negative control vines ($p > 0.05$; Figure 1). No significant differences for lesion length were observed between Botryosphaeriaceae species. All three Botryosphaeriaceae species were re-isolated from their corresponding trunk sections with necrotic lesions, while negative control vines were free of BD pathogens.

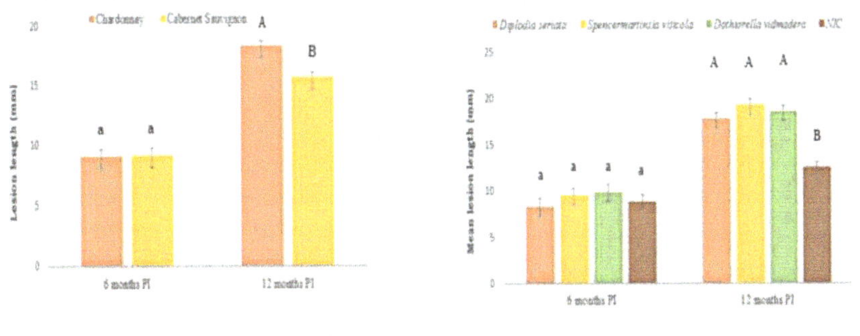

(a) **(b)**

Figure 1. (**a**) Overall mean lesion lengths on Chardonnay and Cabernet Sauvignon at 6- and 12-months (PI). (**b**) Mean lesion caused by *Diplodia seriata* H141a, *Dothiorella vidmadera* DAR78993 and *Spencermartinsia viticola* DAR78870 at 6- and 12-months PI = post inoculation. NIC = non-inoculated control. Lowercase letters refers to 6 months PI vines, uppercase letters refer to 12 months PI vines. Bars with different letters for each inoculation period are significantly different at $p \leq 0.05$ least significant difference (LSD). Error bars are standard error of the means.

At 12 months post-inoculation, vascular staining and discoloration of wood that progressed upward and downward from the inoculation point were also observed from the longitudinal sections. The lesion lengths for all treatments were significantly longer than those vines assessed at 6 months (Figure 1). The overall lesion lengths differed significantly between varieties with Chardonnay being the most susceptible with a mean lesion length of 20.2 ± 0.8 mm, which was significantly longer ($p = 0.01$) than the mean lesion length in Cabernet Sauvignon (17.3 ± 1.0mm) (Figure 1a). The lesion lengths also varied between inoculation treatments with lesions produced by *Dothiorella vidmadera* (20.8 ± 2.7 mm), *Spencermartinsia viticola* (20.4 ± 2.8 mm), and *Diplodia seriata* (19.4 ± 0.60 mm) being significantly longer ($p \leq 0.05$) compared to the negative control vines (13.1 ± 1.9 mm). No significant differences for lesion length were observed between Botryosphaeriaceae species. There were no significant interactions between varieties and inoculation treatments based on lesion lengths ($p > 0.05$) (Figure 1b).

2.1.2. Botryosphaeriaceae DNA in Wood Tissues of Artificially Inoculated Vines

At 6 months post-inoculation, Botryosphaeriaceae DNA was detected by qPCR from wood sections with necrotic lesions near the inoculation point (IP) from vines inoculated with all three Botryosphaeriaceae species (Figure 1). The highest amount of pathogen DNA was detected from Chardonnay vines inoculated with *S. viticola*, with significantly higher amounts of Botryosphaeriaceae DNA than Chardonnay inoculated with *D. seriata* ($p = 0.009$). The amount of Botryosphaeriaceae DNA from Chardonnay inoculated with *Do. vidmadera* was not significantly different from those inoculated with *D. seriata* or *S. viticola* (Figure 2). Botryosphaeriaceae DNA detected from all Cabernet Sauvignon infected vines was significantly lower compared to those from Chardonnay, regardless of species. Significant interaction between variety and species was observed ($p < 0.05$). Botryosphaeriaceae DNA was not detected from any of the non-inoculated vines, while DNA from wood samples inoculated with *Neofusicoccum parvum* from a separate study and included as the positive control for all qPCR assays, all tested positive to the Botryosphaeriaceae DNA (data not shown).

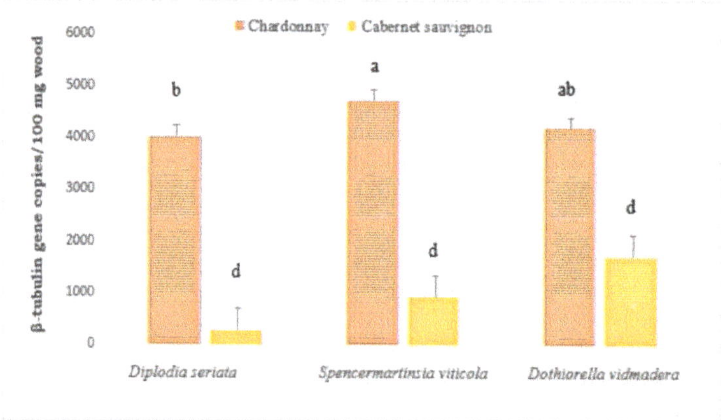

Figure 2. Overall mean of Botryosphaeriaceae β-tubulin gene copies detected from *Vitis vinifera* (cvs. Chardonnay and Cabernet Sauvignon) inoculated with *Diplodia seriata* H141a, *Spencermartinsia viticola* DAR78870 and *Dothiorella vidmadera* DAR78993 at 6 months post inoculation using quantitative PCR. Bars with different letters for each inoculation period are significantly different at $p \leq 0.05$ LSD. Error bars are standard error of the means. All non-inoculated control vines tested negative to Botryosphaeriaceae DNA and were excluded in the graph.

At 6 months, Botryosphaeriaceae DNA was further detected by qPCR from non-necrotic tissues (AA′; Figure 3) for some inoculated vines. For Chardonnay, Botryosphaeriaceae DNA (69 copies) was detected from lesion-free wood sections (AA′; Figure 3) in one out of three replicate vines inoculated with *S. viticola*, while Botryosphaeriaceae DNA was not detected from any of the lesion-free wood sections (AA′; Figure 3) of vines inoculated with *D. seriata* and *Do. vidmadera*. For Cabernet Sauvignon, one out of three replicate vines inoculated with *D. seriata*, *S. viticola* and *Do. vidmadera* showed a different number of copies of Botryosphaeriaceae DNA at 40, 440 and 190 copies, respectively. For those AA′ samples (Figure 3) that were positive to qPCR, none of their subsequent BB′ sample sections (Figure 3) were positive to pathogen DNA (data not presented).

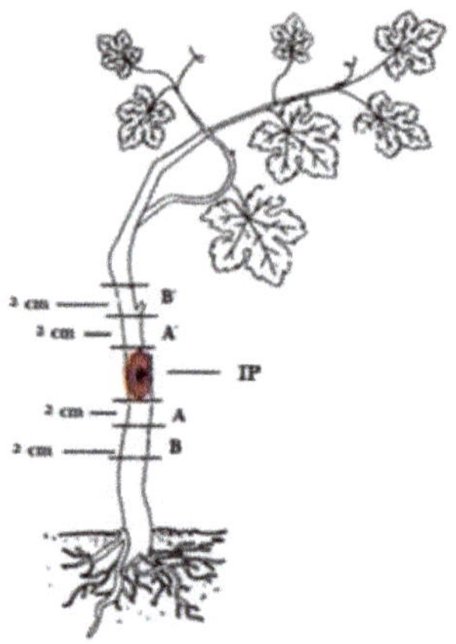

Figure 3. A diagram of an inoculated vine showing the positions and sizes of tissue samples collected and used for analysis. IP: inoculation point. AA′: lesion-free trunk sections subsequent to necrotic lesions; BB′: lesion-free trunk sections subsequent to AA′.

At 12 months post-inoculation, Botryosphaeriaceae DNA was further detected by qPCR from necrotic wood samples (IP; Figure 1) excised from vines inoculated with either of the three species (Figure 4). The amount of pathogen DNA across treatments increased by 10-fold compared to the amount detected at 6 months post-inoculation. The highest amount of pathogen DNA was detected from Chardonnay vines inoculated with *D. seriata* with significantly higher amounts compared to Chardonnay and Cabernet Sauvignon vines inoculated with *S. viticola* (Chardonnay $p = 0.007$; Cabernet Sauvignon $p = 0.000007$) and *Do. Vidmadera* (Chardonnay $p = 0.003$; Cabernet Sauvignon $p = 0.00008$). The lowest amount of pathogen DNA was detected from Cabernet Sauvignon vines inoculated with *D. seriata* and this was significantly lower than for all other inoculated vines. Significant interaction between variety and species was observed ($p < 0.05$) which was associated with the highest amount of Botryosphaeriaceae DNA from Chardonnay and the lowest from Cabernet Sauvignon vines inoculated with *D. seriata*. Furthermore, all asymptomatic tissues (AA′; Figure 3) regardless of treatments were negative to Botryosphaeriaceae DNA by qPCR at 12 months post inoculation.

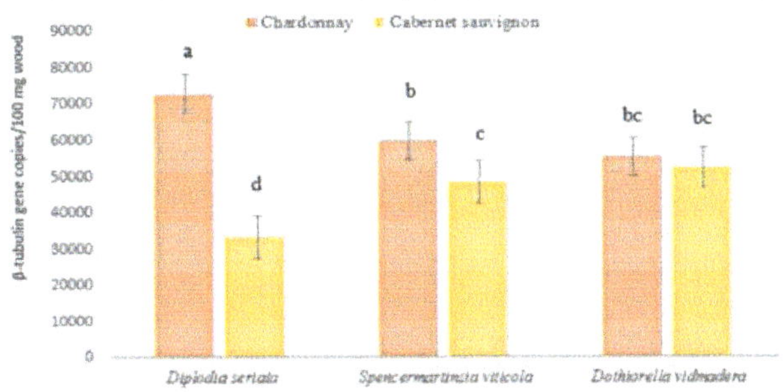

Figure 4. Overall mean of Botryosphaeriaceae β-tubulin gene copies detected from necrotic tissues at the inoculation point of *Vitis vinifera* (cvs. Chardonnay and Cabernet Sauvignon) vines inoculated with *Diplodia seriata* H141a, *Spencermartinsia viticola* DAR78870 and *Dothiorella vidmadera* DAR78993 at 12 months post inoculation using quantitative PCR. Bars with different letters for each inoculation period are significantly different at $p \leq 0.05$ LSD. Error bars are standard error of the means. All non-inoculated vines tested negative to Botryosphaeriaceae DNA and were excluded in the graph.

2.2. Naturally Infected Vines

All wood samples with dieback symptoms, cankers and typical wedge-shaped necrosis were positive to Botryosphaeriaceae species (Figure 5, Table 1). *D. seriata* was the most prevalent species being present in three vineyards and six out of the nine vines sampled. In Hilltops, one vine was positive to *N. parvum*, while the other two vines were positive to *D. seriata* and *D. mutila*. In Tumbarumba, two vines were positive to *D. seriata* while the other vines were positive to both *D. seriata* and *N. parvum*. For the Riverina, two vines were positive to *D. seriata* while one was positive to *B. dothidea*.

2.3. Botryosphaeriaceae DNA in Wood Tissues of Naturally Infected Vines

All necrotic tissue samples tested positive to Botryosphaeriaceae DNA by qPCR (Figure 6). Wood samples from the Riverina vineyard contained the highest number of DNA copies that was 5-fold higher and 1.4-fold higher than the DNA from vines in Hilltops and Tumbarumba vineyards, respectively. The lowest amount of DNA was detected from all vines in Hilltops and Tumbarumba. No Botryosphaeriaceae DNA was detected in any of the non-necrotic wood samples that were collected from the same vines. Statistical analysis was not applied to the naturally infected vines because of differences in the storage and sampling of the collected wood.

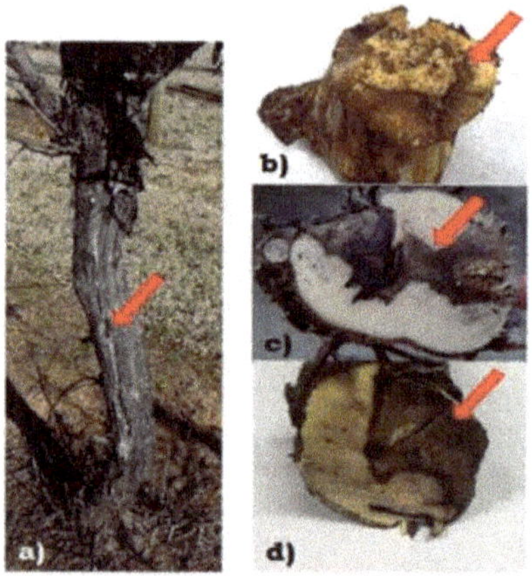

Figure 5. (**a**) Vine with trunk canker (arrow); (**b**) cross-section of the trunk with central necrosis; (**c**) cross-section of the cordon with wedge-shape necrosis; (**d**) cross-section of the trunk with wedge-shape necrosis.

Table 1. Botryosphaeriaceae species isolated from naturally infected vines from three vineyards in New South Wales, Australia.

Location	Variety	Vine Sample	Botryosphaeriaceae Species
Hilltops	Chardonnay	1	*Neofusicoccum parvum*
		2	*Diplodia seriata*
		3	*Diplodia mutila*
Tumbarumba	Chardonnay	1	*D.seriata*
		2	*D. seriata,* *N. parvum*
		3	*D. seriata*
Riverina	Shiraz	1	*D. seriata*
		2	*Botryosphaeria dothidea*
		3	*D. seriata*

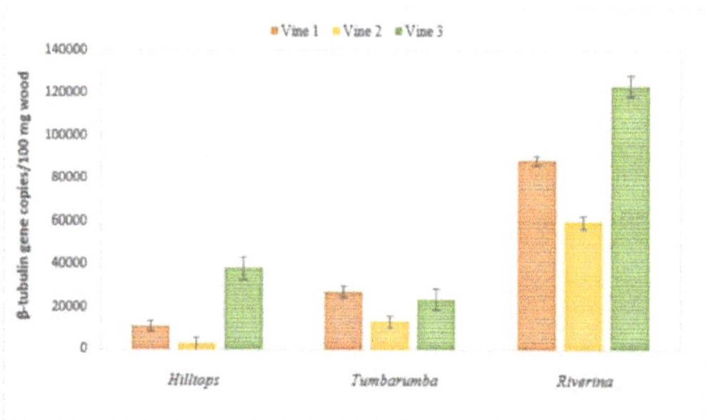

Figure 6. Number of copies of pathogen DNA quantified by qPCR in naturally infected vines from three vineyards in New South Wales, Australia. Error bars are standard deviation of the means.

2.4. Selection of Protocol for Extraction of PMs from Wood

Two different protocols were tested for the extraction of PMs from naturally infected wood material: (A) *n*-hexane/MeOH [34]; and (B) H_2O/MeOH/CH_3Cl [41]. Protocol B was less time consuming and resulted in a higher amount of organic compounds. Analysis of the organic extracts using LC-MS/MS to detect (*R*)-mellein also showed that protocol B yielded a higher amount of the target metabolite. The peak of (*R*)-mellein detected in the extract obtained with Protocol B (Figure S2, red) was 12-fold higher than the organic extract obtained with protocol A (Figure S2, green). Identification of (*R*)-mellein from extracts of symptomatic wood samples was accomplished according to its retention time, precursor ion 179.1 m/z $[M + H]^+$ and fragment ions (Figure S2b). Protocol B was selected for extracting PMs from inoculated vines.

2.5. PMs in Wood Tissues of Naturally Infected Vines

Both symptomatic and asymptomatic wood materials were analysed by LC-MS/MS after extraction, giving different results depending on the vines sampled and vineyard location. For symptomatic wood samples, only one out of three Chardonnay vines from Hilltops, two out of three Chardonnay vines from Tumbarumba and all Shiraz vines from the Riverina were positive to (*R*)-mellein. (*R*)-mellein was not detected in any of the lesion-free wood samples. A comparison of the chromatograms (Figure S3a) led to the detection of a signal at a retention time of 27.44 min in the infected sample (Figure S3a, red, green and purple), which was absent in the asymptomatic sample (Figure S3a, black). The signal in the chromatograms was due to the (*R*)-mellein fragment ions (Figure S3b). Furthermore, the area of (*R*)-mellein signals in the LC-MS/MS analysis was most significant for the Riverina vines with a high number of DNA copies of the pathogen detected in the trunk (Figure S3c), indicating production of (*R*)-mellein by BD pathogens in field vines.

2.6. PMs in Wood Tissues of Artificially Inoculated Vines

At six months post-inoculation, no spencertoxin, protocatechuic acid and (*R*)-mellein were detected in any of the wood sections (IP, Figure 3) from non-inoculated vines. However, (*R*)-mellein was detected from the extracts of necrotic wood samples (IP) of both Chardonnay and Cabernet Sauvignon vines infected with *D. seriata* at 12 months post-inoculation (Figure S4, red) indicating that this pathogen produced (*R*)-mellein during fungal colonization. Furthermore, neither spencertoxin nor protocatechuic acid were detected from necrotic wood samples (IP) infected with *S. viticola* and *Do. vidmadera*, respectively, at 12 months post-inoculation. (*R*)-mellein was further detected from extracts obtained from the Chardonnay vine infected with *N. parvum* (DAR78998) from a separate experiment and included in the test as a positive control (Figure S4, green). The peak of (*R*)-mellein detected from this vine was 2.4-fold higher than the amount detected in vines infected with *D. seriata* 12 months post-inoculation. No target PMs were detected in the negative control vines. Targeted PMs were also not detected in lesion-free wood samples (AA′ and BB′, Figure 3), thus, no further analyses were performed for tissues collected further away from the necrotic wood (IP) including the leaves.

3. Discussion

To the best of our knowledge, this is the first study to investigate the production and translocation of PMs by Botryosphaeriaceae species in BD naturally infected and artificially inoculated vines using a multifaceted approach. However, multidisciplinary approaches for the detection of target PMs in plant tissues have been reported in other pathosystems [42–45].

This current study also represents the first study to use a combination of conventional plant pathology and molecular techniques to detect and quantify Botryosphaeriaceae DNA from artificially inoculated and naturally infected vines. The qPCR primers and probe used in this study [46] are not species-specific, therefore they cannot distinguish the Botryosphaeriaceae pathogen at the species level. However, the isolation of pathogens from naturally infected wood allowed the identification of Botryosphaeriaceae spp. Similarly, re-isolations of the pathogens from artificially inoculated vines resulted in the recovery of *D. seriata*, *S. viticola* and *Do. vidmadera* from previously inoculated vines. Lesion length by itself was insufficient to show differences in disease severity and pathogen virulence at the early stage of infection (6 months post-inoculation). Furthermore, negative control vines also exhibited necrosis near the inoculation point similar to those vines inoculated with pathogens. The qPCR analysis, on the other hand, was able to quantify the amount of pathogen DNA between treatments and assessment periods. Consequently, pathogen DNA was not detected in negative control vines. It is important to note that low levels of pathogen DNA were detected from lesion-free tissues adjacent to the lesions from a few inoculated vines at 6-month post-inoculation, particularly for *S. viticola*. This suggests that the pathogens were able to move endophytically beyond the lesions as latent pathogen, similar to the study reported by Billones-Baaijens et al. [47]. The absence of pathogen DNA in non-necrotic tissues adjacent to the lesions at 12 months post inoculation suggests that as the infection advanced, these pathogens shifted from being latent to necrotrophic.

The qPCR analysis of inoculated vines showed a significant difference in susceptibility between Chardonnay and Cabernet Sauvignon to different Botryosphaeriaceae species, with the former being more susceptible to *D. seriata*. The reduced susceptibility of Cabernet Sauvignon to some Botryosphaeriaceae species may be associated with the amount of stilbene polyphenols that are usually higher in red vine varieties. These compounds have fungistatic activity [48,49], that may assist the plant in limiting infection by BD pathogens. Nevertheless, stilbene polyphenols antimicrobial activity depends on the pathogen infecting the plant [48,49]. For instance, Cabernet Sauvignon is one of the most susceptible cultivars to Esca pathogens: *Phaeomoniella chlamydospora* and *Phaeoacremonium minimum* [10].

The qPCR analysis used in this study was also useful in quantifying pathogen DNA in symptomatic wood samples collected from naturally infected vines. Overall, these results confirm that molecular techniques could be applied to determine and quantify the pathogens in field material.

The LC-MS/MS analysis of the naturally infected and artificially inoculated vines showed that (*R*)-mellein can be detected in infected woody tissues with symptoms of BD. A previous study that investigated the production of Botryosphaeriaceae PMs in planta only used a limited number of naturally infected vines with both BD and Esca symptoms [34]. More recently (*R*)-mellein was also detected in grapevine tissues with symptoms of Esca and Grapevine leaf stripe disease (GLSD) [50]. However, none of the previous studies identified the pathogens or quantified the pathogen DNA in the analysed samples. Furthermore, the application of a multifaceted approach suggests a probable correlation between the amount of pathogen DNA in the wood and the area of the peak of the (*R*)-mellein in the chromatograms. This correlation was more evident in the Riverina vines, which resulted in more intense peaks corresponding to (*R*)-mellein and the highest amount of pathogen DNA in the wood. Moreover, the amount of (*R*)-mellein detected in the wood infected with *N. parvum*, an aggressive Botryosphaeriaceae pathogen [4], 6 months post-inoculation was 2.4-fold higher than the amount detected in vines infected with the lesser aggressive *D. seriata* at 12 months post-inoculation. These data further support those previously reported that the amount of (*R*)-mellein produced by *N. parvum* and *D. seriata* under in vitro conditions is proportional to the aggressiveness of the pathogens [51].

BD pathogens colonize the woody parts of the plants, and during colonization, the fungi can produce and release a series of PMs. The translocation hypothesis suggests that PMs produced by the pathogens can migrate far from the inoculum along the asymptomatic wood, reaching the leaves or the green shoots without being catabolized or without being entirely detoxified [26]. The main goal of our experiment was to verify the movement of targeted PMs along the trunk. For this purpose, symptomatic and adjacent lesion-free functional woody tissues (trunk and branches) were sampled. To ensure that the detection of the target PMs in asymptomatic wood was due to their migration, qPCR analysis was performed. Our results showed that (*R*)-mellein was detected only in the wood samples with necrotic lesions showing a high amount of pathogen DNA. The failure to detect (*R*)-mellein in all the lesion-free wood samples suggests that, at least under these experimental conditions, the translocation of (*R*)-mellein in its native form did not occur along the trunk to the foliage. It is possible that *D. seriata* produces (*R*)-mellein as part of its strategy to breakdown the host cells during infection. In a recent study using in vitro plantlets, Trotel-Aziz et al. showed that *(R)*-mellein strongly suppresses the expression of genes involved in plant defence and that *(R)*-mellein may be accumulated in planta in its native chemical form. The detection of this compound in symptomatic wood during our experiment supports this hypothesis [52].

However, the mere presence of PMs in the wood may not be enough to induce foliar symptoms in grapevines with BD. The development of symptoms may be more complicated than previously thought, and they may arise from interactions between biotic and abiotic stresses (water stress, drought, heat stress), which require more in-depth studies to understand. Fungi involved in GTDs can act as endophytes for several years before becoming pathogens, and many have hypothesised that the abiotic conditions, in particular thermal and water stress, can weaken the plant defence and therefore result in the development of GTD foliar symptom [9,10,22,53–56]. For instance, previous investigation, showed that GLSD leaf symptoms increased when the rainfall was abundant in June–July [54].

More comprehensive field studies on the influence of climatic conditions on foliar symptoms associated with GTD, including the role of PMs produced in planta, are fundamental for elucidating the relationship between fungal PMs and physiological changes in the vine which result in the expression of foliar symptoms.

Another result arising from this study was the failure to detect protocatechuic acid and spencertoxin in the LC-MS/MS analysis of vines inoculated with *D. vidmadera* and *S. viticola*, respectively. These two species were shown to produce these PMs in vitro [35,39]. The lack of detection of these PMs may be due to various reasons: (i) they may not be produced in planta; (ii) they could be detoxified by the plant; (iii) they may form toxin derivatives with other compounds; and/or (iv) they may be irreversibly bound to the wood contributing to lesion expression. The latter hypothesis was already suggested for other phytotoxins produced by the ED pathogen, *E. lata* [27–29]. Overall, the general conclusion may be that not all the secondary metabolites produced in vitro can be detected in planta since their fate mainly depends on the biological role played in the interaction of the pathogen with the host. This can be further validated by applying our approach to different experimental conditions. For instance, plant materials with reported BD foliar symptoms could be analyzed to further investigate the role of these PMs in foliar symptom development. Comparing data obtained from various vine-growing regions can be valuable to validate our results and help to explain the lack of BD foliar symptoms observed in Australian vineyards.

PMs are also known to interact with different cellular targets or can inhibit the activity of plant enzymes [57]. All these processes can result in the formation of toxin conjugates or derivatives that could explain the failure to detect target PMs in their native form. Metabolomics approaches can be applied to investigate the formation of PM derivatives and their role in symptoms development or plant-pathogen interaction. Indeed, untargeted, and targeted metabolomics have become fundamental tools in plant science and chemical ecology [58,59]. Recently, metabolic changes in grapevine wood infected with *N. parvum* have been reported [60]. This was accompanied by an accumulation of a number of unknown metabolites in the infected wood samples. Their structural elucidation will be crucial to better understand the response of the plant to GTDs and to also identify if any of these unknown metabolites could be potential derivatives of PMs.

4. Materials and Methods

4.1. Artificially Inoculated Vines

4.1.1. Planting Materials

Vitis vinifera cvs. Chardonnay and Cabernet Sauvignon, two of the most commonly grown varieties in Australian vineyards [61] were selected for a glasshouse experiment. Twenty- eight dormant, apparently healthy cuttings for each variety were collected in a commercial vineyard in Hilltops, New South Wales (NSW), Australia in Winter (June 2017) and stored at 4 °C for 4 weeks until rooting. All cuttings were surface-sterilised with 0.5% sodium hypochlorite for 1 min, rinsed twice with tap water and rooted in plastic trays containing perlite. The trays were placed on heat mats at 30 °C for 4 weeks to facilitate rooting. The rootlings were planted in 10 L pots containing commercial garden mix (60% compost, 20% wash sand, 20% screen loam). All vines were maintained in a glasshouse (17–27 °C) and watered every 12 h for 5 min (8 L/h) with an automatic dripper system for 6–12 months until assessment.

4.1.2. Fungal Isolates

The three Botryosphaeriaceae species: (a) *Diplodia seriata* H141a; (b) *Spencermartinsia viticola* DAR78870 and; (c) *Dothiorella vidmadera* DAR78993 from the National Wine and Grape Industry Centre (Charles Sturt University, Wagga Wagga, NSW, Australia) culture collection, which produced PMs in vitro and were previously characterised [33,34,37], were used for inoculating the glasshouse vines in Spring (November 2017). For inoculations, wounds were created in the middle internode of the trunk for each vine using a flame-sterilised 4 mm sterile cork borer. Mycelial plugs (4 mm) cut from the margins of 4- day-old cultures of the selected fungal species grown on potato dextrose agar supplemented with Chloramphenicol (100 mg/L) (PDA-C), were inserted into the wounds and sealed with Parafilm (Bemis, USA). Sterile non-colonised plugs of PDA-C were used as negative controls. The inoculated vines for each variety were arranged in a randomised complete block design (RCBD) at 7 replications per inoculum per variety combination.

4.1.3. Sampling of Artificially Inoculated Vines

The trunks of randomly selected vines were cut at the base and shoots were trimmed off. The bark surrounding the inoculation point was removed and lesions were measured using a digital caliper (Workzone, Australia). Tissue samples from each vine were collected as shown in Figure 3. Three different samples were collected from each vine: (a) trunk sections with visible lesions including the inoculation point (IP); (b) 2 cm lesion-free trunk sections cut above and below the necrotic lesions (labelled as AA'); (c) 2 cm trunk sections above and below section A and A' (labelled as BB'). The samples were surface-sterilised for 2 min in 70% ethanol and rinsed three times with sterile deionised water (SDW) before cutting longitudinally to obtain four quarters of each section. One quarter of the section was used for isolation of the pathogen, while the remaining sections were stored at −80 °C and used for DNA and toxin extractions.

4.1.4. Fungal Isolation from Artificially Inoculated Vines

One quarter of surface-sterilised trunk sections collected from the inoculated vines (Figure 1) were placed onto PDA-C. Plates were incubated at 25 °C in the dark and observed for growth of Botryosphaeriaceae species for 4–7 days. The re-isolated pathogens were identified using morphological and molecular methods. Three vines per treatment were assessed at 6 months post inoculation in Autumn (May 2018). Four vines per treatment were further assessed at 12 months post inoculation which occurred in Spring (November 2018).

4.2. Fungal Isolation from Naturally Infected Vines

Wood samples (cordons and trunks) exhibiting BD cankers were collected from 20–24 years old grapevines from three wine regions in NSW, Australia in Winter (June–August 2017). The Tumbarumba region has a cool climate with an altitude of 700 m.

The Hilltops region has a continental climate with an altitude of 450 m. The Riverina region has a semiarid climate with an altitude of 66–540 m [62]. Four to five wood pieces (10–20 cm) with necrotic lesions and non-necrotic tissues 20–30 cm away from the necrotic ones were cut from three different vines in Hilltops (cv. Shiraz), Riverina (cv. Shiraz) and Tumbarumba (cv. Chardonnay) and stored at 4 °C for 2 weeks until processed. Each sample were processed and analysed individually. For isolations, the bark was removed to expose the necrotic lesions and the wood was cut into ~1 cm sections (20–24 pcs) with each section containing necrotic and healthy wood. All sections were surface-sterilised following the methods described for the inoculated vines. Approximately 10–12 pieces of the surface-sterilised tissues were stored at −80 °C for DNA and toxin extractions. The remaining sections were plated onto PDA-C with four sections per plate for a total of three plates per vine. Plates were incubated at 25 °C for 4–7 days and observed for growth of Botryosphaeriaceae species. All Botryosphaeriaceae isolates were identified by morphological and molecular methods.

4.3. DNA Extractions from Fungal Mycelia

DNA samples were extracted from the mycelium for all the isolated fungi using PrepMan Ultra (Applied Biosystems, UK) and following the manufacturer's instructions. Mycelium (~100 mg) was scraped from the edge of the colony for each isolate using a sterile pipette tip and transferred into a sterile 1.5 mL tube containing 100 µL of the PrepMan Ultra (Applied Biosystems, UK) preparation reagent. Mycelial suspensions were vortexed for 30 s and incubated at 95 °C in a heat block for 10 min. The tubes were centrifuged for 2 min at 3220× *g* and 50 µL of the supernatant was transferred to a new sterile 1.5 mL tube and stored at −20 °C until required for PCR.

4.4. DNA Extraction from Grapevine Wood

Wood samples stored at −80 °C were freeze-dried (Christ, John Morris Scientific, USA) for 24–36 h. Dried samples were homogenized at 20 Hz for 2 min 30 s using 10 mL grinding jars attached to a TissueLyser II (Qiagen, Hilden, Germany). The ground wood (100 mg) was transferred into a sterile 2 mL tube for DNA extraction while the remaining ground wood was stored in a separate tube for the extraction of toxins. DNA was extracted from wood samples using the methods described by Pouzoulet et al. [63] with some modifications. The CTAB (Cetyl Trimethyl Ammonium Bromide) extraction buffer was prepared according to Doyle and Doyle [64], and 1 ml was added to each tube containing 100 mg of ground wood and gently mixed by pipetting. The mixture was incubated at 65 °C for 1 h using a heat block. After incubation, 500 µL of chloroform/isoamyl alcohol (24:1, Sigma Aldrich, St. Louis, MI, USA) was added and the tube was inverted 10×, incubated on ice for 5 min and then centrifuged at 4 °C for 10 min at 2300× *g*. Approximately 420 µL of the lysate was pipetted into a QIAshredder spin column placed in a 2 mL collection tube from the Qiagen DNeasy Plant DNA extraction kit (Qiagen, Hilden, Germany) and centrifuged for 2 min at 20,000× *g*. The subsequent steps were performed using the buffers, materials and protocol from the DNeasy Plant DNA extraction kit. All DNA samples were eluted to a final volume of 100 µL using the Qiagen AE buffer. All DNA samples were quantified using a Quantus™ Fluorometer (Promega, Madison, WI, USA) prior to qPCR.

DNA was further extracted from the necrotic tissues collected from potted vines (cv. Chardonnay) inoculated six months prior with *N. parvum* DAR78998 following the methods described above. This isolate was found to be highly virulent in a separate experiment [18] and is known to produce (*R*)-mellein in vitro [35]. The DNA extracted from these vines was included in the qPCR analysis and toxin analyses.

4.5. Identification of Isolated Botryosphaeriaceae by PCR

All Botryosphaeriaceae recovered from the artificially inoculated and naturally infected vines were identified using PCR and DNA sequencing. To amplify the internal transcribed spacer (ITS) region of the ribosomal DNA of the pathogens, PCR was performed using universal primers ITS1 and ITS4 [65]. Each 25 µL PCR reaction contained 1x PCR buffer (Bioline, Memphis, TN, USA), 0.4 µL for each of primer, 1.25 U of My TaqRed DNA polymerase (Bioline, Memphis, USA) and approximately 1–5 ng of DNA template. PCRs were performed using a thermal cycler (C100 Thermal cycler, Biorad Laboratories, Pty, Ltd., Hercules, CA, USA) under the following conditions: initial denaturation at 95 °C for 5 min, 35 cycles of 30 s at 94 °C, 45 s at 55 °C, and 90 s at 72 °C, with a final extension of 5 min at 72 °C. Following amplification, the PCR products were visualized by gel electrophoresis. PCR products were purified with FavorPrep Gel/PCR purification kit (Favorgen Biotech Corp, Taiwan) and sequenced at the Australian Genome Research Facility (AGRF; Sydney, NSW, Australia). All DNA sequences and chromatographs were analyzed using the DNAMAN 5.2 (Lynnon Biosoft©, San Ramon, CA, USA) and Chromas Lite 2.1© (Technelysium PTY Ltd, Brisbane, Australia) software. All trimmed DNA sequences were analysed using the Basic Local Alignment Search Tool (BLAST) in GenBank (https://www.ncbi.nlm.nih.gov/genbank/ (accessed on 18 April 2021)).

4.6. Quantification of Botryosphaeriaceae spp. from Wood Samples by qPCR

The qPCR assay using Botryosphaeriaceae multi-species primers and hydrolysis probe developed by Billones-Baaijens et al. [46] were used to detect and quantify Botryosphaeriaceae spp. from artificially inoculated and naturally infected vines. All qPCR assays were performed with the RotorGene 6000 system (Corbett Life Science, Qiagen, Hilden, Germany) using Botryosphaeriaceae multi-species primers Bot-BtF1 (5′-GTATGGCAATCTTCTGAACG-3′) and Bot-BtR1 (5′-CAGTTGTTACCGGCRCCRGA-3′), and a hydrolysis probe, Taq-Bot probe 5′-/56-FAM/TCGAGCCCG/ZEN/GCACSATGGAT/3IBkFQ/-3′) [41]. For each assay, three controls were included: (1) non-template control (H_2O); (2) standard (500 pg) Bot-Btub gBlock [41]; (3) DNA from vine inoculated with *N. parvum* DAR78998 at four technical replicates each.

For the artificially inoculated vines, all necrotic tissue samples collected from the IP were first analysed followed by the tissues collected above and below the necrotic lesions (AA′). For AA′ tissues which tested positive to Botryosphaeriaceae DNA, their subsequent BB samples were further analysed for a total of 32 and 36 samples for 6- and 12- months PI, respectively. For naturally infected vines, wood samples containing necrotic tissues were first analysed by qPCR. When the necrotic tissue sample was positive to qPCR, the healthy wood samples away from the necrotic tissue were further analysed for a total of 18 samples overall.

To determine the amount of pathogen DNA that was amplified by each qPCR assay, previously developed standard curves [46] were imported in the Rotor-Gene Q software (Version 2.3.1). The standard (Bot-Btub gBlock, 500 pg) that was included in each qPCR assay was used to calibrate the imported standard curve and the resulting regression equations were used to quantify the number of Botryosphaeriaceae β-tubulin gene copies in each reaction as previously described by Billones-Baaijens et al. [46] following the MIQE guidelines [66]. To calculate the number of copies of the Botryosphaeriaceae β-tubulin gene in each wood sample, the following formula was used:

$$N = g\ (d \times c)/t \times c, \quad (1)$$

where N: calculated number of β-tubulin gene copies in one wood sample, g: the mean number of gene copies detected by qPCR; d: total gDNA extracted from 100 mg of wood (100 µL), c: DNA concentration (µL); T: the amount of DNA template (5 µL) in one reaction.

4.7. Statistical Analysis

Data arising from the glasshouse experiment were analysed using IBM SPSS 24 software. All data were tested for homogeneity using Levene's test at $p \leq 0.05$. For inoculated vines, univariate analysis of variance (ANOVA) was used to assess differences in lesion lengths and pathogen copies between varieties, inoculated pathogens and their interactions ($p \leq 0.05$). All means were separated by pairwise comparison using Fischer's least significant differences (LSD) test at 5% significance level. Statistical analysis was not applied to the naturally infected vines because of differences in the storage and sampling of the collected wood.

4.8. Chemicals and Standards for LC-MS/MS

CH_3Cl, MeOH and *n*-hexane were analytical grade (Sigma-Aldrich, St. Louis, USA). H_2O, 0.1% HCOOH and MeCN were LC-MS grade (Sigma-Aldrich, St. Louis, USA). The (*R*)-mellein [35], protocatechuic alcohol [36] and spencertoxin [39] that were used as standards were isolated from in vitro cultures of *D. seriata*, *Do. vidmadera* and *S. viticola* as previously reported.

4.9. Testing of Protocols for Extraction of PMs from Wood

Two published extraction protocols were compared for their suitability to extract toxins from wood samples [34,41]. Freeze-dried wood samples collected from the naturally infected vines (36 in total) were extracted using both protocols and were analysed by LC-MS/MS. Freeze-dried wood samples (100 mg) were used for each extraction. The protocol by Saviano et al. [41] was subsequently used to extract PMs from the wood samples (100 mg) collected from inoculated vines (100 in total).

4.10. LC-MS/MS Analysis of Targeted PMs from Wood

Analyses were carried out using a 1290 Infinity II LC system (Agilent) hyphenated to an Agilent 6470 triple quadrupole (QqQ). The UPLC system included a binary pump and a cooled autosampler maintained at 15 °C. Mass Hunter software was used to control the instruments and to acquire the data which were then processed for analysis. The chromatographic separation was performed using a reverse phase column Phenomenex Luna 5.4 µm 250 × 4 mm i.d., protected by a security guard column Phenomenex maintained at 30 °C. The mobile phase consisted of H_2O 1% (*v*/*v*) HCOOH (Phase A) and MeCN (Phase B). The flow rate was 0.7 mL/min, the gradient system was initiated with 10% of Phase B for 2 min and reached 30% at 15 min, 80% at 25 min, isocratic until 27 min and 95% at 40 min. Samples were injected into the column with an injection volume of 20 µL.

The Agilent 6470 triple quadrupole (QqQ) was used as the detector in MRM mode with electrospray ionization (ESI) in positive ionization mode. The source and desolvation temperatures were respectively set at 350 °C, Nebulizer, 40 psi; N_2 flow, 12 L min^{-1}. Capillary voltage was set at 3.5 kV in positive mode. The MRM transitions (precursor ion → daughter ions), fragmentor energy and collision energy for (*R*)-mellein, spencertoxin and protocatechuic acid were optimized using the Agilent Optimizer Software, and the optimized parameters were shown in Table 2. For the parameter optimization, standards of (*R*)-mellein, spencertoxin and protocatechuic acid were used.

Table 2. Toxins used as standards for the LC-MS/MS and their corresponding optimized parameters.

Toxin	Precursor Ion m/z	Fragment Ion m/z	Fragmentor Voltage *	CV **	Retention Time (Min)
(*R*)-mellein 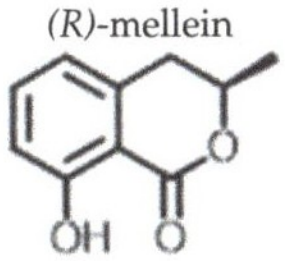	179.1 $[M + H]^+$	161.0 133.0 105.0	90	12 16 24	27.44

Table 2. *Cont.*

Toxin	Precursor Ion m/z	Fragment Ion m/z	Fragmentor Voltage *	CV **	Retention Time (Min)
Protocatechuic alcohol 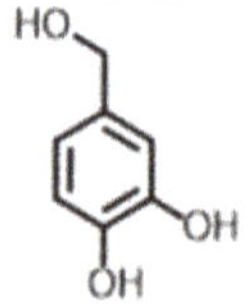	123.1 [M-H2O + H]$^+$	67.1 55.1 51.1	90	16 24 40	6.94
Spencertoxin 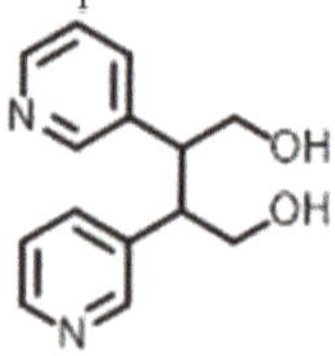	283.1 [M + K]$^+$	177.8 118	90	44 50	18.11

* Fragmentor voltage: controls the speed at which the ions pass through a medium pressure capillary between the electrospray chamber and the mass spectrometer. ** CV: collision energy voltage.

5. Conclusions

Detection of PMs in infected plant tissue may provide insights into the involvement of PMs in the pathogenicity of fungal isolates and symptom development of GTDs. This multidisciplinary approach was appropriate for investigating the production and translocation of (*R*)-mellein in vines showing BD symptoms. Our results highlighted a possible correlation between the amount of (*R*)-mellein and the amount of pathogen DNA copies in the wood samples. The development of a robust quantitative LC-MS/MS method for the detection of (*R*)-mellein can assist further investigations regarding this correlation. These data may provide valuable information on the infection strategies of the pathogens and may clarify the role played by (*R*)-mellein in the development of the disease.

Our study did not find any evidence that PMs are translocated into healthy tissues of the vine. Thus, the targeted PMs we investigated in this study were most likely produced by BD pathogens for penetrating and invading the hosts. The migration of PMs in their native form into the woody tissues of the plant should not be considered a generalised process for all pathosystems.

Supplementary Materials: The following are available online at https://www.mdpi.com/article/10.3390/plants10040802/s1, Figure S1: Representative photos of the inoculated vines showing internal staining of the wood, Figure S2: Representative LC-MS/MS chromatograms of naturally infected wood extract, Figure S3: LC-MS/MS analysis of Riverina vines, Figure S4: MRM Chromatograms of Chardonnay vine inoculated with *D. seriata*.

Author Contributions: Conceptualisation, S.S., R.B.-B., P.R. and A.E.; methodology, P.R., S.S. and R.B.-B.; validation, S.S. and R.B.-B.; formal analysis, S.S. and R.B.-B.; investigation, P.R. and J.M.N.; resources, S.S., R.B.-B. and A.E.; writing—original draft preparation, P.R.; writing—review and editing, P.R., R.B.-B., J.M.N., M.M., A.C., A.E. and S.S.; visualization, P.R., S.S. and R.B.-B.; supervision, S.S., R.B.-B. and A.E.; project administration, S.S., R.B.-B. and A.E.; funding acquisition, S.S. and A.E. All authors have read and agreed to the published version of the manuscript.

Funding: This work was supported by academic grants from the National Wine Grape Industry Centre, Charles Sturt University, Australia; Dipartimento di Scienze Chimiche, Università di Napoli Federico II, Italy and a Wine Australia PhD Student Support Grant.

Institutional Review Board Statement: Not applicable.

Informed Consent Statement: Not applicable.

Data Availability Statement: The data presented in this study are contained within the article and Supplementary Material.

Conflicts of Interest: The authors declare no conflict of interest.

References

1. Popa, E.O.; Roşca, I. Main trends of the pests management in agroecosystems of grapevine plantations. *Sci. Pap. Ser. Manag. Econ. Eng. Agric. Rural. Dev.* **2011**, *11*, 146–150.
2. Halleen, F.; Fourie, P.H.; Crous, P.W. A review of black foot disease of grapevine. *Phytopathol. Med.* **2006**, *45*, 55–67.
3. Gessler, C.; Pertot, I.; Perazzolli, M. *Plasmopara viticola*: A review of knowledge on downy mildew of grapevine and effective disease management. *Phytopathol. Med.* **2011**, *50*, 3–44.
4. Úrbez-Torres, J.R. The status of Botryosphaeriaceae species infecting grapevines. *Phytopathol. Med.* **2011**, *50*, 5–45.
5. Fontaine, F.; Pinto, C.; Vallet, J.; Clément, C.; Gomes, A.C.; Spagnolo, A. The effects of grapevine trunk diseases (GTDs) on vine physiology. *Eur. J. Plant Pathol.* **2016**, *144*, 707–721. [CrossRef]
6. Mondello, V.; Songy, A.; Battiston, E.; Pinto, C.; Coppin, C.; Trotel-Aziz, P.; Clément, C.; Mugnai, L.; Fontaine, F. Grapevine trunk diseases: A review of fifteen years of trials for their control with chemicals and biocontrol agents. *Plant Dis.* **2018**, *102*, 1189–1217. [CrossRef]
7. Baránek, M.; Armengol, J.; Holleinová, V.; Pečenka, J.; Calzarano, F.; Peňázová, E.; Vachůn, M.; Eichmeier, A. Incidence of symptoms and fungal pathogens associated with grapevine trunk diseases in Czech vineyards: First example from a north-eastern European grape-growing region. *Phytopathol. Mediterr.* **2018**, *57*, 449–458.
8. Calzarano, F.; D'Agostino, V.; Pepe, A.; Osti, F.; Della Pelle, F.; De Rosso, M.; Flamini, R.; Di Marco, S. Patterns of phytoalexins in the grapevine leaf stripe disease (esca complex)/grapevine pathosystem. *Phytopathol. Mediterr.* **2016**, *55*, 410–426.
9. Calzarano, F.; Di Marco, S. Further evidence that calcium, magnesium and seaweed mixtures reduce grapevine leaf stripe symptoms and increase grape yield. *Phytopathol. Mediterr.* **2018**, *57*, 459–471.
10. Claverie, M.; Notaro, M.; Fontaine, F.; Wéry, J. Current knowledge on grapevine trunk diseases with complex etiology: A systemic approach. *Phytopathol. Med.* **2020**, *59*, 29–53.
11. Hrycan, J.; Hart, M.; Bowen, P.; Forge, T.; Úrbez-Torres, J.R. Grapevine trunk disease fungi: Their roles as latent pathogens and stress factors that favour disease development and symptom expression. *Phytopathol. Med.* **2020**, *59*, 395–424.
12. Bruez, E.; Vallance, J.; Gautier, A.; Laval, V.; Compant, S.; Maurer, W.; Sessitsch, A.; Lebrun, M.H.; Rey, P. Major changes in grapevine wood microbiota are associated with the onset of esca, a devastating trunk disease. *Environ. Microbiol.* **2020**, *22*, 5189–5206. [CrossRef]
13. Sosnowski, M.R.; Ayres, M.; Scott, E. The influence of water deficit stress on the grapevine trunk disease pathogens Eutypa lata and Diplodia seriata. *Plant Dis.* **2020**. [CrossRef]
14. Bertsch, C.; Ramírez-Suero, M.; Magnin-Robert, M.; Larignon, P.; Chong, J.; Abou-Mansour, E.; Spagnolo, A.; Clément, C.; Fontaine, F. Grapevine trunk diseases: Complex and still poorly understood. *Plant Pathol.* **2013**, *62*, 243–265. [CrossRef]
15. Billones-Baaijens, R.; Savocchia, S. A review of Botryosphaeriaceae species associated with grapevine trunk diseases in Australia and New Zealand. *Austral. Plant Pathol.* **2019**, *48*, 3–18. [CrossRef]
16. Guerin-Dubrana, L.; Fontaine, F.; Mugnai, L. Grapevine trunk disease in European and Mediterranean vineyards: Occurrence, distribution and associated disease-affecting cultural factors. *Phytopathol. Med.* **2019**, *58*, 49–71.
17. Larignon, P.; Dubos, B.; Cere, L.; Fulchic, R. Observation on black dead arm in French vineyards. *Phytopathol. Med.* **2001**, *40*, 336–342.
18. Pitt, W.M.; Huang, R.; Steel, C.; Savocchia, S. Pathogenicity and epidemiology of Botryosphaeriaceae species isolated from grapevines in Australia. *Austral. Plant Pathol.* **2013**, *42*, 573–582. [CrossRef]
19. Carter, M. The Status of Eutypa lata as a Pathogen. In *Monograph–Phytopathological Paper No. 32*; International Mycological Institute: Surrey, UK, 1991.
20. Mugnai, L.; Graniti, A.; Surico, G. Esca (black measles) and brown wood-streaking: Two old and elusive diseases of grapevines. *Plant Dis.* **1999**, *83*, 404–418. [CrossRef]
21. Surico, G.; Mugnai, L.; Marchi, G. The esca disease complex. In *Integrated Management of Diseases Caused by Fungi, Phytoplasma and Bacteria*; Springer: Berlin/Heidelberg, Germany, 2008; pp. 119–136.
22. Sosnowski, M.; Shtienberg, D.; Creaser, M.; Wicks, T.; Lardner, R.; Scott, E. The influence of climate on foliar symptoms of Eutypa dieback in grapevines. *Phytopathology* **2007**, *97*, 1284–1289. [CrossRef]
23. Reis, P.; Magnin-Robert, M.; Nascimento, T.; Spagnolo, A.; Abou-Mansour, E.; Fioretti, C.; Clement, C.; Rego, C.; Fontaine, F. Reproducing Botryosphaeria dieback foliar symptoms in a simple model system. *Plant Dis.* **2016**, *100*, 1071–1079. [CrossRef] [PubMed]
24. Larignon, P.; Dubos, B. The villainy of black dead arm. *Wines Vines* **2001**, *82*, 86–89.
25. Masi, M.; Cimmino, A.; Reveglia, P.; Mugnai, L.; Surico, G.; Evidente, A. Advances on fungal phytotoxins and their role in grapevine trunk diseases. *J. Agric. Food Chem.* **2018**, *66*, 5948–5958. [CrossRef]

26. Tey-Rulh, P.; Philippe, I.; Renaud, J.-M.; Tsoupras, G.; de Angelis, P.; Fallot, J.; Tabacchi, R. Eutypine, a phytotoxin produced by *Eutypa lata* the causal agent of dying-arm disease of grapevine. *Phytochemistry* **1991**, *30*, 471–473. [CrossRef]
27. Mahoney, N.; Molyneux, R.J.; Smith, L.R.; Schoch, T.K.; Rolshausen, P.E.; Gubler, W.D. Dying-arm disease in grapevines: Diagnosis of infection with *Eutypa lata* by metabolite analysis. *J. Agric. Food Chem.* **2005**, *53*, 8148–8155. [CrossRef] [PubMed]
28. Lardner, R.; Mahoney, N.; Zanker, T.P.; Molyneux, R.J.; Scott, E.S. Secondary metabolite production by the fungal pathogen *Eutypa lata*: Analysis of extracts from grapevine cultures and detection of those metabolites in planta. *Aust. J. Grape Wine Res.* **2006**, *12*, 107. [CrossRef]
29. Rolshausen, P.; Greve, L.; Labavitch, J.; Mahoney, N.; Molyneux, R.; Gubler, W. Pathogenesis of *Eutypa lata* in grapevine: Identification of virulence factors and biochemical characterization of cordon dieback. *Phytopathology* **2008**, *98*, 222–229. [CrossRef] [PubMed]
30. Andolfi, A.; Mugnai, L.; Luque, J.; Surico, G.; Cimmino, A.; Evidente, A. Phytotoxins produced by fungi associated with grapevine trunk diseases. *Toxins* **2011**, *3*, 1569–1605. [CrossRef]
31. Martos, S.; Andolfi, A.; Luque, J.; Mugnai, L.; Surico, G.; Evidente, A. Production of phytotoxic metabolites by five species of Botryosphaeriaceae causing decline on grapevines, with special interest in the species *Neofusicoccum luteum* and *N. parvum*. *Eur. J. Plant Pathol.* **2008**, *121*, 451–461. [CrossRef]
32. Djoukeng, J.D.; Polli, S.; Larignon, P.; Abou-Mansour, E. Identification of phytotoxins from *Botryosphaeria obtusa*, a pathogen of black dead arm disease of grapevine. *Eur. J. Plant Pathol.* **2009**, *124*, 303–308. [CrossRef]
33. Evidente, A.; Punzo, B.; Andolfi, A.; Cimmino, A.; Melck, D.; Luque, J. Lipophilic phytotoxins produced by *Neofusicoccum parvum*, a grapevine canker agent. *Phytopathol. Med.* **2010**, *49*, 74–79.
34. Abou-Mansour, E.; Débieux, J.-L.; Ramírez-Suero, M.; Bénard-Gellon, M.; Magnin-Robert, M.; Spagnolo, A.; Chong, J.; Farine, S.; Bertsch, C.; L'Haridon, F. Phytotoxic metabolites from *Neofusicoccum parvum*, a pathogen of Botryosphaeria dieback of grapevine. *Phytochemistry* **2015**, *115*, 207–215. [CrossRef]
35. Reveglia, P.; Savocchia, S.; Billones-Baaijens, R.; Masi, M.; Cimmino, A.; Evidente, A. Phytotoxic metabolites by nine species of Botryosphaeriaceae involved in grapevine dieback in Australia and identification of those produced by *Diplodia mutila, Diplodia seriata, Neofusicoccum australe* and *Neofusicoccum luteum*. *Nat. Prod. Res.* **2019**, *33*, 2223–2229. [CrossRef]
36. Reveglia, P.; Savocchia, S.; Billones-Baaijens, R.; Cimmino, A.; Evidente, A. Isolation of phytotoxic phenols and characterization of a new 5-hydroxymethyl-2-isopropoxyphenol from *Dothiorella vidmadera*, a causal agent of grapevine trunk disease. *J. Agric. Food Chem.* **2018**, *66*, 1760–1764. [CrossRef]
37. Masi, M.; Reveglia, P.; Baaijens-Billones, R.; Górecki, M.; Pescitelli, G.; Savocchia, S.; Evidente, A. Phytotoxic metabolites from three *Neofusicoccum* species causal agents of Botryosphaeria dieback in Australia, luteopyroxin, neoanthraquinone, and luteoxepinone, a disubstituted furo-α-pyrone, a hexasubstituted anthraquinone, and a trisubstituted oxepi-2-one from *Neofusicoccum luteum*. *J. Nat. Prod.* **2020**, *83*, 453–460.
38. Masi, M.; Reveglia, P.; Femina, G.; Baaijens-Billones, R.; Savocchia, S.; Evidente, A. Luteoethanones A and B, two phytotoxic 1-substituted ethanones produced by *Neofusicoccum luteum*, a causal agent of Botryosphaeria dieback on grapevine. *Nat. Prod. Res.* **2020**, 1–8. [CrossRef]
39. Reveglia, P.; Savocchia, S.; Billones-Baaijens, R.; Masi, M.; Evidente, A. Spencertoxin and spencer acid, new phytotoxic derivatives of diacrylic acid and dipyridinbutan-1,4-diol produced by *Spencermartinsia viticola*, a causal agent of grapevine Botryosphaeria dieback in Australia. *Arab. J. Chem.* **2020**, *13*, 1803–1808. [CrossRef]
40. Reveglia, P.; Savocchia, S.; Billones-Baaijens, R.; Masi, M.; Cimmino, A.; Evidente, A. Diploquinones A and B, two new phytotoxic tetrasubstituted 1, 4-naphthoquinones from *Diplodia mutila*, a causal agent of grapevine trunk disease. *J. Agric. Food Chem.* **2018**, *66*, 11968–11973. [CrossRef]
41. Saviano, G.; Paris, D.; Melck, D.; Falasca, A.; Trupiano, D.; Iorizzi, M.; Scippa, G.S.; Motta, A. Monitoring spatial and temporal metabolic dynamics of woody poplar root under mechanical stress conditions by NMR-based metabolomics. *Metabolomics* **2016**, *12*, 65. [CrossRef]
42. Deighton, N.; Muckenschnabel, I.; Colmenares, A.J.; Collado, I.G.; Williamson, B. Botrydial is produced in plant tissues infected by *Botrytis cinerea*. *Phytochemistry* **2001**, *57*, 689–692. [CrossRef]
43. Zhao, Y.; Jones, W.; Sutherland, P.; Palmer, D.; Mitchell, R.; Reynolds, P.; Damicone, J.; Bender, C. Detection of the phytotoxin coronatine by ELISA and localization in infected plant tissue. *Physiol. Mol. Plant. Path.* **2001**, *58*, 247–258. [CrossRef]
44. Ghosh, M.; Amudha, R.; Jayachandran, S.; Sakthivel, N. Detection and quantification of phytotoxic metabolites of *Sarocladium oryzae* in sheath rot-infected grains of rice. *Lett. Appl. Microbiol.* **2002**, *34*, 398–401. [CrossRef] [PubMed]
45. Buckel, I.; Andernach, L.; Schüffler, A.; Piepenbring, M.; Opatz, T.; Thines, E. Phytotoxic dioxolanones are potential virulence factors in the infection process of *Guignardia bidwellii*. *Sci. Rep.* **2017**, *7*, 1–7. [CrossRef] [PubMed]
46. Billones-Baaijens, R.; Úrbez-Torres, J.R.; Liu, M.; Ayres, M.; Sosnowski, M.; Savocchia, S. Molecular methods to detect and quantify Botryosphaeriaceae inocula associated with grapevine dieback in Australia. *Plant Dis.* **2018**, *102*, 1489–1499. [CrossRef]
47. Billones-Baaijens, R.; Jones, E.; Ridgway, H.; Jaspers, M. Virulence affected by assay parameters during grapevine pathogenicity studies with Botryosphaeriaceae nursery isolates. *Plant Pathol.* **2013**, *62*, 1214–1225. [CrossRef]
48. Del Río, J.A.; Gómez, P.; Báidez, A.; Fuster, M.D.; Ortuño, A.; Frías, V. Phenolic compounds have a role in the defence mechanism protecting grapevine against the fungi involved in Petri disease. *Phytopathol. Med.* **2004**, *43*, 87–94.

49. Lambert, C.; Bisson, J.; Waffo-Téguo, P.; Papastamoulis, Y.; Richard, T.; Corio-Costet, M.-F.; Mérillon, J.-M.; Cluzet, S.P. Phenolics and their antifungal role in grapevine wood decay: Focus on the Botryosphaeriaceae family. *J. Agric. Food Chem.* **2012**, *60*, 11859–11868. [CrossRef]
50. Magnin-Robert, M.; Spagnolo, A.; Boulanger, A.; Joyeux, C.; Clément, C.; Abou-Mansour, E.; Fontaine, F. Changes in plant metabolism and accumulation of fungal metabolites in response to esca proper and apoplexy expression in the whole grapevine. *Phytopathology* **2016**, *106*, 541–553. [CrossRef]
51. Ramírez-Suero, M.; Bénard-Gellon, M.; Chong, J.; Laloue, H.; Stempien, E.; Abou-Mansour, E.; Fontaine, F.; Larignon, P.; Mazet-Kieffer, F.; Farine, S. Extracellular compounds produced by fungi associated with Botryosphaeria dieback induce differential defence gene expression patterns and necrosis in *Vitis vinifera* cv. Chardonnay cells. *Protoplasma* **2014**, *251*, 1417–1426. [CrossRef]
52. Trotel-Aziz, P.; Abou-Mansour, E.; Courteaux, B.; Rabenoelina, F.; Clément, C.; Fontaine, F.; Aziz, A. Bacillus subtilis PTA-271 counteracts Botryosphaeria dieback in grapevine, triggering immune responses and detoxification of fungal phytotoxins. *Front. Plant. Sci.* **2019**, *10*, 25. [CrossRef]
53. Larignon, P.; Fontaine, F.; Farine, S.; Clement, C.; Bertsch, C. Esca and black dead arm: Two major actors of grapevine trunk diseases. *Compt. Rendus Biol.* **2009**, *332*, 765–783. [CrossRef]
54. Calzarano, F.; Osti, F.; Baranek, M.; Di Marco, S. Rainfall and temperature influence expression of foliar symptoms of grapevine leaf stripe disease (esca complex) in vineyards. *Phytopathol. Med.* **2018**, *57*, 488–505.
55. Serra, S.; Ligios, V.; Schianchi, N.; Prota, V.A.; Scanu, B. Expression of grapevine leaf stripe disease foliar symptoms in four cultivars in relation to grapevine phenology and climatic conditions. *Phytopathol. Med.* **2019**, *57*, 557–568.
56. Songy, A.; Fernandez, O.; Clément, C.; Larignon, P.; Fontaine, F. Grapevine trunk diseases under thermal and water stresses. *Planta* **2019**, *249*, 1655–1679. [CrossRef]
57. Möbius, N.; Hertweck, C. Fungal phytotoxins as mediators of virulence. *Curr. Opin. Plant Biol.* **2009**, *12*, 390–398. [CrossRef]
58. Ghatak, A.; Chaturvedi, P.; Weckwerth, W. Metabolomics in plant stress physiology. In *Plant Genetics and Molecular Biology*; Springer: Berlin/Heidelberg, Germany, 2018; pp. 187–236.
59. Poulin, R.X.; Pohnert, G. Simplifying the complex: Metabolomics approaches in chemical ecology. *Anal. Bioanal. Chem.* **2019**, *411*, 13–19. [CrossRef]
60. Labois, C.; Wilhelm, K.; Laloue, H.; Tarnus, C.; Bertsch, C.; Goddard, M.-L.; Chong, J. Wood metabolomic responses of wild and cultivated grapevine to infection with *Neofusicoccum parvum*, a trunk disease pathogen. *Metabolites* **2020**, *10*, 232. [CrossRef]
61. WineAustralia. *Wine Australia Annual Report*; WineAustralia: Kent Town, Australia, 2018.
62. WineAustralia.com. Available online: https://www.wineaustralia.com/growing-making/environment-and-climate (accessed on 31 March 2020).
63. Pouzoulet, J.; Mailhac, N.; Couderc, C.; Besson, X.; Daydé, J.; Lummerzheim, M.; Jacques, A. A method to detect and quantify *Phaeomoniella chlamydospora* and *Phaeoacremonium aleophilum* DNA in grapevine-wood samples. *Appl. Microbiol. Biotechnol.* **2013**, *97*, 10163–10175. [CrossRef]
64. Doyle, J.J.; Doyle, J.L. A rapid DNA isolation procedure for small quantities of fresh leaf tissue. *Phytochem. Bul.* **1987**, *19*, 11–15.
65. White, T.J.; Bruns, T.; Lee, S.; Taylor, J. Amplification and direct sequencing of fungal ribosomal RNA genes for phylogenetics. *PCR Protoc. A Guide Methods Appl.* **1990**, *18*, 315–322.
66. Bustin, S.A.; Benes, V.; Garson, J.A.; Hellemans, J.; Huggett, J.; Kubista, M.; Mueller, R.; Nolan, T.; Pfaffl, M.W.; Shipley, G.L. The MIQE guidelines: Minimum information for publication of quantitative real-time PCR experiments. *Clin. Chem.* **2009**, *55*, 611–622. [CrossRef] [PubMed]

plants

MDPI

Article

Factors Involved on Tiger-Stripe Foliar Symptom Expression of Esca of Grapevine

Francesco Calzarano [1,*], Giancarlo Pagnani [1], Michele Pisante [1], Mirella Bellocci [2], Giuseppe Cillo [3], Elisa Giorgia Metruccio [4] and Stefano Di Marco [4]

1 Faculty of Bioscience and Technologies for Food, Agriculture and Environment, University of Teramo, Via Renato Balzarini, 1, 64100 Teramo, Italy; gpagnani@unite.it (G.P.); mpisante@unite.it (M.P.)
2 Veterinary Public Health Institute of Abruzzo and Molise Regions (IZSAM), Campo Boario, 64100 Teramo, Italy; m.bellocci@izs.it
3 Department of Agronomy, Food, Natural Resources, Animal and Environment, University of Padova, 16-35020 Legnaro, Italy; giuseppe.cillo@studenti.unipd.it
4 CNR, IBE, Via Gobetti 101, 40129 Bologna, Italy; elisa.metruccio@ibe.cnr.it (E.G.M.); stefano.dimarco@ibe.cnr.it (S.D.M.)
* Correspondence: fcalzarano@unite.it

Abstract: Esca of grapevine causes yield losses correlated with incidence and severity symptom expression. Factors associated with leaf symptom mechanisms are yet to be fully clarified. Therefore, in 2019 and 2020, macro and microelement analyses and leaf reflectance measurements were carried out on leaves at different growth stages in a vineyard located in Abruzzo, central Italy. Surveys were carried out on leaves of both never leaf-symptomatic vines and different categories of diseased vine shoots. Never leaf-symptomatic and diseased vines were also treated with a fertilizer mixture that proved to be able to limit the symptom expression. Results showed that untreated asymptomatic diseased vines had high calcium contents for most of the vegetative season. On the contrary, treated asymptomatic diseased vines showed higher contents of calcium, magnesium, and sodium, at berries pea-sized, before the onset of symptoms. These vines had better physiological efficiency showing higher water index (WI), normalized difference vegetation index (NDVI), and green normalized difference vegetation index (GNDVI) values, compared to untreated asymptomatic vines, at fruit set. Results confirmed the strong response of the plant to symptom expression development and the possibility of limiting this response with calcium and magnesium applications carried out before the symptom onset.

Keywords: grapevine; esca complex; wood disease control; leaf symptoms

Citation: Calzarano, F.; Pagnani, G.; Pisante, M.; Bellocci, M.; Cillo, G.; Metruccio, E.G.; Di Marco, S. Factors Involved on Tiger-Stripe Foliar Symptom Expression of Esca of Grapevine. *Plants* **2021**, *10*, 1041. https://doi.org/10.3390/plants10061041

Academic Editor: Carlos Agustí-Brisach

Received: 25 March 2021
Accepted: 17 May 2021
Published: 21 May 2021

1. Introduction

Esca of grapevine is a complex, destructive and widely spread disease. The disease includes the involvement of several microorganisms producing different types of wood deterioration at different age of the plant, from nursery to ageing vineyards. The disease has been commonly associated with tracheomycotic pathogens as *Phaeomoniella chlamydospora*, *Phaeoacremonium minimum* (or another species of *Phaeoacremonium*), and with the basidiomycete *Fomitiporia mediterranea*, or other species recently isolated in United States [1–5]. The complexity of the disease, especially for the role of pathogens, i.e., of *Botryosphaeriaceae* species [2], led to the proposal of a classification, although the fail to fulfill Koch's postulates. In particular, Esca of grapevine was defined Esca complex and divided into five syndromes [6]. Three diseases, Brown wood streaking, Petri disease and grapevine leaf stripe disease (GLSD) were grouped as "grapevine phaeo-tracheomycosis complex". White rot, the fourth disease of Esca complex, was mainly caused by *Fomitiporia mediterranea* [7]. The fifth disease, named as Esca proper, was considered the concomitant occurrence of tracheomycotic and white rot pathogens and relative wood alterations in the same plant, traditionally reported as "Esca of grapevine".

Some studies where tiger-stripe symptom expression was associated with infections of tracheomycotic fungi, without the occurrence of white-rot necrosis, as for "young esca" in Australia, somehow supported this classification [8,9]. On the other hand, and more recently, a correlation between the amount of white-rot necrotic tissues and leaf symptoms was recorded in different studies [10–13].

The leaf blade of diseased plants showed light green chlorosis that can expand and coalesce in interveinal yellow red-brown stripes, displaying the so-called tiger-stripe symptoms [1,3]. In any case, incidence and severity of the foliar symptom expression demonstrated to be correlated with qualitative–quantitative yield losses [14,15]. Moreover, the symptoms occurrence and the consequent yield losses may vary from one season to another, regardless of the rate of woody tissue deterioration, and in relation to physiological, cultural, and environmental factors not completely cleared yet [9]. Meteorological factors can influence the occurrence of foliar symptoms in a given area and season as demonstrated for June and July rainfall, which appeared to be correlated to the foliar symptom expression [16,17]. Cultural factors also proved to play a role in foliar symptoms expression [18].

Studies hypothesized that foliar symptoms occurrence was based on the toxic metabolites produced by fungi in the wood [19–21]. Toxins can reach the canopy through the transpiration stream and contribute to induce leaf responses that leads to the formation of interveinal necrosis as the results of a hypersensitivity reaction [2,22,23]. However, further hypotheses on the occurrence of foliar symptoms were recently considered [24]. A functional disorder of sap flow, as sap disruption caused by the wood altered by pathogens [25], nongaseous embolisms or occlusions of vessels by gels and/or tyloses [26,27], and/or a role of annual infections on annual shoots [28] were postulated. These studies once again highlight the number and complex interaction of parameters associated with the outburst of foliar symptoms.

Phytoalexins, mainly present in symptomatic leaves at pre-bunch closure, were synthesized in symptomatic leaves as a consequence of the appearance of necrosis, and therefore did not seem effective in reducing symptom development [29]. This hypothesis seemed to be confirmed by the proportional increasing of phytoalexins with leaf symptoms increase [30]. Therefore, plants might react to the toxic metabolites with a kind of hypersensitivity reaction, in agreement with studies on the formation of anti-microbial compounds after the occurrence of symptoms [31].

The response of plant to the occurrence of symptoms seemed also associated with the nutritional status of the infected vines, and in particular with the calcium content [32,33]. Diseased vines treated with a fertilizer mixture based on calcium, magnesium, and seaweed significantly reduced foliar symptoms, synthesizing as well higher amounts of *trans*-resveratrol in the leaves earlier in the season, compared to untreated diseased vines. Therefore, an early increase of phytoalexins might contribute to the reduction of symptoms [34,35].

Precision agriculture is a site-specific informed management system that analyzes the factors that can vary over space and time in the production process, minimizing inputs (as water, fertilizers, plant protection products). The aim is to reduce the impact of agricultural practices on the environment, to increase the quality of products and the profitability of agriculture [36]. In this context, remote sensing techniques allow to detect physical or chemical characteristics of soil or plant organs either from proximal platforms (sensors) or remotely via satellite, drone, and aircraft [37].

Assessments are carried out in relation to the electromagnetic radiation emitted in a certain range of different wavelengths, called spectrum [38]. The bands of a spectrum in the optical domain make it possible to investigate phenomena connected with the photosynthetic capacity of crops. The optical domain [39] includes the visible, also called photosynthetically active radiation (PAR), with wavelengths between 400 and 700 nm, the infrared (IR) which in turn includes: near infrared (NIR, 700–1300 nm); short wave infrared (SWIR, 1300–2500 nm); midwave infrared (MWIR, 3000–8000 nm); thermal infrared (TIR, 7000–20,000 nm).

The most used vegetation indices to estimate the photosynthetic capacities of a crop based on the spectral response are the normalized difference vegetation index (NDVI), [40] and the green normalized difference vegetation index (GNDVI), [41]. These indices evaluated the leaf reflectance, expressed by the ratio between intensity of the reflected radiant flux and intensity of the incident radiant flux. NDVI considers the reflectance on wavelengths (λ) of the NIR (λ = 770 nm) and of the visible red (RED) (λ = 660 nm). GNDVI is calculated from the NDVI index by substituting the reflectance in the red with that in the green.

Through leaf reflectance measurements it is possible to detect different indices including the water index (WI), [42], as response of plant regarding specific infrared bands (R900/R970) so that water concentration in plants (PWC) and consequently the water stress can be estimated.

The aim of this study was to investigate parameters and mechanisms involved in complex processes of tiger-stripe symptoms formation, through leaf reflectance measurements and assessment of the main macro and microelements in leaves of both vines with different disease expressions and never leaf-symptomatic vines. The same analysis was carried out in diseased and never leaf-symptomatic vines treated with a fertilizer mixture capable of interfering with the foliar symptom expression, to obtain further information on the factors that regulate foliar symptoms expression.

2. Results

2.1. *Leaf Fertilizer Applications and Foliar Symptom Surveys*

Both years of survey on incidence and severity of foliar symptoms in the Controguerra vineyard confirmed the dynamics of symptom expression, characterized by a remarkable increasing from the berries developing color stage. The effect of applications of the fertilizer mixture was noticeable in treated vines, in particular since the assessments carried out at majority of berries touching stage (31 July 2019 and 31 July 2020), and in the following growth stages, with the increasing of symptom expression in the vineyard (Figures 1 and 2). Both in 2019 and 2020, the differences in symptoms of incidence and severity percentages between treated and untreated vines, evaluated at harvest (13 September 2019 and 12 September 2020), were statistically significant for both parameters, by means of Chi-square test for p = 0.05.

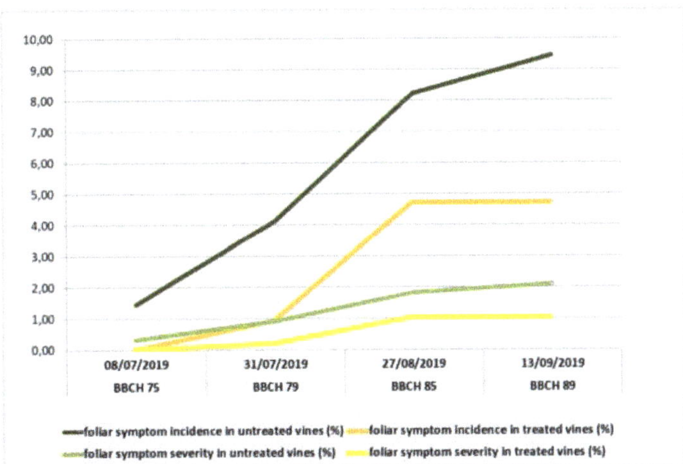

Figure 1. Incidence and severity of tiger-stripe foliar symptoms in vines treated with the fertilizer mixture and in untreated vines, in 2019, in Controguerra vineyard.

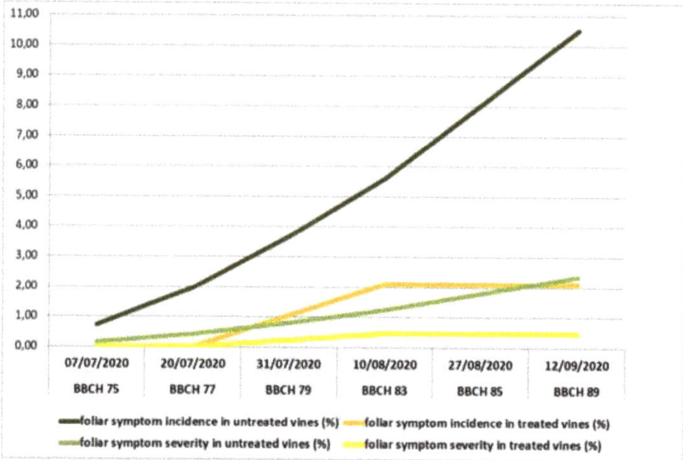

Figure 2. Incidence and severity of tiger-stripe foliar symptoms in vines treated with the fertilizer mixture and in untreated vines, in 2020, in Controguerra vineyard.

2.2. *Leaf Reflectance Measurements*

2.2.1. Comparisons between Treated and Untreated Never Leaf-Symptomatic and Diseased Asymptomatic Vine Leaves

The results of reflectance measurements, carried out on leaves of both never leaf-symptomatic (NLS) and asymptomatic vines (AS), treated and untreated, showed higher values in leaves east facing and in full light (side A), compared to the leaves on side B, west facing and in shadow (Tables 1 and 2).

Table 1. NDVI and GNDVI in leaves of treated and untreated never leaf-symptomatic vines and diseased asymptomatic vines in the side A rows of Controguerra vineyard.

					NDVI				
	66 BBCH	**69 BBCH**	**71 BBCH**	**75 BBCH**	**77 BBCH**	**79 BBCH**	**83 BBCH**	**85 BBCH**	**89 BBCH**
Treatment	**14/06/2019**	**21/06/2019**	**27/06/2019**	**08/07/2019**	**16/07/2019**	**31/07/2019**	**12/08/2019**	**27/08/2019**	**13/09/2019**
NLSnt	0.914 a	0.866 a	0.820 ab	0.806 a	0.838 a	0.783 a	0.810 a	0.823 a	0.819 a
ASnt	0.897 a	0.857 a	0.808 b	0.834 a	0.823 a	0.799 a	0.808 a	0.836 a	0.856 a
NLSt	0.804 b	0.867 a	0.798 b	0.788 a	0.842 a	0.826 a	0.814 a	0.844 a	0.844 a
ASt	0.777 b	0.868 a	0.886 a	0.791 a	0.834 a	0.813 a	0.824 a	0.827 a	0.844 a
Treatment	**15/06/2020**	**22/06/2020**	**27/06/2020**	**09/07/2020**	**20/07/2020**	**31/07/2020**	**10/08/2020**	**27/08/2020**	**12/09/2020**
NLSnt	0.902 a	0.853 a	0.848 b	0.873 a	0.832 a	0.833 a	0.853 a	0.854 a	0.834 a
ASnt	0.885 a	0.861 a	0.812 b	0.858 a	0.812 a	0.831 a	0.856 a	0.861 a	0.822 a
NLSt	0.798 b	0.859 a	0.833 b	0.830 a	0.856 a	0.848 a	0.855 a	0.831 a	0.806 a
ASt	0.792 b	0.878 a	0.894 a	0.852 a	0.870 a	0.834 a	0.824 a	0.835 a	0.818 a
					GNDVI				
Treatment	**14/06/2019**	**21/06/2019**	**27/06/2019**	**08/07/2019**	**16/07/2019**	**31/07/2019**	**12/08/2019**	**27/08/2019**	**13/09/2019**
NLSnt	0.727 a	0.624 a	0.596 ab	0.519 a	0.568 a	0.486 a	0.547 a	0.542 a	0.590 a
ASnt	0.698 a	0.590 a	0.551 b	0.567 a	0.540 a	0.477 a	0.533 a	0.549 a	0.673 a
NLSt	0.479 b	0.646 a	0.599 ab	0.532 a	0.608 a	0.574 a	0.571 a	0.539 a	0.651 a
ASt	0.424 b	0.639 a	0.663 a	0.489 a	0.562 a	0.529 a	0.567 a	0.568 a	0.634 a
Treatment	**15/06/2020**	**22/06/2020**	**27/06/2020**	**09/07/2020**	**20/07/2020**	**31/07/2020**	**10/08/2020**	**27/08/2020**	**12/09/2020**
NLSnt	0.712 a	0.625 a	0.558 b	0.612 a	0.522 b	0.553 a	0.615 a	0.600 a	0.540 a
ASnt	0.701 a	0.636 a	0.574 b	0.604 a	0.517 b	0.557 a	0.635 a	0.637 a	0.552 a
NLSt	0.542 b	0.667 a	0.648 a	0.632 a	0.685 a	0.658 a	0.654 a	0.642 a	0.511 a
ASt	0.568 b	0.656 a	0.623 a	0.639 a	0.699 a	0.648 a	0.643 a	0.667 a	0.575 a

NDVI = normalized difference vegetation index. GNDVI = green normalized difference vegetation index. NLS = never leaf-symptomatic vine leaves; AS = diseased but asymptomatic for the entire season vine leaves; t = treated with fertilizer mixture; nt = untreated. For each column of each year, values followed by the same letter do not differ statistically according to Tukey's honest significant difference (HSD) test at $p = 0.05$.

Table 2. NDVI and GNDVI in leaves of treated and untreated never leaf-symptomatic vines and diseased asymptomatic vines in the side B rows of Controguerra vineyard.

					NDVI				
	66 BBCH	**69 BBCH**	**71 BBCH**	**75 BBCH**	**77 BBCH**	**79 BBCH**	**83 BBCH**	**85 BBCH**	**89 BBCH**
Treatment	**14/06/2019**	**21/06/2019**	**27/06/2019**	**08/07/2019**	**16/07/2019**	**31/07/2019**	**12/08/2019**	**27/08/2019**	**13/09/2019**
NLSnt	0.928 a	0.904 a	0.938 b	0.950 a	0.951 a	0.934 a	0.950 a	0.936 a	0.935 a
ASnt	0.907 a	0.915 a	0.958 b	0.971 a	0.957 a	0.947 a	0.957 a	0.946 a	0.959 a
NLSt	0.810 c	0.934 a	0.978 a	0.974 a	0.967 a	0.927 a	0.913 a	0.944 a	0.950 a
ASt	0.867 b	0.936 a	0.981 a	0.971 a	0.975 a	0.933 a	0.943 a	0.927 a	0.940 a
Treatment	**15/06/2020**	**22/06/2020**	**27/06/2020**	**09/07/2020**	**20/07/2020**	**31/07/2020**	**10/08/2020**	**27/08/2020**	**12/09/2020**
NLSnt	0.981 a	0.913 a	0.946 b	0.979 a	0.979 a	0.971 a	0.964 a	0.971 a	0.938 a
ASnt	0.977 a	0.922 a	0.943 b	0.977 a	0.976 a	0.974 a	0.971 a	0.977 a	0.931 a
NLSt	0.838 b	0.945 a	0.984 a	0.965 a	0.942 b	0.969 a	0.970 a	0.974 a	0.948 a
ASt	0.849 b	0.955 a	0.984 a	0.975 a	0.951 b	0.967 a	0.973 a	0.973 a	0.940 a
					GNDVI				
Treatment	**14/06/2019**	**21/06/2019**	**27/06/2019**	**08/07/2019**	**16/07/2019**	**31/07/2019**	**12/08/2019**	**27/08/2019**	**13/09/2019**
NLSnt	0.764 a	0.713 a	0.794 b	0.764 a	0.787 a	0.737 a	0.799 a	0.735 a	0.757 a
ASnt	0.730 a	0.745 a	0.831 b	0.850 a	0.814 a	0.766 a	0.829 a	0.768 a	0.847 a
NLSt	0.504 b	0.790 a	0.864 b	0.880 a	0.847 a	0.770 a	0.752 a	0.839 a	0.833 a
ASt	0.546 b	0.781 a	0.877 a	0.864 a	0.859 a	0.789 a	0.809 a	0.768 a	0.808 a
Treatment	**15/06/2020**	**22/06/2020**	**27/06/2020**	**09/07/2020**	**20/07/2020**	**31/07/2020**	**10/08/2020**	**27/08/2020**	**12/09/2020**
NLSnt	0.812 a	0.823 a	0.865 ab	0.872 a	0.864 a	0.853 a	0.819 b	0.848 a	0.715 a
ASnt	0.807 a	0.845 a	0.833 b	0.864 a	0.859 a	0.855 a	0.845 ab	0.871 a	0.730 a
NLSt	0.622 b	0.812 a	0.921 a	0.878 a	0.838 a	0.878 a	0.868 ab	0.886 a	0.800 a
ASt	0.676 b	0.833 a	0.915 a	0.895 a	0.842 a	0.875 a	0.883 a	0.892 a	0.765 a

NDVI = normalized difference vegetation index. GNDVI = green normalized difference vegetation index. NLS = never leaf-symptomatic vine leaves; AS = diseased but asymptomatic for the entire season vine leaves; t = treated with fertilizer mixture; nt = untreated. For each column of each year values followed by the same letter do not differ statistically according to Tukey's honest significant difference (HSD) test at $p = 0.05$.

Conversely, the trend of reflectance measurements carried out from growth stage 66 to 89, were similar in the two sides (A and B) in both years of survey (Tables 1 and 2).

The NDVI and GNDVI values showed very similar trends and were always significantly lower in leaves of treated compared to untreated NLS and AS vines, in the first survey, at growth stage 66 (Tables 1 and 2). On the contrary, in the third survey, at growth stage 71, the NDVI and GNDVI values of treated AS leaves appeared significantly higher than those of AS untreated leaves. In the same survey, the treated NLS leaves showed higher or similar values than those of untreated NLS leaves (Tables 1 and 2). From the growth stage 75 to the last survey, growth stage 89, no differences were found between NLS and AS leaves. Few exceptions were observed only in 2020, at growth stage 77, for GNDVI (side A), and for NDVI (side B), with values of treated NLS and AS leaves significantly higher and lower than untreated NLS and AS leaves, respectively (Tables 1 and 2).

In the first three surveys (growth stages 66, 69, and 71), values of the WI were higher in leaves of treated NLS and AS vines, compared to those of the corresponding untreated vines, both in side A and in side B. From the survey at growth stage 75 to the survey at growth stage 83 any difference between treated and untreated NLS and AS values was detected (Table 3). In growth stages 85 and 89 surveys, WI values were lower, in most of the cases significantly, in treated compared to untreated AS leaves (Table 3).

Table 3. Water Index in leaves of treated and untreated never leaf-symptomatic vines and diseased asymptomatic vines in the Controguerra vineyard.

					Side A WI				
	66 BBCH	69 BBCH	71 BBCH	75 BBCH	77 BBCH	79 BBCH	83 BBCH	85 BBCH	89 BBCH
Treatment	**14/06/2019**	**21/06/2019**	**27/06/2019**	**08/07/2019**	**16/07/2019**	**31/07/2019**	**12/08/2019**	**27/08/2019**	**13/09/2019**
NLSnt	1.042 b	1.056 b	1.039 c	1.069 ab	1.046 a	1.020 a	1.041 a	1.013 a	1.018 ab
ASnt	1.086 b	1.072 b	1.064 bc	1.091 a	1.049 a	1.036 a	1.065 a	1.028 a	1.037 a
NLSt	1.114 a	1.100 a	1.078 ab	1.048 b	1.011 a	1.027 a	1.041 a	1.000 a	0.995 ab
ASt	1.110 a	1.103 a	1.107 a	1.068 ab	1.086 a	1.007 a	1.046 a	0.995 a	0.987 b
Treatment	**15/06/2020**	**22/06/2020**	**27/06/2020**	**09/07/2020**	**20/07/2020**	**31/07/2020**	**10/08/2020**	**27/08/2020**	**12/09/2020**
NLSnt	1.036 b	1.023 b	1.043 b	1.059 a	1.050 a	1.029 b	1.039 a	1.038 a	1.032 ab
ASnt	1.056 b	1.044 b	1.062 b	1.082 a	1.056 a	1.051 ab	1.047 a	1.063 a	1.051 a
NLSt	1.112 a	1.114 a	1.081 ab	1.060 a	1.058 a	1.081 a	1.048 a	1.029 a	1.017 ab
ASt	1.116 a	1.110 a	1.106 a	1.058 a	1.058 a	1.047 b	1.056 a	1.044 a	1.013 b
					Side B WI				
Treatment	**14/06/2019**	**21/06/2019**	**27/06/2019**	**08/07/2019**	**16/07/2019**	**31/07/2019**	**12/08/2019**	**27/08/2019**	**13/09/2019**
NLSnt	1.054 c	1.084 b	1.102 c	1.153 a	1.116 a	1.053 a	1.112 a	1.043 ab	1.070 ab
ASnt	1.082 bc	1.116 b	1.130 bc	1.199 a	1.131 a	1.080 a	1.175 a	1.091 a	1.107 a
NLSt	1.123 ab	1.166 a	1.257 ab	1.179 a	1.180 a	1.056 a	1.098 a	0.965 b	1.020 ab
ASt	1.140 a	1.180 a	1.286 a	1.140 a	1.193 a	1.033 a	1.096 a	0.969 b	1.007 b
Treatment	**15/06/2020**	**22/06/2020**	**27/06/2020**	**09/07/2020**	**20/07/2020**	**31/07/2020**	**10/08/2020**	**27/08/2020**	**12/09/2020**
NLSnt	1.088 b	1.101 b	1.111 b	1.076 a	1.105 a	1.066 b	1.067 b	1.087 b	1.078 a
ASnt	1.098 b	1.086 b	1.157 b	1.129 a	1.111 a	1.142 a	1.129 a	1.186 a	1.129 a
NLSt	1.122 a	1.143 a	1.203 a	1.106 a	1.094 a	1.119 ab	1.104 ab	1.157 ab	1.040 a
ASt	1.151 a	1.148 a	1.208 a	1.106 a	1.089 a	1.144 a	1.127 a	1.102 ab	1.052 a

WI = water index. NLS = never leaf-symptomatic vine leaves; AS = diseased but asymptomatic for the entire season vine leaves; t = treated with fertilizer mixture; nt = untreated. For each column of each year values followed by the same letter do not differ statistically according to Tukey's honest significant difference (HSD) test at $p = 0.05$.

2.2.2. Comparisons among Untreated Never Leaf-Symptomatic, Diseased Asymptomatic, and Diseased Symptomatic Vine Leaves

The leaf reflectance measurements carried out from the first leaf symptom appearance (8 July 2019 and 9 July 2020, growth stage 75) to the harvest (growth stage 89) on leaves of untreated never leaf-symptomatic (NLS) and asymptomatic (AS) vines, and on leaves of the different shoots of untreated symptomatic diseased vines (PRE-S, ASTIGR, and TIGR), did not show differences for both NDVI and GNDVI, except for significantly lower values in symptomatic leaves (TIGR) (Table 4).

Table 4. NDVI and GNDVI in never leaf-symptomatic vine leaves and in different categories of leaves of diseased vines in the side A rows of Controguerra vineyard.

			NDVI			
	75 BBCH	77 BBCH	79 BBCH	83 BBCH	85 BBCH	89 BBCH
Treatment	**08/07/2019**	**16/07/2019**	**31/07/2019**	**12/08/2019**	**27/08/2019**	**13/09/2019**
NLS	0.806 a	0.838 a	0.783 a	0.810 a	0.823 a	0.819 a
AS	0.834 a	0.823 a	0.799 a	0.808 a	0.836 a	0.856 a
PRE-S	0.806 a	0.844 a	0.812 a	0.786 a	0.830 a	0.821 a
ASTIGR	0.870 a	0.880 a	0.778 a	0.778 a	0.829 a	0.810 a
TIGR	0.723 b	0.622 b	0.570 b	0.759 a	0.636 b	0.635 b
Treatment	**09/07/2020**	**20/07/2020**	**31/07/2020**	**10/08/2020**	**27/08/2020**	**12/09/2020**
NLS	0.873 a	0.832 a	0.833 a	0.853 a	0.854 a	0.834 a
AS	0.858 a	0.812 a	0.831 a	0.856 a	0.861 a	0.822 a
PRE-S	0.852 a	0.798 a	0.820 a	0.859 a	0.875 a	0.784 a
ASTIGR	0.875 a	0.806 a	0.825 a	0.831 a	0.833 a	0.816 a
TIGR	0.743 b	0.658 b	0.678 b	0.634 b	0.600 b	0.542 b

Table 4. *Cont.*

	GNDVI					
Treatment	**08/07/2019**	**16/07/2019**	**31/07/2019**	**12/08/2019**	**27/08/2019**	**13/09/2019**
NLS	0.519 a	0.568 a	0.486 a	0.547 a	0.542 ab	0.590 a
AS	0.567 a	0.540 a	0.477 a	0.533 a	0.549 ab	0.673 a
PRE-S	0.523 a	0.550 a	0.485 a	0.521 a	0.608 a	0.633 a
ASTIGR	0.569 a	0.648 a	0.462 a	0.496 a	0.563 a	0.601 a
TIGR	0.465 b	0.451 b	0.303 b	0.448 a	0.415 b	0.487 b
Treatment	**09/07/2020**	**20/07/2020**	**31/07/2020**	**10/08/2020**	**27/08/2020**	**12/09/2020**
NLS	0.612 a	0.522 a	0.553 a	0.615 a	0.600 a	0.540 a
AS	0.604 a	0.517 a	0.557 a	0.635 a	0.637 a	0.552 a
PRE-S	0.522 ab	0.502 a	0.466 ab	0.573 a	0.589 a	0.425 ab
ASTIGR	0.625 a	0.559 a	0.553 a	0.564 a	0.635 a	0.552 a
TIGR	0.448 b	0.505 a	0.392 b	0.402 b	0.424 b	0.365 b

NDVI = normalized difference vegetation index. GNDVI = green normalized difference vegetation index. NLS = never leaf-symptomatic vine leaves; AS = diseased but asymptomatic for the entire season vine leaves; PRE-S = pre-symptomatic shoot leaves of symptomatic vines; ASTIGR = asymptomatic shoot leaves of symptomatic vines; TIGR = tiger-striped shoot leaves of symptomatic vines. For each column of each year values followed by the same letter do not differ statistically according to Tukey's honest significant difference (HSD) test at $p = 0.05$.

In both years at growth stage 75 values of WI were significantly higher in AS leaves than in PRE-S, ASTIGR and TIGR leaves, but not different from values of NLS leaves. From surveys at growth stage 79 to the last surveys (growth stage 89), AS leaves showed higher values than the other leaves, in some cases significantly compared to those of PRE-S, ASTIGR, and TIGR (Table 5).

Table 5. Water Index in never leaf-symptomatic vine leaves and in different categories of leaves of diseased vines in the side A rows of Controguerra vineyard.

	WI					
	75 BBCH	**77 BBCH**	**79 BBCH**	**83 BBCH**	**85 BBCH**	**89 BBCH**
Treatment	**08/07/2019**	**16/07/2019**	**31/07/2019**	**12/08/2019**	**27/08/2019**	**13/09/2019**
NLS	1.069 ab	1.046 a	1.020 a	1.041 ab	1.013 a	1.018 a
AS	1.091 a	1.049 a	1.036 a	1.065 a	1.028 a	1.037 a
PRE-S	1.059 b	1.057 a	1.027 a	1.020 bc	1.023 a	1.017 a
ASTIGR	1.008 c	1.108 a	1.013 a	0.987 c	1.020 a	1.004 a
TIGR	1.001 c	1.033 a	1.013 a	0.983 c	1.013 a	1.003 a
Treatment	**09/07/2020**	**20/07/2020**	**31/07/2020**	**10/08/2020**	**27/08/2020**	**12/09/2020**
NLS	1.059 a	1.050 a	1.029 ab	1.039 a	1.038 ab	1.032 ab
AS	1.082 a	1.056 a	1.051 a	1.047 a	1.063 a	1.051 a
PRE-S	1.010 b	1.033 a	0.995 b	1.020 a	1.028 abc	1.011 b
ASTIGR	1.021 b	1.048 a	0.995 b	1.011 a	1.007 bc	1.012 b
TIGR	1.009 b	1.042 a	1.001 b	1.002 a	0.993 c	1.002 b

WI = water index. NLS = never leaf-symptomatic vine leaves; AS = diseased but asymptomatic for the entire season vine leaves; PRE-S = pre-symptomatic shoot leaves of symptomatic vines; ASTIGR = asymptomatic shoot leaves of symptomatic vines; TIGR = tiger-striped shoot leaves of symptomatic vines. For each column of each year values followed by the same letter do not differ statistically according to Tukey's honest significant difference (HSD) test at $p = 0.05$.

2.3. *Analysis of Macro and Microelements in the Leaf*

2.3.1. Macroelements

In untreated vines, only some differences in P content, not confirmed in both years of the study, were observed between leaves of the different categories. Only at harvest (growth stage 89), the P content was significantly higher in AS leaves than in leaves of the other categories of untreated shoots (Table S1).

P contents did not differ between leaves of treated plants, both in 2019 and 2020, except for TIGR leaves which showed higher values at growth stage 89 (Table S1).

In the comparison between treated and untreated vines, P contents were always lower (often significantly) in the treated AS leaves compared to untreated AS ones. Furthermore, in both years, at growth stage 85, lower values of treated compared to untreated ASTIGR leaves were recorded, although significantly only in 2020 (Table S1).

Contents of K did not show remarkable differences among leaves of the different categories of shoots. Some sporadic differences, such as those observed in 2019, between untreated PRE-S and TIGR leaves and treated NLS leaves, at growth stage 75, were not confirmed the following year (Table S1).

In both years, in untreated vines, Ca and Mg contents were higher in leaves of AS vines compared to NLS ones, from growth stage 75 to growth stage 85. In these stages the differences between NLS and AS were always significant for Ca. At growth stage 71 and 89 no significant differences were observed between the two categories of leaves. In leaves of ASTIGR shoots, Ca and Mg contents were similar in most of the cases to those of AS leaves. At growth stage 75, leaves of PRE-S shoots always had Ca and Mg contents lower than those of the other untreated leaves, significantly compared to AS leaves. In the following growth stages, Ca and Mg contents of PRE-S leaves increased and generally were not different from those of the other untreated leaves. Higher contents of Ca were recorded in treated compared to untreated TIGR leaves. This difference was significant at growth stage 75 in both years (Table 6).

Table 6. Calcium and magnesium content in never leaf-symptomatic vine leaves and in different categories of leaves of diseased vines.

	BBCH 71	**BBCH 75**		**BBCH 85**		**BBCH 89**	
	27/06/2020	**11/07/2019**	**09/07/2020**	**27/08/2019**	**27/08/2020**	**12/09/2019**	**12/09/2020**
Treatment	**Calcium (mg g^{-1})**						
NLSnt	5.98 bc	4.62 cd	4.80 de	7.67 c	7.23 c	12.21 ab	7.93 b
NLSt	6.37 ab	5.38 bc	5.27 cde	10.73 abc	7.80 bc	6.92 bc	9.58 ab
ASnt	5.15 c	7.06 b	6.23 bc	12.58 ab	10.53 a	12.51 ab	7.73 b
ASt	7.02 a	9.49 a	7.67 d	8.74 c	7.55 c	8.03 bc	9.87 ab
PRE-Snt	n.p.	3.22 d	3.98 e	10.28 abc	10.57 a	11.94 ab	9.75 ab
PRE-St	n.p.	n.p.	n.p.	9.18 bc	8.28 abc	12.00 ab	9.53 ab
ASTIGRnt	n.p.	6.59 b	6.7 bc	11.13 abc	9.63 abc	12.82 a	7.07 b
ASTIGRt	n.p.	n.p.	n.p.	9.38 bc	8.77 abc	13.71 a	8.42 ab
TIGRnt	n.p.	8.91 a	10.03 a	13.17 a	10.07 ab	14.03 a	10.10 ab
TIGRt	n.p.	n.p.	n.p.	13.17 a	10.42 a	13.73 a	11.48 a
Treatment	**Magnesium (mg g^{-1})**						
NLSnt	0.76 ab	0.73 cd	0.66 cd	1.21 bc	0.90 b	1.65 a	1.01 abc
NLSt	0.85 a	1.01 bc	1.13 a	1.52 bc	1.56 a	1.57 a	1.35 a
ASnt	0.63 b	1.27 b	0.76 c	1.73 b	1.03 b	1.62 a	0.86 bc
ASt	0.86 a	1.66 a	1.01 ab	1.35 bc	0.87 b	1.14 a	1.22 ab
PRE-Snt	n.p.	0.55 d	0.57 d	1.10 c	1.07 b	1.39 a	1.23 ab
PRE-St	n.p.	n.p.	n.p.	1.41 bc	0.98 b	1.92 a	0.99 abc
ASTIGRnt	n.p.	0.77 cd	0.68 cd	1.41 bc	0.96 b	1.63 a	0.71 c
ASTIGRt	n.p.	n.p.	n.p.	1.48 bc	1.16 ab	2.05 a	1.09 ab
TIGRnt	n.p.	1.00 bc	0.93 b	1.57 bc	1.00 b	1.55 a	1.01 abc
TIGRt	n.p.	n.p.	n.p.	2.35 a	1.23 ab	1.45 a	1.18 ab

NLS = never leaf-symptomatic vine leaves; AS = asymptomatic diseased vine leaves; PRE-S = pre-symptomatic leaves of symptomatic vines; ASTIGR = asymptomatic leaves of symptomatic vines; TIGR = tiger-striped leaves of symptomatic vines. t = treated with fertilizer mixture; nt = untreated; n.p. = not present category. For each column of each year values followed by the same letter do not differ statistically according to Tukey's honest significant difference (HSD) test at $p = 0.05$.

In treated vines, Ca content was higher in AS leaves than in NLS ones, from 71 to 75 growth stage, to a significant extent in 2019, at growth stage 75, when Mg contents were also significantly higher in the AS leaves. Conversely, at growth stage 85, for both elements and in each of the two years of sampling, higher contents were observed in NLS leaves compared to AS ones, although significantly only for Mg in 2020. At this growth stage, Ca and Mg contents in leaves of the other categories of treated shoots were not different to those of NLS and AS ones, except for higher contents in TIGR leaves. At growth stage 89, no difference in Ca and Mg contents was detected among the various categories of treated leaves, with the exception of higher Ca levels in TIGR and ASTIGR leaves compared to NLS and AS leaves, recorded in 2019 (Table 6).

In both years of sampling, Ca and Mg contents were significantly higher in treated compared to untreated AS leaves, at growth stages 71 and 75. In the following sampling, at growth stage 85, a clear inversion of these results was observed, since significantly higher contents of Ca and Mg were recorded in untreated compared to treated AS leaves. In 2019, at growth stage 89, higher Ca and Mg contents in untreated than treated AS leaves were still detected. Unlike AS vines, Ca and Mg contents in leaves of NLS treated vines were not different compared to untreated NLS ones at growth stages 85 and 89 (Table 6).

2.3.2. Microelements

In untreated vines, in 2019, at growth stage 75, Na content was significantly higher in NLS leaves, compared to AS, PRE-S, ASTIGR, and TIGR ones. In these categories of diseased vine, Na contents increased during the growing season. Therefore, in the following growth stages, a progressive leveling of Na content in the different categories was observed, which no longer differed from each other, with the exception of TIGR leaves. In both years TIGR leaves showed generally higher Na content than that of the other categories. Except for TIGR leaves, in 2020, at growth stage 75, the leaves of the different categories did not differ in Na content. However, as in 2019, in the following growth stages 85 and 89, no differences were noticed among NLS, AS, PRE-S, and ASTIGR leaves (Table 7).

Table 7. Sodium content in never leaf-symptomatic vine leaves and in different categories of leaves of diseased vines.

	BBCH 71	BBCH 75		BBCH 85		BBCH 89	
	27/06/2020	11/07/2019	09/07/2020	27/08/2019	27/08/2020	12/09/2019	12/09/2020
Treatment	**Sodium (mg Kg^{-1})**						
NLSnt	32.8 b	112.6 a	57.7 c	106.3 ab	64.2 de	84.3 d	72.5 a
NLSt	52.7 a	83.3 ab	110.0 a	86.5 ab	127.0 a	180.0 ab	113.3 a
ASnt	29.5 b	53.7 bc	61.6 bc	111.7 ab	101.7 abcd	135.5 bcd	109.7 a
ASt	55.5 a	112.4 a	124.0 a	90.5 ab	73.5 cde	112.0 bcd	111.2 a
PRE-Snt	n.p.	34.2 c	72.3 bc	55.8 b	60.8 e	100.2 cd	91.5 a
PRE-St	n.p.	n.p.	n.p.	134.5 ab	85.3 bcde	168.3 bc	91.7 a
ASTIGRnt	n.p.	28.3 c	68.7 bc	108.6 ab	66.2 de	87.5 d	67.3 a
ASTIGRt	n.p.	n.p.	n.p.	120.7 ab	76.0 cde	106.2 bcd	71.7 a
TIGRnt	n.p.	35.6 bc	96.3 ab	147.5 a	108.0 abc	102.3 bcd	138.3 a
TIGRt	n.p.	n.p.	n.p.	99.4 ab	117.3 ab	250.2 a	134.8 a

NLS = never leaf-symptomatic vine leaves; AS = asymptomatic diseased vine leaves; PRE-S = pre-symptomatic leaves of symptomatic vines; ASTIGR = asymptomatic leaves of symptomatic vines; TIGR = tiger-striped leaves of symptomatic vines. t = treated with fertilizer mixture; nt = untreated; n.p. = not present category. For each column of each year values followed by the same letter do not differ statistically according to Tukey's honest significant difference (HSD) test at $p = 0.05$.

In treated vines, except for some sporadic cases, no differences were recorded among NLS, AS, PRE-S, and ASTIGR leaves. Only TIGR leaves showed high Na contents especially recorded at growth stages 89, in 2019. At growth stages 85 and 89, AS, PRE-S, and ASTIGR leaves of treated diseased vines never differed from each other in both years of survey (Table 7).

In 2019 and 2020, Na contents appeared significantly higher in treated compared to untreated AS leaves, at growth stages 71 and 75. On the contrary, in the following growth stages, higher Na contents were recorded in untreated AS leaves. In 2020, treated NLS leaves had significantly higher content of Na at growth stages 71, 75, and 85, than untreated NLS leaves. At these growth stages, in 2019, no difference was noticed.

No significant differences in Na levels were observed between treated and untreated leaves of the other categories, except for significantly higher contents in treated compared to untreated TIGR leaves, at growth stage 89, in 2019 (Table 7).

Fe contents were similar in the different categories of untreated leaves at the first two growth stages, in both years of study. Later in the season, significantly higher contents in AS, ASTIGR, and TIGR leaves compared to NLS ones, at growth stage 85, in 2020, were noticed (Table S2). Furthermore, in both years, Fe contents of untreated TIGR leaves were significantly higher than that recorded in NLS and AS leaves, at growth stages 89.

Treated vines did not show differences among shoot categories, with the exception of TIGR leaves in 2019 at growth stage 85, with Fe content significantly higher than NLS, AS, and PRE-S leaves (Table S2).

No difference in Fe levels was generally recorded between treated and untreated leaves (Table S2).

In untreated vines, no significant differences in Cu content were noticed among the leaves of the various shoot categories, apart from very few exceptions only related to higher NLS leaf content at growth stage 75 in 2019, in ASTIGR at growth stage 75 in 2020 and in AS at growth stage 85 in 2020 (Table S2).

In treated vines, leaves from the different categories of shoots never differed from each other, except for significantly higher Cu contents in TIGR compared to NLS leaves, at growth stage 85, in 2019 (Table S2).

In the comparison between treated and untreated vines, no differences were generally noticed in Cu content among the different investigated categories (Table S2).

In untreated vines, in 2020, Mn contents were significantly higher: in ASTIGR than PRE-S leaves, at growth stage 75; in AS than NLS, PRE-S and TIGR leaves, at growth stage 85. These results were not observed at the corresponding growth stages in 2019 (Table S3).

No differences in Mn content were found among leaves of treated shoots, except for significantly higher contents in PRE-S than NLS and ASTIGR leaves, at growth stage 85 of 2019 (Table S3).

Mn contents did not show any difference between the leaves of the same shoot categories of treated and untreated vines.

No differences were generally recorded in Zn leaf content (Table S3).

3. Discussion

The grapevine trunk diseases (GTDs) are increasing in almost all grape growing areas [43]. Although Esca complex is the most widespread and studied GTDs in European vineyards, some etiological and epidemiological aspects still remained unexplained as well as for the development of an effective control strategy [3]. Losses in quantity and quality of yield in diseased vineyards appeared to be correlated with incidence and severity of plant symptom expression [14,15]. However, mechanisms and factors associated with the occurrence of symptoms are yet to be fully clarified. Therefore, the present study aimed to evaluate some aspects of the physiology of diseased vines involved with the appearance and development of the disease symptom expression.

3.1. Macro and Microelements in the Leaf

The analysis of macro and microelements carried out in the present study on samples of leaves collected at different stages of growing season highlighted the dynamics of each element both in never leaf-symptomatic and in the different categories of diseased vine shoots. Therefore, indications have been obtained on a possible role played by these elements in the complex mechanisms of symptom expression.

Levels of calcium and magnesium differed between leaves of never leaf-symptomatic and diseased vines, but also between leaves of different diseased vine shoots. In particular, for most of the growing season, in the parcel not treated with the fertilizer mixture, the leaves of asymptomatic vines, AS, and, at berries pea-sized, the leaves of asymptomatic shoots of symptomatic diseased vines, ASTIGR, had higher levels of calcium and magnesium compared to untreated leaves of never leaf-symptomatic vines, NLS. On the contrary, at pea-sized berries, still green leaves at the base of shoots that began to show symptoms in the leaves of distal part, PRE-S, had lower contents of calcium and magnesium, maybe also due to a shoot malfunction, compared to the leaves of the other shoots, and in particular compared to AS leaves. As observed in a previous study [32], accumulations of calcium, magnesium, sodium, and microelements were recorded in symptomatic leaves, TIGR, probably because of their impaired functionality. Other remarkable differences were in low sodium contents in PRE-S shoot leaves compared to AS leaves, at softening of berries. These results provided a first evidence on the role of calcium and magnesium in the expression of leaf symptoms, notable in both AS and ASTIGR leaves, which remained asymptomatic throughout the season [34,44]. The involvement of calcium and magnesium in the symptom occurrence could be confirmed by low levels of both elements in leaves of PRE-S, compared to AS shoots, which then showed a progressive development of symptoms. Furthermore, the lower sodium contents in PRE-S compared to AS leaves, indicated a possible function of this element in the dynamics of foliar symptom expression. The correlation between sodium and the expression of leaf symptoms is in agreement with what was found in the decay of kiwifruit, a disease similar for many aspects to Esca of grapevine [45].

Fertilizer mixture applications confirmed to be effective in reducing foliar symptom expression [34,44], but also contributed to obtain further information on leaf symptom expression dynamics. Leaves of treated AS vines had higher contents of calcium and magnesium, compared to NLS treated leaves, at berries pea-sized, while at softening of berries, higher calcium and magnesium contents in NLS compared to AS leaves were recorded. AS vines probably responded better to the treatment than NLS vines, because most likely NLS vines did not require calcium and magnesium. Therefore, it may not be excluded that AS vines, due to infection, could have a better physiological attitude to calcium and magnesium supply. Unlike what was observed in untreated AS leaves, treated AS leaves showed sodium dynamics similar to that found for calcium and magnesium, providing further indications of a possible involvement of sodium in foliar symptom expression. Also treated TIGR leaves, as for the untreated, accumulated calcium, magnesium, sodium, and microelements.

The comparison of elements content between treated and untreated leaves gave further information on their dynamics in leaf symptom development. The same dynamics observed for calcium, magnesium, and sodium in treated AS and NLS leaves appeared clearer in the comparison between treated and untreated AS leaves. In fact, contents of the three elements were higher in treated compared to untreated AS leaves, at fruit set and berries pea-sized stages, while at softening of berries contents reversed and were higher in untreated compared to treated AS leaves.

Calcium and magnesium contents were not different between treated and untreated NLS leaves, indicating how the applications of calcium and magnesium, at fruit set and berries pea-sized stages, before the onset of symptoms in the vineyard, were useful in AS vines particularly, by reducing the symptom expression.

Comparisons between leaves of treated and untreated shoots of the same category did generally not show any difference for the other investigated elements.

3.2. *Role of Calcium in Leaf Symptom Expression*

The overall analysis of results on macro and microelements content in leaves of the different categories confirmed the involvement of calcium in the expression of foliar symptoms, hypothesizing a possible role in reducing the symptom expression, as shown in a previous study [34].

The role of calcium was evident both in untreated AS vines, which showed leaf calcium increases for most of the season, and in treated AS vines, whose leaves accumulated calcium up to pea-sized berries, at the onset of symptoms in the vineyard. In treated vines, including the ones who most likely would show symptoms if not treated, calcium accumulations assessed in the first part of the season and due to fertilizer applications, indicated a possible role of the element in the modulation of the plant response, which was hypothesized for leaf symptom expression [46]. This plant response can be associated with phytotoxic substances produced by fungi in infected woody tissues and translocated to the leaf via transpiration current [20,21,47]. Therefore, the development of necrosis in symptomatic leaves was associated with a strong response of the plant to such phytotoxic substances, comparable to a hypersensitive response [48].

To support the hypothesis mentioned, previous studies showed that at pre-bunch closure, tiger-stripe leaves, compared to never leaf-symptomatic and asymptomatic vine leaves, had higher contents of *trans*-resveratrol and other phytoalexins, which are synthesized after the hypersensitive response of the plant [29,31].

Furthermore, previous studies demonstrated that vines without foliar symptoms as a consequence of fertilizer mixture applications had higher contents of *trans*-resveratrol, *trans*-ε-viniferin, *trans*-δ-viniferin, and flavonoids, as well as calcium oxalate druse, at pre-bunch closure, with respect to both untreated never leaf-symptomatic and asymptomatic diseased vines [34,35].

In the subsequent growth stages phytoalexins contents progressively decreased in treated vines and increased in untreated vines which showed higher contents than treated vines [35], reproducing the same dynamics of calcium, magnesium, and sodium observed in the present study.

These results strengthened the hypothesis of a defense plant response involved in the foliar symptom expression, indicating as well the role of the fertilizer mixture in this response, by reducing symptoms.

Moreover, what assessed on both untreated and treated asymptomatic plants suggested that calcium activity could be linked to the reduction of plant oxidative response [48]. The penetration of calcium into leaves, facilitated by the seaweed, can increase the synthesis of calmodulin. Calmodulin regulates salicylic acid and, consequently, plant response to fungal toxic metabolites, reducing effects of the hypersensitive response, associated with leaf symptom expression [49].

Moreover, the reduction of symptom expression could also be due to the ability of calcium to strengthen cell walls, as calcium oxalate crystal accumulation, assessed in leaves of treated asymptomatic vines [34,50,51]. Furthermore, the increasing of calcium content in extracellular spaces can increase phytoalexin synthesis [52,53].

3.3. Role of Magnesium and Sodium in Leaf Symptom Expression

Applications of the fertilizer mixture were particularly effective in diseased plants, likely for a plant health improvement stimulated by magnesium [54]. It therefore might not be excluded an involvement of magnesium as fundamental constituent of chlorophyll, in reducing or delaying its degradation, limiting as well leaf symptom expression in synergy with calcium [34,44].

The role of magnesium associated with disease control proved to be much more controversial, as it can reduce, but also favor, disease severity [55]. Magnesium can have indirect beneficial effects on diseases with the possibility to withstand disease outbreak, because of the involvement of this element in several physiological processes associated with plant health [54].

In GTDs, studies demonstrated that Mg^{++} and Mn^{++} detoxified eutypin, the toxin produced by *Eutypa lata*, causal agent of Eutypa dieback [56]. It might not be excluded similar effects on toxic metabolites produced by the pathogens involved in Esca of grapevine. Although the reduction of symptom expression was demonstrated in the present study mainly as effect of calcium and magnesium, further studies are needed to give a specific role to magnesium.

The significant sodium increasing at berries pea-sized in treated compared to untreated AS vine leaves could be related to phytoalexin increasing observed in a previous study at the same growth stage. Regulatory mechanisms associated with second messenger metabolites and sodium uptake have been studied; in particular, fast and temporary increases of reactive oxygen species (ROS) involved in hypersensitive response and secondary metabolites synthesis, soon after increases of Na^{+}, were recorded [57].

Given the structural similarity of sodium and potassium ions [58], sodium can replace potassium but only in conditions of its deficiency [59,60]. Therefore, a peculiar function of sodium in the leaf symptom expression might even be possible, since the uptake of this element did not seem linked to a leaf potassium deficiency never found in the present study in AS leaves.

3.4. *Leaf Reflectance Measurements*

At berries pea-sized, leaves of never leaf-symptomatic and asymptomatic diseased vines had WI values higher than leaves of the different categories of symptomatic shoots, indicating a higher plant water concentration (PWC). In the first three surveys carried out at full flowering, end of flowering, and fruit set stages, never leaf-symptomatic and asymptomatic diseased vines, treated with the fertilizer mixture, showed WI leaves values higher than those of leaves of corresponding untreated vines. The higher WI values in untreated AS vine leaves, recorded at berries pea-sized, highlighted better physiological efficiency of AS vines, further improved by fertilizer applications.

On the contrary, at full flowering, NDVI and GNDVI values were lower in treated NLS and AS leaves, compared to the untreated ones, while at fruit set, both indices were higher in AS treated compared to untreated, but similar between treated and untreated NLS leaves. These results demonstrated the activity of fertilizer mixture applications carried out until the fruit set. Higher WI and lower NDVI and GNDVI values, observed at full flowering in treated compared to the untreated AS leaves, might be related to the 'energy' required for plant defense response modulated by the components of fertilizer mixture.

However, decreasing NDVI and GNDVI values in treated leaves was transitory; in fact, at fruit set, near the beginning of first symptom appearance, these values increased in treated compared to untreated AS leaves, because of an increased photosynthetic activity probably due to fertilizer mixture applications.

At fruit set, higher contents of calcium and magnesium were also recorded on treated compared to untreated AS leaves. In a previous study, at the first days of symptom appearance, *trans*-resveratrol, was higher in treated compared to untreated AS leaves [26]. Thus, the mixture applications seemed to act with a further mechanism of action, by increasing photosynthetic activity, probably favoring increases of the two elements and *trans*-resveratrol.

On the other hand, at berries pea-sized low contents of calcium and magnesium, as well lower values of WI, were detected in untreated PRE-S leaves, confirming the role played by the two elements in limiting foliar symptoms.

4. Materials and Methods

4.1. *Vineyard*

Field trials were carried out in a 42-year-old vineyard, cv. Trebbiano d'Abruzzo on 420A rootstock, trained as Geneva Double Courtain (GDC). The vineyard, located in Controguerra (TE), Abruzzo, a central Italy Region, consisted of 700 vines, planted on an area of 5984 m^2 on a clayey-limestone soil, with a 2×4 m planting. The average yield per vine ranged from 13 to 16.5 kg. This vineyard is being surveyed since 1994 for the incidence and severity of tiger-stripe foliar symptoms.

4.2. *Leaf Applications*

Two parcels were identified in the vineyard; one of them was treated with the fertilizer mixture (Algescar ®, Natural Development Group, Castelmaggiore, Bologna, Italy) based on $CaCl_2$ (33.7%), $Mg(NO_3)_2$ (29.1%), and *Fucales* seaweed extract (6.1%), the second parcel represented the untreated control. Each parcel included three plots of 90 vines, each consisting of a replicate. In 2019 and 2020, vines of a parcel were subjected to six foliar applications with the fertilizer mixture. The mixture was applied with an air blast sprayer at a dose of 5 L, in a water volume of 800 L ha^{-1}, at 10–15 days interval, from 53 "inflorescences clearly visible" to 77 "berries beginning to touch" BBCH growth stages [61]. In 2019, applications were performed on 14 May; 3, 12, 21 and 26 June; and 15 July; in 2020 on 13 May; 3, 13, 22 and 26 June; and 15 July.

4.3. Leaf Symptom Surveys

The foliar symptom incidence and severity were recorded every year for 25 consecutive seasons. Thus, during the survey of the present study carried out in 2019 and 2020, it was possible to identify the asymptomatic infected vines, which did not show symptoms in the years of survey but have shown symptoms in at least one of the previous years of survey. These asymptomatic vines were distinguished from the never leaf-symptomatic vines, which did not show symptoms in any of the 25 years of survey.

In 2019, four surveys on foliar symptoms were carried out on 8 and 31 July, 27 August and 13 September; in 2020, six surveys were carried out on 7, 20, and 31 July, 10 and 27 August; and 12 September, both in treated and untreated parcels. Surveys in July were performed from 75 "berries pea sized" to 79 "majority of berries touching" BBCH growth stages; in August at 83 "berries developing color" and 85 "softening of berries" stages; in September at 89 "berries ripe for harvest" stage [61].

In each survey, both the disease incidence and severity were recorded on each single plant. The incidence was calculated by dividing the number of plants with symptoms on the total number of diseased vines (that are plants that showed symptoms in at least one of the 25 years of survey) and multiplying by 100. The severity was calculated using the formula $SN \times 100/(Y \times Z)$, where SN = sum of symptom severity values; Y = number of the monitored plants; Z = maximum value of the symptom scale [62]. The symptom severity was calculated using an arbitrary 0–5 scale, where 0 = absence of symptom; 1 = 1–10%; 2 = 11–30%; 3 = 31–50%; 4 = 51–70%; 5 = 71–100% of foliar symptoms on the canopy.

4.4. Leaf Reflectance Measurements

For each plant category (never leaf-symptomatic, diseased but asymptomatic for the entire season and symptomatic diseased) leaf reflectance measurements were carried out in both years of the study on leaves of shoots classified as follows: never leaf-symptomatic (NLS) from never leaf-symptomatic vines and asymptomatic (AS) from diseased but asymptomatic for the entire season vines. In the symptomatic diseased vine group measurements were carried out on: i) pre-symptomatic shoots, namely shoots that showed leaves with early symptoms at the bottom of the shoot, and leaves still without symptoms in the remaining part (PRE-S); ii) shoots with only asymptomatic leaves for the entire season (ASTIGR); and iii) shoots with only symptomatic leaves (TIGR). For each category of shoot, measurements, performed with a portable spectroradiometer mod. Fieldspec Pro®, Malvern Panalytical Ltd. (Malvern, UK), were carried out on leaves located in the median part of the primary shoots, opposite to a cluster, in order to avoid variability along the shoot.

The spectroradiometer sensor is equipped with high sensitivity detector array, low straylight, built-in shutter, background current compensation system, and second-order filter that allows detections with a high signal-to-noise ratio in less than a second. Spectrum detection consists in the acquisition of discrete measurements, recorded by the instrument through the internal software and shown on the integrated display. It is then possible to connect the instrument to a computer for processing and exporting the detected spectra, using a special program supplied with the instrument.

For each of categories never leaf-symptomatic vines (NLS) and diseased but asymptomatic for the entire season vines (AS), six plants were identified. In each plant, two shoots were considered, one facing east (side A) and one facing west (side B). In each of the two shoots, five measurements were made on three leaves, finally obtaining an average. Also in the symptomatic diseased vines, six vines were chosen, but for each vine the measurements were performed on three shoots, one for each PRE-S, ASTIGR, and TIGR treatments; in this case, measurements were carried out only on the shoots of side A, given the lack of symptomatic shoots in the part of the canopy facing west (side B).

In NLS and AS vines data of both treated and control parcels were recorded. On the contrary, measurements on the five types of shoot (NLS, AS, PRE-S, ASTIGR, and TIGR) were performed only in the untreated parcel, due to the low expression of foliar symptoms, probably because of the combination of the fertilizer mixture activity and seasonal meteorological conditions.

Reflectance measurements of NLS and AS shoots were carried out in each of the 9 surveys, from 66 "full flowering" to 89 "berries ripe for harvest" BBCH growth stages, whereas measurements of the categories of shoot in the untreated parcel were carried out in six surveys, from 75 "berries pea-sized", at the onset of symptoms, to 89 "berries ripe for harvest" BBCH growth stages.

NDVI, GNDVI, and WI values were recorded at each reflectance measurement.

4.5. Analysis of Macro and Microelements in the Leaf

4.5.1. Leaf Sampling

In order to assess the content of macro and microelements in leaves of the different categories of plants, in 2019 leaf samples were taken on 11 July, 27 August, and 12 September, corresponding respectively to the following BBCH growth stages: 77 "berries beginning to touch"; 85 "softening of berries"; and 89 "berries ripe for harvest". In 2020, samples were taken on July 9, August 27, and September 12, with an extra sampling on June 27, at 71 "fruit set" [61].

For each plant category (never leaf-symptomatic, diseased but asymptomatic for the entire season and symptomatic diseased) samples were taken on six vines. Eight leaves were taken from each never leaf-symptomatic vine (NLS) and all season asymptomatic diseased vine (AS). From each symptomatic diseased vine, eight leaves were taken for each of the three types of shoot—PRE-S, ASTIGR, and TIGR. For each type of shoot, six samples were collected, each consisting of eight leaves. Leaves were always collected from the median portion of primary shoots and in the opposite position to a cluster.

4.5.2. Macro and Microelements Analysis

Reagents and Standards

Mix elements stock standard solution of calcium (Ca), magnesium (Mg), potassium (K), sodium (Na) at 2000, 400, 200 e 1000 mg L^{-1} respectively and single element stock solution of yttrium (Y) at 1000 mg L^{-1} were purchased from Sigma-Aldrich (St. Louis, MO, USA). Mix elements stock standard solution of manganese (Mn), copper (Cu), iron (Fe), zinc (Zn), and single element stock solution of phosphorus (P) at 100 mg L^{-1} were provided by Panreac Química SLU (Castellar del Vallès, Barcelona, Spain) and CPAchem (Stara Zagora, Bulgaria) respectively.

Solutions were prepared with high-purity water of 18.2 $M\Omega \cdot cm$ resistivity obtained from a PURELAB® (ELGA LabWater, High Wycombe, United Kingdom). Reagents used for the sample digestion were nitric acid at concentration $\geq$67%, (CHEM-LAB NV, Zedelgem, Belgium), and hydrogen peroxide at 30% (Merck, Darmstadt, Germany). Argon gas of 99.9995% purity was supplied by Sapio (Monza, Italy).

The reference material BCR 1573a, tomato leaves (National Institute of Standard & Technology, Gaithersburg, Maryland, USA) has certified values of concentration of all investigated elements.

Analytical Method—Sample Preparation

Each sample was analyzed to determine levels of calcium, magnesium, potassium, sodium, manganese, copper, iron, zinc, and phosphorus.

The sample preparative step was made applying two official methods UNI EN 13804:2013 (sample homogenization) and UNI EN 13805:2014 (sample microwave digestion).

The instrumental analysis of all investigated elements was conducted by inductively coupled plasma–atomic emission spectrometry (ICP-AES) applying official method UNI EN 15621:2017.

The method provides for a complete destruction of organic matter with nitric acid and hydrogen peroxide at high temperatures and pressure, in a closed vessel, applying microwave assisted heating, prior to ICP-AES analysis.

At each session, a certified reference material (BCR 1573a) as quality control for all investigated elements was analyzed.

Prior to analysis, all the apparatus intended to come into direct contact with the sample and glassware were treated with nitric acid solution (1 ÷ 2%) and then rinsed with high-purity water.

Leaf samples, with own foil and petiole, after sampling in the field, were stored at the temperature of −20 °C.

At the time of analysis, samples were homogenized with cutting mill (Grindomix GM-200, Retsch, Germany) at 9000 rpm for (30 ÷ 60) seconds.

Homogenized sample (500 ± 50) mg L^{-1} was weighed into PTFE vessels and dissolved in 5 mL of concentrated nitric acid and 1 mL of hydrogen peroxide at 30%. Mineralization was performed in a Multiwave 3000 microwave digestion system (Anton Paar, Graz, Austria) according to the program shown in Table 8.

Table 8. Instrumental parameters of microwave digestion oven.

Step	Power (watt)	Ramp (°C/min)	Duration (min)
1	450	1	4
2	800	5	8
3	1000	5	15
4	0	-	15

After cooling, the resulting clear solutions (samples, blank and reference material) were quantitatively transferred and diluted exactly to a volume of 15 mL with high-purity water.

Analytical Method—Instrumental Analysis

Analysis of investigated elements was carried out with an inductively coupled plasma atomic emission spectrometer Optima 7000 (PerkinElmer, Waltham, MA, USA). Measurements were performed applying the instrumental conditions mentioned below (Tables 9 and 10).

Table 9. Instrumental conditions of ICP-AES.

Parameter	Value
Power radiofrequency (W)	1300 ÷ 1500
Nebulizer gas flow (L min^{-1})	0.6 ÷ 0.7
Plasma gas flow (L min^{-1})	15
Auxiliary gas flow (L min^{-1})	0.2
Nebulizer	Mira Mist (in peek)
Spray chamber	Cyclonic

Quantitative determination was performed by an external calibration with Y as internal standard. The linearity range of all calibration curves are reported in Table 11.

At each analytical session an aliquot of certified reference material BCR 1573a was analyzed. In Table 12 certified values and achieved recovery average were reported.

Table 10. Wavelength and mode view.

Element	Wavelenght (nm)	Plasma Mode View
Fe	238.204	Axial
Cu	327.393	
Zn	213.857	
Mn	257.610	Attenuated Axial
P	213.617	Radial
Na	589.592	
Ca	317.933	
Mg	285.213	
K	766.490	
Y	361.104–371.029	-

Table 11. Linearity range of calibration curve of elements.

Element	Linearity Range (mg L^{-1})
Fe	0.0050 ÷ 1.0
Cu	
Zn	
Mn	
P	0.25 ÷ 10
Na	0.13 ÷ 2.0
Ca	0.25 ÷ 10
Mg	0.050 ÷ 2.0
K	0.025 ÷ 1.0

Table 12. Certified values of BCR1573a and achieved recovery average.

Element	Certified Values (mg Kg^{-1})	Recovery (%) (n = 5)
Fe	368 ± 7	100
Cu	4.70 ± 0.14	109
Zn	30.9 ± 0.7	97
Mn	246 ± 8	98
P *	0.216 ± 0.004	102
Na	136 ± 4	95
Ca *	5.05 ± 0.09	101
Mg *	1.2 **	93
K *	2.70 ± 0.05	91

* measurement unit (%); ** information value; n = number of analytical session.

4.6. Statistical Analysis

In both years of the study, the incidence and severity of foliar symptoms recorded in the parcel treated with the fertilizer mixture, were compared with values of the untreated parcel. The comparison was carried out at harvest (13 September 2019 and 12 September 2020) using Chi-square tests at $p = 0.05$, following what reported in our previous study [25]. A one-way analysis of variance (ANOVA) was applied for each survey or sampling values, respectively of NDVI, GNDVI, WI or macro and microelements, recorded in leaves of NLS and AS vines and in leaves of shoots of symptomatic diseased vines ASTIGR, PRE-S, and TIGR. When significant differences emerged, means separation was performed by Tukey's honest significant difference (HSD) test at $p = 0.05$. Statistical analysis was performed using XLSTAT 2016 (Addinsoft, Paris, France).

5. Conclusions

Dynamics of macro and microelements and vegetation indices highlighted in this study, as for the dynamics of phytoalexins discussed in our recent studies, suggested the role of calcium and magnesium in tiger-stripe foliar symptoms expression. These results also reinforced the hypothesis on the triggering of a complex response of the plant to the occurrence and development of leaf symptoms induced by toxic substances promoted by pathogens.

Calcium could play a role in modulating the plant's response to toxic fungal metabolites, reducing the effects of an uncontrolled reaction associated with the expression of foliar symptoms in diseased vines.

The activity of calcium was indicated both in untreated vines and in the vines treated with the fertilizer mixture, which significantly reduced the symptom expression, although with different dynamics.

Calcium, magnesium, and sodium contents were particularly high in leaves of treated asymptomatic diseased vines at berries pea-sized, when foliar symptoms began to appear in the vineyard. In these leaves, the high content of the three elements was accompanied by high values of WI, preceded by high NDVI and GNDVI values recorded at fruit set, and followed by an early synthesis of phytoalexins, demonstrated in previous studies.

Therefore, increased availability of calcium and magnesium up to pea-sized berries reduced foliar symptom expression just when they begin to appear.

Further evidence carried out in different grape growing areas and infection conditions are needed to strengthen the hypothesis discussed in the present study.

Supplementary Materials: The following are available online at https://www.mdpi.com/article/10.3390/plants10061041.../s1, Table S1: Phosphorus and potassium content in never leaf-symptomatic vine leaves and in different categories of leaves of diseased vines, Table S2: Iron and Copper content in never leaf-symptomatic vine leaves and in different categories of leaves of diseased vines, Table S3: Manganese and Zinc content in never leaf-symptomatic vine leaves and in different categories of leaves of diseased vines.

Author Contributions: Conceptualization, F.C. and S.D.M.; methodology, F.C., M.B. and G.C.; software, G.P.; validation, F.C., S.D.M. and M.P.; formal analysis, G.P.; investigation, F.C. and E.G.M.; data curation, F.C. and S.D.M.; writing—original draft preparation, F.C.; writing—review and editing, F.C. and S.D.M. All authors have read and agreed to the published version of the manuscript.

Funding: This research received no external funding.

Institutional Review Board Statement: Not applicable.

Informed Consent Statement: Not applicable.

Data Availability Statement: Data are reported within the Article and Supplementary materials.

Acknowledgments: The authors thank Francesco Pitocco for much help given during this study.

Conflicts of Interest: The authors declare no conflict of interest.

References

1. Mugnai, L.; Graniti, A.; Surico, G. Esca (black measles) and brown wood streaking: Two old and elusive diseases of grapevines. *Plant Dis.* **1999**, *83*, 404–418. [CrossRef] [PubMed]
2. Bertsch, C.; Ramírez-Suero, M.; Magnin-Robert, M.; Larignon, P.; Chong, J.; Abou-Mansour, E.; Spagnolo, A.; Clèment, C.; Fontaine, F. Grapevine trunk diseases: Complex and still poorly understood. *Plant Pathol.* **2013**, *62*, 243–265. [CrossRef]
3. Gramaje, D.; Urbez-Torres, J.R.; Sosnowski, M.R. Managing grapevine trunk diseases with respect to etiology and epidemiology: Current strategies and future prospects. *Plant Dis.* **2018**, *102*, 12–39. [CrossRef]

4. Brown, A.A.; Lawrence, D.P.; Baumgartner, K. Role of basidiomycete fungi in the grapevine trunk disease esca. *Plant Pathol.* **2020**, *69*, 205–220. [CrossRef]
5. Baránek, M.; Armengol, J.; Holleinová, V.; Pečenka, J.; Calzarano, F.; Peňázová, E.; Vachůn, M.; Eichmeier, A. Incidence of symptoms and fungal pathogens associated with grapevine trunk diseases in Czech vineyards: First example from a north-eastern European grape-growing region. *Phytopathol. Mediterr.* **2018**, *57*, 449–458. [CrossRef]
6. Surico, G. Towards a redefinition of the diseases within the esca complex. *Phytopathol. Mediterr.* **2009**, *48*, 5–10.
7. Fischer, M. Biodiversity and geographic distribution of basidiomycetes causing esca-associated white rot in grapevine: A worldwide perspective. *Phytopathol. Mediterr.* **2006**, *45*, S30–S42.
8. Edwards, J.; Marchi, G.; Pascoe, I. Young esca in Australia. *Phytopathol. Mediterr.* **2001**, *40*, S303–S310.
9. Calzarano, F.; Di Marco, S. Wood discoloration and decay in grapevines with esca proper and their relationship with foliar symptoms. *Phytopathol. Mediterr.* **2007**, *46*, 96–101.
10. Lecomte, P.; Darrieutort, G.; Liminana, J.M.; Louvet, G.; Tandonnet, J.P. Eutypiose et esca: I: Eléments de réflexion pour mieux appréhender ces phénomènes de dépérissement. *Phytoma* **2008**, *615*, 42–48.
11. Ouadi, L.; Bruez, E.; Bastien, S.; Vallance, J.; Lecomte, P.; Domec, J.C.; Rey, P. Ecophysiological impacts of Esca, a devastating grapevine trunk disease, on *Vitis vinifera* L. *PloS ONE* **2019**, *14*, e0222586. [CrossRef] [PubMed]
12. Maher, N.; Piot, J.; Bastien, S.; Vallance, J.; Rey, P.; Guérin-Dubrana, L. Wood necrosis in esca-affected vines: Types, relationships and possible links with foliar symptom expression. *J. Int. Sci. Vigne Vin.* **2012**, *46*, 15–27. [CrossRef]
13. Bruez, E.; Vallance, J.; Gerbore, J.; Lecomte, P.; Da Costa, J.P.; Guerin-Dubrana, L.; Rey, P. Analyses of the temporal dynamics of fungal communities colonizing the healthy wood tissues of esca leaf-symptomatic and asymptomatic vines. *PLoS ONE* **2014**, *9*, e95928. [CrossRef] [PubMed]
14. Calzarano, F.; Seghetti, L.; Del Carlo, M.; Cichelli, A. Effect of esca on the quality of berries, musts and wines. *Phytopathol. Mediterr.* **2004**, *43*, 125–135.
15. Lorrain, B.; Ky, I.; Pasquier, G.; Jourdes, M.; Guerin Dubrana, L.; Gény, L.; Rey, P.; Donèche, B.; Teissedre, P.L. Effect of Esca disease on the phenolic and sensory attributes of Cabernet Sauvignon grapes, musts and wines. *Aust. J. Grape Wine R.* **2012**, *18*, 64–72. [CrossRef]
16. Marchi, G.; Peduto, F.; Mugnai, L.; Di Marco, S.; Calzarano, F.; Surico, G. Some observations on the relationship on manifest and hidden esca to rainfall. *Phytopathol. Mediterr.* **2006**, *45*, 117–126.
17. Calzarano, F.; Osti, F.; Baránek, M.; Di Marco, S. Rainfall and temperature influence expression of foliar symptoms of grapevine leaf stripe disease (esca complex) in vineyards. *Phytopathol. Mediterr.* **2018**, *57*, 488–505. [CrossRef]
18. Lecomte, P.; Diarra, B.; Carbonneau, A.; Rey, P.; Chevrier, C. Esca of grapevine and training practices in France: Results of a 10-year survey. *Phytopathol. Mediterr.* **2018**, *57*, 472–487.
19. Sparapano, L.; Bruno, G.; Graniti, A. Effects on plants of metabolites produced in culture by *Phaeoacremonium chlamydosporum*, *P. aleophilum* and *Fomitiporia punctata*. *Phytopathol. Mediterr.* **2000**, *39*, 169–177.
20. Evidente, A.; Sparapano, L.; Andolfi, A.; Bruno, G. Two naphthalenone pentaketides from liquid cultures of *Phaeoacremonium aleophilum*, a fungus associated with esca of grapevine. *Phytopathol. Mediterr.* **2000**, *39*, 162–168.
21. Tabacchi, R.; Fkyerat, A.; Poliart, C.; Dubin, G. Phytotoxins from fungi of esca of grapevine. *Phytopathol. Mediterr.* **2000**, *39*, 156–161.
22. Andolfi, A.; Mugnai, L.; Luque, J.; Surico, G.; Cimmino, A.; Evidente, A. Phytotoxins produced by fungi associated with grapevine trunk diseases. *Toxins* **2011**, *3*, 1569–1605. [CrossRef]
23. Magnin-Robert, M.; Letousey, P.; Spagnolo, A.; Rabenoelina, F.; Jacquens, L.; Mercier, L.; Clément, C.; Fontaine, F. Leaf stripe form of esca induces alteration of photosynthesis and defence reactions in presymptomatic leaves. *Funct. Plant Biol.* **2011**, *38*, 856–866. [CrossRef] [PubMed]
24. Claverie, M.; Notaro, M.; Fontaine, F.; Wery, J. Current knowledge on Grapevine Trunk Diseases with complex etiology: A systemic approach. *Phytopathol. Mediterr.* **2020**, *59*, 29–53.
25. Lecomte, P.; Darrieutort, G.; Liminana, J.M.; Comont, G.; Muruamendiaraz, A.; Legorburu, F.J.; Choueiri, E.; Jreijiri, F.; El Amil, R.; Fermaud, M. New insights into esca of grapevine: The development of foliar symptoms and their association with xylem discoloration. *Plant Dis.* **2012**, *96*, 924–934. [CrossRef] [PubMed]
26. Bortolami, G.; Gambetta, G.A.; Delzon, S.; Lamarque, L.J.; Pouzoulet, J.; Badel, E.; Burlett, R.; Charrier, G.; Cochard, H.; Dayer, S.; et al. Exploring the hydraulic failure hypothesis of esca leaf symptom formation. *Plant Physiol.* **2019**, *181*, 1163–1174. [CrossRef]
27. Pouzoulet, J.; Scudiero, E.; Schiavon, M.; Santiago, L.S.; Rolshausen, P.E. Modeling of xylem vessel occlusion in grapevine. *Tree Physiol.* **2019**, *39*, 1438–1445. [CrossRef]
28. Larignon, P. Effect of sodium arsenite on the life cycles of the pathogenic agents involved in wood grapevine diseases. *Phytopathol. Mediterr.* **2017**, *56*, 537.
29. Calzarano, F.; D'Agostino, V.; Pepe, A.; Osti, F.; Della Pelle, F.; de Rosso, M.; Flamini, R.; Di Marco, S. Patterns of phytoalexins in the grapevine leaf stripe disease (esca complex)/grapevine pathosystem. *Phytopathol. Mediterr.* **2016**, *55*, 410–426. [CrossRef]
30. Calzarano, F.; Osti, F.; D'Agostino, V.; Pepe, A.; Della Pelle, F.; de Rosso, M.; Flamini, R.; Di Marco, S. Levels of phytoalexins in vine leaves with different degrees of grapevine leaf stripe disease symptoms (Esca complex of diseases). *Phytopathol. Mediterr.* **2017**, *56*, 494–501. [CrossRef]
31. Heath, M.C. Hypersensitive response-related death. *Plant Mol. Biol.* **2000**, *44*, 321–334. [CrossRef] [PubMed]

32. Calzarano, F.; Amalfitano, C.; Seghetti, L.; Cozzolino, V. Nutritional status of vines affected with esca proper. *Phytopathol. Mediterr.* **2009**, *48*, 20–31.
33. Di Marco, S.; Osti, F. Effect of biostimulant sprays on *Phaeomoniella chlamydospora* and esca proper infected vines under greenhouse and field conditions. *Phytopathol. Mediterr.* **2009**, *48*, 47–58.
34. Calzarano, F.; Di Marco, S.; D'Agostino, V.; Schiff, S.; Mugnai, L. Grapevine leaf stripe disease (esca complex) are reduced by a nutrients and seaweed mixture. *Phytopathol. Mediterr.* **2014**, *53*, 543–558.
35. Calzarano, F.; Osti, F.; D'Agostino, V.; Pepe, A.; Di Marco, S. Mixture of calcium, magnesium and seaweed affects leaf phytoalexin contents and grape ripening on vines with grapevine leaf stripe disease. *Phytopathol. Mediterr.* **2017**, *56*, 445–457. [CrossRef]
36. Bhakta, I.; Phadikar, S.; Majumder, K. State-of-the-art technologies in precision agriculture: A systematic review. *J. Sci. Food Agric.* **2019**, *99*, 4878–4888. [CrossRef]
37. Vincini, M.; Calegari, F.; Casa, R. Sensitivity of leaf chlorophyll empirical estimators obtained at Sentinel-2 spectral resolution for different canopy structures. *Precis. Agric.* **2015**, *17*, 313–331. [CrossRef]
38. Wójtowicz, M.; Wójtowicz, A.; Piekarczyk, J. Application of remote sensing methods in agriculture. *Commun. Biom. Crop. Sci.* **2016**, *11*, 31–50.
39. Daughtry, C. Estimating corn leaf chlorophyll concentration from leaf and canopy reflectance. *Remote Sens. Environ.* **2000**, *74*, 229–239. [CrossRef]
40. Rouse, J.W.; Haas, R.H.; Schell, J.A.; Deering, D.W. Monitoring Vegetation Systems in the Great Plains with ERTS. In *Third Earth Resources Technology Satellite–1 Symposium*; Freden, S.C., Mercanti, E.P., Becker, M., Eds.; NASA Special Publication-351; Nasa: Washington, DC, USA, 1974; Volume I: Technical Presentations, pp. 309–317.
41. Gitelson, A.A.; Kaufman, Y.; Merzlyak, M.N. Use of a green channel in remote sensing of global vegetation from EOS-MODIS. *Remote Sens. Environ.* **1996**, *58*, 289–298. [CrossRef]
42. Penuelas, J.; Pinol, J.; Ogaya, R.; Filella, I. Estimation of plant water concentration by the reflectance Water Index WI (R900/R970). *Int. J. Remote Sens.* **1997**, *18*, 2869–2875. [CrossRef]
43. Mondello, V.; Songy, A.; Battiston, E.; Pinto, C.; Coppin, C.; Trotel-Aziz, P.; Clément, C.; Mugnai, L.; Fontaine, F. Grapevine trunk diseases: A review of fifteen years of trials for their control with chemicals and biocontrol agents. *Plant Dis.* **2018**, *102*, 1189–1217. [CrossRef]
44. Calzarano, F.; Di Marco, S. Further evidence that calcium, magnesium and seaweed mixtures reduce grapevine leaf stripe symptoms and increase grape yield. *Phytopathol. Mediterr.* **2018**, *57*, 459–471. [CrossRef]
45. Di Marco, S.; Osti, F. Foliar symptom expression of wood decay in *Actinidia deliciosa* in relation to environmental factors. *Plant Dis.* **2008**, *92*, 1150–1157. [CrossRef]
46. Lecourieux, D.; Ranjeva, R.; Pugin, A. Calcium in plant defence-signalling pathways. *New Phytol.* **2006**, *171*, 249–269. [CrossRef]
47. Andolfi, A.; Cimmino, A.; Evidente, A.; Iannaccone, M.; Capparelli, R.; Mugnai, L.; Surico, G. A new flow cytometry technique to identify *Phaeomoniella chlamydospora* exopolysaccharides and study mechanisms of esca grapevine foliar symptoms. *Plant Dis.* **2009**, *93*, 680–684. [CrossRef] [PubMed]
48. Lima, M.R.M.; Ferreres, F.; Dias, A.C.P. Response of *Vitis vinifera* cell cultures to *Phaeomoniella chlamydospora*: Changes in phenolic production, oxidative state and expression of defence-related genes. *Eur. J. Plant Pathol.* **2012**, *132*, 133–146. [CrossRef]
49. Du, L.; Ali, G.S.; Simons, K.A.; Hou, J.; Yang, T.; Reddy, A.S.N.; Poovaiah, B.W. Ca^{2+}/ calmodulin regulate salicylic-acid-mediated plant immunity. *Nature* **2009**, *457*, 1154–1158. [CrossRef]
50. Conway, W.S.; Sams, C.E.; Abbott, J.A.; Bruton, B.D. Postharvest calcium treatment of apple fruit to provide broad-spectrum protection against postharvest pathogens. *Plant Dis.* **1991**, *75*, 620–622. [CrossRef]
51. Kratzke, M.G. Study of Mechanism of Calcium Uptake by Potato Tubers and of Cellular Properties Affecting Soft Rot. Ph.D. Thesis, Department of Horticulture, University of Wisconsin, Wisconsin, MD, USA, 1988; p. 321.
52. Stäb, M.R.; Ebel, J. Effects of Ca^{2+} on phytoalexin induction by fungal elicitor in soybean cells. *Arch. Biochem. Biophys.* **1987**, *257*, 416–423. [CrossRef]
53. Tavernier, E.; Wendehenne, D.; Blein, J.P.; Pugin, A. Involvement of free calcium in action of cryptogein, a proteinaceous elicitor of hypersensitive reaction in tobacco cells. *Plant Physiol.* **1995**, *109*, 1025–1031. [CrossRef]
54. Datnoff, L.E.; Elmer, W.H.; Huber, D.M. *Mineral Nutrition and Plant Disease*; The American Phytopathological Society: St. Paul, MN, USA, 2007; p. 278.
55. Huber, D.M.; Jones, J.B. The role of magnesium in plant disease. *Plant Soil* **2013**, *368*, 73–85. [CrossRef]
56. Colrat, S.; Deswarte, C.; Latché, A.; Klaébe, K.; Bouzayen, M.; Fallot, J.; Roustan, J.P. Enzymatic detoxification of eutypine, a toxin from *Eutypa lata*, by *Vitis vinifera* cells: Partial purification of an NADPH-dependent aldehyde reductase. *Planta* **1999**, *207*, 544–550. [CrossRef]
57. Maathuis, F.J. Sodium in plants: Perception, signalling, and regulation of sodium fluxes. *J. Exp. Bot.* **2014**, *65*, 849–858. [CrossRef]
58. Amtmann, A.; Sanders, D. Mechanisms of Na (+) uptake by plant cells. *Adv. Bot. Res.* **1999**, *29*, 75–112.
59. Fageria, V.D. Nutrient interactions in crop plants. *J. Plant Nutr.* **2001**, *24*, 1269–1290. [CrossRef]
60. Garciadeblas, B.; Senn, M.E.; Banuelos, M.A.; Rodriguez-Navarro, A. Sodium transport and HKT transporters: The rice model. *Plant J.* **2003**, *34*, 788–801. [CrossRef] [PubMed]

61. Lorenz, D.H.; Eichhorn, K.W.; Bleiholder, H.; Close, R.; Meier, U.; Weber, E. Phenological growth stages of the grapevine (*Vitis vinifera* L. ssp. vinifera). Encoding and description of the phenological stages of the grapevine according to the extended BBCH scheme. *Aust. J. Grape Wine Res.* **1995**, *1*, 100–103.
62. McKinney, H.H. Influence of soil temperature and moisture on infection of wheat seedlings by *Helminthosporium sativum*. *J. Agr. Res.* **1923**, *26*, 195–218.

plants

MDPI

Article

Cultivar Susceptibility to Natural Infections Caused by Fungal Grapevine Trunk Pathogens in La Mancha Designation of Origin (Spain)

Juan L. Chacón-Vozmediano [1,*], David Gramaje [2], Maela León [3], Josep Armengol [3], Juan Moral [4], Pedro M. Izquierdo-Cañas [1] and Jesús Martínez-Gascueña [1]

[1] Institute for Agri-food and Forestry Research and Development of Castilla-La Mancha (IRIAF), Tomelloso, 13700 Ciudad Real, Spain; pmizquierdo@jccm.es (P.M.I.-C.); jmartinezg@jccm.es (J.M.-G.)
[2] Institute of Grapevine and Wine Sciences (ICVV), Spanish National Research Council (CSIC), University of La Rioja and Government of La Rioja, 26007 Logroño, Spain; david.gramaje@icvv.es
[3] Instituto Agroforestal Mediterráneo, Universitat Politècnica de València, 46022 Valencia, Spain; maela.leon@uv.es (M.L.); jarmengo@eaf.upv.es (J.A.)
[4] Department of Agronomy, María de Maeztu Unit of Excellence, Campus of Rabanales, University of Córdoba, 14071 Córdoba, Spain; juan.moral@uco.es
* Correspondence: jlchacon@jccm.es

Abstract: Grapevine trunk diseases (GTDs) are one of the main biotic stress factors affecting this crop. The use of tolerant grapevine cultivars would be an interesting and sustainable alternative strategy to control GTDs. To date, most studies about cultivar susceptibility have been conducted under controlled conditions, and little information is available about tolerance to natural infections caused by GTD fungi. The objectives of this study were: (i) to identify tolerant cultivars to GTD fungi within a Spanish germplasm collection, based on external symptoms observed in the vineyard; and (ii) to characterize the pathogenic mycoflora associated with symptomatic vines. For this purpose, a grapevine germplasm collection including 22 white and 25 red cultivars was monitored along three growing seasons, and their susceptibility for esca foliar symptoms was assessed. Fungi were identified by using morphological and molecular methods. Cultivars such as, 'Monastrell', 'Graciano', 'Cabernet Franc', 'Cabernet Sauvignon', 'Syrah', 'Moscatel de Alejandría', 'Sauvignon Blanc', and 'Airén' displayed high susceptibility to GTDs, whereas others such as 'Petit Verdot', 'Pinot Noir', 'Chardonnay', and 'Riesling' were considered as tolerant. The prevalent fungal species isolated from symptomatic vines were *Phaeomoniella chlamydospora* (27.9% of the fungal isolates), *Cryptovalsa ampelina* (24.6%), and *Dothiorella sarmentorum* (21.3%).

Keywords: fungal pathogens; grapevine; grapevine trunk diseases; natural infections; pathogenicity

Citation: Chacón-Vozmediano, J.L.; Gramaje, D.; León, M.; Armengol, J.; Moral, J.; Izquierdo-Cañas, P.M.; Martínez-Gascueña, J. Cultivar Susceptibility to Natural Infections Caused by Fungal Grapevine Trunk Pathogens in La Mancha Designation of Origin (Spain). *Plants* **2021**, *10*, 1171. https://doi.org/10.3390/plants10061171

Academic Editor: Gabriella Cirvilleri

Received: 12 April 2021
Accepted: 4 June 2021
Published: 9 June 2021

1. Introduction

Grapevine trunk diseases (GTDs) are currently considered one of the main types of biotic stress of this crop due to reducing both yield and lifespan of vineyards, which results in substantial economic losses to the grape and wine industry worldwide [1]. GTDs are characterized by presenting a broad diversity of internal wood and foliar symptoms, resulting in an overall decline and eventual death of the affected plants [2]. These diseases are as old as vine cultivation; however, their impact and significance have only been recognized in the early 1990s, when wine growers and the wine industry began to worry about the economic losses that they caused [1]. This emergence is thought to be correlated with several factors, including changes in viticulture practices and vineyard management, and the prohibition of effective fungicides against GTD fungi [1,3,4]. The increasing incidence of GTDs over recent decades is probably related to a sum of pathogen, hostplant, environmental (i.e., abiotic stresses), and cultural factors [5].

The aggressiveness and symptoms caused by fungal pathogens associated with GTDs differ significantly between grapevine-growing regions and vary depending on cultivars [6]. In field trials, one of the main problems to diagnose GTDs is related to the variability in external symptom expression, whereby symptoms on leaves and berries may be obvious one year but are not apparent in another [5]. Furthermore, it is common for several GTDs to overlap in the same grapevine simultaneously [6]. Therefore, it is difficult to associate visual symptoms with causal agents.

A complex of fungal genera and species of taxonomically unrelated—ascomycetous and basidiomycetous—fungi are associated with GTDs [1], which can cause more than one disease [2]. Fungal GTD complex currently includes six main different diseases affecting both grapevine planting material in nurseries, as well as young and mature vineyards. These diseases are: black-foot, Petri and esca diseases, and Botryosphaeria, Eutypa, and Diaporthe diebacks [1,2,5]. Black-foot and Petri disease affect planting material and young vineyards of up to 8-years-old, whereas esca disease, and Botryosphaeria, Eutypa, and Phomopsis diebacks predominantly affect mature grapevines that are more than eight years old. Among these diseases, Botryosphaeria dieback caused by several species in the Botryosphaeriaceae family is the most widespread worldwide [7,8].

GTDs pathogens can be propagated using infected planting material in nurseries [1]. In mature vineyards, infection of grapevines by these fungi primarily occurs through pruning wounds. Air-borne spores are spread by rain splashes, wind, or arthropods, coming in contact with and colonizing exposed wood vessels [9–11]. Grapevines have the highest risk of infection during the pruning period because of the high number of wounds made on a single grapevine and the frequency of rain events that occur during that period. Grapevine wounds remain susceptible to infection for several weeks [9,10].

Currently, there are no effective strategies to control GTDs. Thus, the use of tolerant cultivars could be considered an interesting and sustainable alternative strategy to minimize their incidence. This approach would be the least expensive, and the most effective means of controlling them [1]. Phenotyping assays to determine the susceptibility of grapevines to GTDs fungi have mainly focused on two directions: (i) mechanical artificial inoculations of the fungi on plant material—cultivar cuttings or canes—under laboratory, greenhouse, and field conditions [12–24], and (ii) field observations of natural fungal infections [14,20,25–27]. In the latter case, the undetermined latency period (asymptomatic status) and the "erratic" behavior of the foliar symptoms displayed for these diseases, especially the esca complex, make the implementation of long-term studies under field conditions necessary.

La Mancha Designation of Origin (DO) (Central Spain) is the largest delimited viticultural area in Europe (157,449 ha) and one of the most important wine-growing regions in the world. In this work, a vineyard with 47 cultivars authorized in this DO has been monitored for three growing seasons to characterize their susceptibility to GTDs, based on visual assessment of external symptoms observed in grapevines, complemented at the end of the third year with the isolation of fungi. The main objectives were: (i) to identify tolerant cultivars to GTDs fungi in a Spanish germplasm collection, based on external symptoms observed in the vineyard; and (ii) to characterize the pathogenic mycoflora associated with symptomatic vines. In the latter case, problematic aspects related to the indeterminate latency period (asymptomatic state) and the "erratic" behavior of the foliar symptoms that these diseases present, especially the Esca complex, make it necessary for studies to be carried out over several years This is the first study carried out about the susceptibility of grapevine cultivars to GTD infections in the La Mancha region. Knowledge on cultivar resistance to fungal trunk pathogens is critical for growers who plant to establish or replant vineyards and wish to reduce their reliance on fungicides and costs for controlling GTDs. This study will also provide information about the prevalent fungal species associated with GTDs in La Mancha DO.

2. Results

2.1. GTD Assessment

Of the 47 cultivars studied, only 18 of them (38.3%) showed symptoms associated with GTDs in at least one vine. The number of vines showing external symptoms was 37 (0.57% of the vines in the vineyard). Figure 1 shows the percentage of symptomatic vines with respect to the total vines of each cultivar ($n = 139$). The highest values corresponded to the cultivars 'Monastrell', 'Moscatel de Alejandría', and 'Sauvignon Blanc' with values of 4.32%, 3.60%, and 2.88%, respectively, on the total vines of each cultivar.

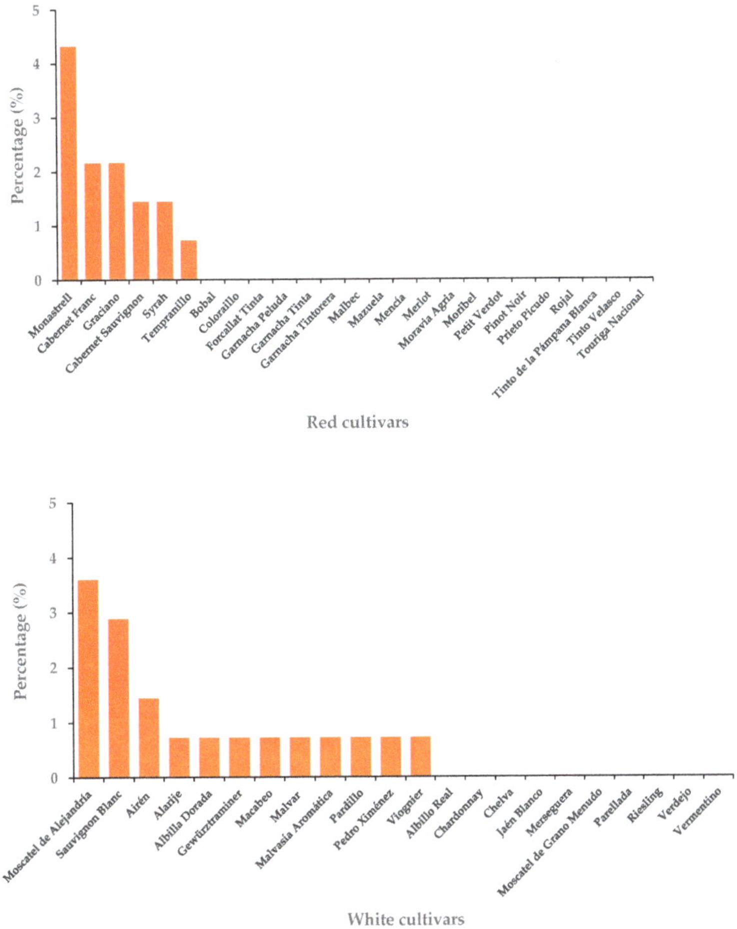

Figure 1. Percentage of grapevines showing foliar/external symptoms associated with GTDs in the different cultivars along with the growing seasons 2016, 2017, and 2018.

According to the Friedman's test, both the season and the grapevine cultivar significantly ($p < 0.05$) influenced the severity of GTDs symptoms. Regarding disease severity, the cultivars were grouped into six homogeneous groups with the cultivars 'Monastrell' and 'Moscatel de Alejandría' being the most susceptible (Table 1). However, when the cultivars were classified considering the presence or absence of the disease (i.e., the disease incidence) during the last evaluation, the cultivars that did have no symptomatic plants (29 cultivars) were classified as more resistant than those cultivars that had at least one replicated plant showing symptoms, which formed a homogeneous group.

Table 1. Homogeneous groups of grapevine cultivars according to severity of grapevine trunk diseases (GTDs) symptoms showed under field conditions in La Mancha Designation of Origin (DO), Central Spain.

Group	n	Cultivars	Mean Rank	Homogeneous Group
1	2	'Monastrell'	10,122.03	A
		'Moscatel de Alejandría'	10,097.48	
2	1	'Sauvignon Blanc'	10,028.47	AB
3	2	'Cabernet Franc'	9958.05	ABC
		'Graciano'	9933.76	
4	3	'Syrah'	9887.52	BCD
		'Airén'	9887.01	
		'Cabernet Sauvignon'	9887.01	
5	10	'Macabeo'	9816.83	CD
		'Gewürztraminer'	9816.69	
		'Alarije'	9816.69	
		'Pardillo'	9816.69	
		'Albilla Dorada'	9816.44	
		'Malvasía Aromática'	9816.44	
		'Malvar'	9816.31	
		'Tempranillo'	9816.26	
		'Pedro Ximénez'	9816.16	
		'Viognier'	9816.16	
6	29	'Albillo Real'	9746.00	D
		'Bobal'	9746.00	
		'Chardonnay'	9746.00	
		'Chelva'	9746.00	
		'Coloraillo'	9746.00	
		'Forcallat Tinta'	9746.00	
		'Garnacha Peluda'	9746.00	
		'Garnacha Tinta'	9746.00	
		'Garnacha Tintorera'	9746.00	
		'Jaén Blanco'	9746.00	
		'Malbec'	9746.00	
		'Mazuela'	9746.00	
		'Mencía'	9746.00	
		'Merlot'	9746.00	
		'Merseguera'	9746.00	
		'Moravia Agria'	9746.00	
		'Moribel'	9746.00	
		'Moscatel de Grano Menudo'	9746.00	
		'Parellada'	9746.00	
		'Petit Verdot'	9746.00	
		'Pinot Noir'	9746.00	
		'Prieto Picudo'	9746.00	
		'Riesling'	9746.00	
		'Rojal'	9746.00	
		'Tinto de la Pámpana Blanca'	9746.00	
		'Tinto Velasco'	9746.00	
		'Touriga Nacional'	9746.00	
		'Verdejo'	9746.00	
		'Vermentino'	9746.00	

Significant differences according to Friedman's test at $p = 0.05$. Homogeneous. groups were formed according to Dunn's test corrected by Bonferroni.

2.2. Fungal Isolation and Identification

Fungi were mainly isolated from the central part of the wood of shoots and arms, collected from the grapevines showing external symptoms. Based on colony morphology, conidial characteristics, molecular approaches, and phylogenetic analyses, 61 fungal isolates had 99–100% identity with reference isolates of seven species belonging to the genera *Cryptovalsa, Diaporthe, Diplodia, Dothiorella, Phaeoacremonium, Phaeomoniella*, and *Phellinus* (Tables 2 and 3) (Supplementary Figure S1). The prevalent species were *Phaeomoniella chlamydospora* (27.9% of the fungal isolates), *Cryptovalsa ampelina* (24.6%), and *Dothiorella sarmentorum* (21.3%). The remaining isolates were identified as *Diplodia seriata* (11.5%), *Phaeoacremonium minimum* (8.2%), *Diaporthe* sp. (4.9%), and *Phellinus mori* (1.6%). Regarding the family, the species belonging to Botryosphaeriaceae—*D. sarmentorum* and *D. seriata*—were the most prevalent fungi isolated from symptomatic vines (32.8%).

Table 2. Fungal trunk pathogens isolated from red cultivars showing foliar/external symptoms in 2018.

Cultivar	Identification Number	Foliar/External Symptom Incidence	Fungal Species	Trunk Disease
'Cabernet Franc'	95	3	*Pa. chlamydospora* *C. ampelina* *D. sarmentorum*	Esca
	101	5	*D. sarmentorum*	Botryosphaeria dieback
	123	5	*D. sarmentorum*	Botryosphaeria dieback
'Cabernet Sauvignon'	75	3	*C. ampelina* *D. seriata*	Eutypa dieback
	101	3	*D. seriata*	Botryosphaeria dieback
'Graciano'	12	1	*Pa. chlamydospora* *C. ampelina* *D. sarmentorum*	Esca
	21	5	*C. ampelina*	Eutypa dieback
	84	1	*C. ampelina*	Eutypa dieback
'Monastrell'	19	3	*Pa. chlamydospora*	Esca
	79	3	*Pa. chlamydospora* *Ph. mori*	Esca
	87	3	*Pa. chlamydospora* *C. ampelina*	Esca
	105	3	*Pa. chlamydospora* *Pm. minimum*	Esca
	109	3	*Pm. minimum* *D. sarmentorum*	Esca
	116	4	*Pa. chlamydospora* *Pm. minimum* *C. ampelina*	Esca

Table 2. *Cont.*

Cultivar	Identification Number	Foliar/External Symptom Incidence	Fungal Species	Trunk Disease
'Syrah'	39	4	*Pa. chlamydospora* *C. ampelina*	Esca
	51	5	*Pa. chlamydospora* *D. sarmentorum* *D. seriata*	Esca
'Tempranillo'	35	2	None	Not detected

Table 3. Fungal trunk pathogens isolated from white cultivars showing foliar/external symptoms in 2018.

Cultivar	Identification Number	Foliar/External Symptom Incidence	Fungal Species	Trunk Disease
'Airén'	16	3	*Pa. chlamydospora* *C. ampelina* *D. sarmentorum*	Esca
	79	3	*D. sarmentorum*	Botryosphaeria dieback
'Alarije'	126	4	*Pa. chlamydospora* *C. ampelina*	Esca
'Albilla Dorada'	36	3	*C. ampelina* *D. seriata*	Eutypa dieback
'Gewürztraminer'	22	4	*Diaporthe* sp.	Phomopsis dieback
'Macabeo'	42	5	*D. sarmentorum* *D. seriata*	Botryosphaeria dieback
'Malvar'	101	2	*Pa. chlamydospora* *D. sarmentorum*	Esca
'Malvasía Aromática'	13	3	*D. seriata*	Botryosphaeria dieback
'Moscatel de Alejandría'	45	2	*Pa. chlamydospora*	Esca
	46	2	*Pa. chlamydospora*	Esca
	47	2	*Pa. chlamydospora* *C. ampelina*	Esca
	48	2	*Pa. chlamydospora* *D. sarmentorum*	Esca
	50	2	*Pm. minimum* *D. sarmentorum*	Esca
'Pardillo'	96	4	*C. ampelina*	Eutypa dieback
'Pedro Ximénez'	91	1	*Pm. minimum*	Esca

Table 3. *Cont.*

Cultivar	Identification Number	Foliar/External Symptom Incidence	Fungal Species	Trunk Disease
'Sauvignon Blanc'	37	5	*Diaporthe* sp.	Phomopsis dieback
	89	4	*D. sarmentorum* *D. seriata*	Botryosphaeria dieback
	92	2	*Diaporthe* sp.	Phomopsis dieback
	107	4	*Pa. chlamydospora* *C. ampelina*	Esca
'Viognier'	96	1	*C. ampelina*	Eutypa dieback

Esca was the most prevalent disease observed during the experiment. It was detected in 20 plants (54.1% of the total symptomatic vines), followed by Botryosphaeria, Eutypa, and Phomopsis diebacks with seven (18.9%), six 16.2%), and three (8.1%) affected vines, respectively. There was only one vine (2.7%) of the 'Tempranillo' cultivar showing GTD-external symptoms; however, no GTDs-fungi were isolated from this vine.

3. Discussion

This is the first study aimed at assessing the cultivar susceptibility to natural infections caused by fungal GTDs pathogens in a grapevine germplasm collection, in a DO in Spain. To date, no evidence of qualitative resistance to any of the most common GTDs fungi has been found. Several infection assays have reported varying GTD resistance of grapevine cultivars to these pathogens [13,17,20,23,24,27], clones [27,28], and rootstocks [29–33], but the vine defense mechanisms underlying those observations, which would explain the tolerance or susceptibility of the different cultivars, have not yet been completely elucidated. Among the different reasons that may cause the difference in susceptibility between cultivars, small xylem vessel diameter and high lignin content in the wood of shoots and arms have been hypothesized to explain tolerance toward fungal vascular pathogens [34,35].

According to GTD symptoms severity observed in our study, six homogeneous groups of cultivars were established. The cultivars in which the symptoms were more severely expressed were coincident with those with the highest number of infected plants. The most severe symptoms were observed in cultivars such as 'Monastrell', 'Moscatel de Alejandría', 'Sauvignon Blanc', 'Cabernet Franc', 'Graciano', 'Syrah', 'Airén', and 'Cabernet Sauvignon', whereas cultivars 'Macabeo', 'Gewürztraminer', 'Alarije', 'Pardillo', 'Albilla Dorada', 'Malvasía Aromática', 'Malvar', 'Tempranillo', 'Pedro Ximénez', and 'Viognier' showed less severe GTDs symptomatology. The remaining 29 cultivars did not show any symptoms.

Previous reports on cultivar susceptibility to esca disease displayed varying results depending on whether the infection occurred artificially or naturally, and also on the environment (in vitro, greenhouse or field) in which the assays were carried out. In studies performed by artificial inoculation of GTDs pathogens, 'Cabernet Sauvignon' was shown to be a highly tolerant genotype to *Pa. chlamydospora* in assays performed in vitro [16], and to *Pa. chlamydospora* and *Pm. minimun* in greenhouse [17] and field [12] conditions. In contrast, 'Cabernet Sauvignon' was considered a susceptible cultivar to esca natural infection under field conditions in Italy [27,36–38], and Australia [20], which is in agreement with the results of our study. 'Tempranillo' cultivar has also been widely evaluated to esca disease susceptibility. In contrast with our results, 'Tempranillo' was considered as susceptible to *Pa. chlamydospora* infection in assays performed in Portugal [22] and

Spain [23] in greenhouse conditions, and to GTD natural infections under field conditions [20,37]. Although cultivar Tempranillo showed GTD external symptoms, no GTD fungi were isolated from this cultivar. This could be due to the sampling methodology and the selection of specific pieces of wood for fungal isolation. In our study, 'Sauvignon Blanc' and 'Syrah' were considered as susceptible cultivars to esca disease, whereas 'Merlot', 'Chardonnay', and 'Riesling' were considered as tolerant. These findings are in agreement with those obtained by other authors when evaluating GTD natural infections under field conditions, who also considered 'Sauvignon Blanc' as susceptible [27,38] and 'Merlot' [26,38], and 'Chardonnay' [27] as tolerant cultivars. By contrast, other researchers considered 'Syrah' and 'Riesling' as tolerant and susceptible cultivars, respectively, to GTDs natural infections [27].

Regarding Botryosphaeria dieback, several inconsistences were found between the results of our study and other research carried out worldwide. For example, in a previous study, the severity of internal wood symptoms caused by *Neofusicoccum parvum* differed amongst several cultivars belonging to the germplasm collection evaluated here, being 'Monastrell' one of the most tolerant cultivars [24]. In vitro studies showed that 'Cabernet Sauvignon' was tolerant to artificial inoculation by *D. seriata*, while 'Gewürztraminer' was considered susceptible [14,19]. Further research under field conditions reported a high tolerance of 'Cabernet Sauvignon' to Botryosphaeria dieback natural infection, whereas 'Syrah' was considered susceptible to this disease [14], which disagrees with the results of the present study. Conversely, in assays performed in field conditions, 'Syrah' and 'Sauvignon Blanc' were considered as susceptible cultivars, whereas 'Gewürztraminer' was considered as moderately susceptible to artificial inoculation by *D. seriata* [20], which partially agree with our results.

In this study, 'Petit Verdot', 'Merlot', 'Tempranillo', 'Chardonnay', and 'Gewürztraminer' displayed more tolerance to infection caused by *C. ampelina* than 'Graciano', 'Monastrell', 'Syrah', 'Cabernet Franc', and 'Sauvignon Blanc'. These results are generally consistent with those obtained earlier by other researchers [14,20,26], who assessed the tolerance of several cultivars to natural GTD infection in field conditions, and considered 'Merlot', 'Petit Verdot', and 'Gewürztraminer' among the most tolerant cultivars to Eutypa dieback, whereas 'Chardonnay', 'Tempranillo', and 'Sauvignon Blanc' were considered as susceptible cultivars.

Regarding the tolerance to the genus *Phellinus*, in a study performed by artificial inoculation in greenhouse conditions, 'Cabernet Sauvignon' and 'Merlot' performed as genotypes more tolerant to *Phellinus* sp. and *Ph. punctatus* than 'Garnacha' [13].

In studies performed by artificial inoculation in greenhouse, 'Cabernet Sauvignon' and 'Merlot' performed as more tolerant cultivars to *D. ampelina* than 'Cabernet Franc', 'Chardonnay', and 'Riesling' [17]. These results are consistent with those obtained in this study, in which the only cultivars showing susceptibility to *Diaporthe* sp. were 'Sauvignon Blanc' and 'Gewürztraminer', whereas the remaining cultivars such as, 'Cabernet Sauvignon', 'Merlot' 'Cabernet Franc', 'Chardonnay', and 'Riesling' performed as tolerant.

Cultivar susceptibility based on visual assessment of external symptoms [39], mainly foliar symptomatology associated with esca disease [25–27,40], has the limitation that the GTD pathogens often occur in mixed infections within the same vine [1]. In contrast with previous studies on natural GTD infection assessment, symptomatic plants were inspected for GTD fungal incidence at the end of the experiment. Isolations from symptomatic vines revealed several pathogens associated with esca disease (*Pa. chlamydospora*, *Pm. minimum*), Eutypa dieback (*C. ampelina*), Botryosphaeria dieback (*D. seriata*, *D. sarmentorum*) and Phomopsis dieback (*Diaporthe* sp.), being *Pa. chlamydospora* the most frequent fungal species, followed by *C. ampelina* and *D. sarmentorum*. In general, these results are consistent with those obtained by other authors in Italy [3] and Spain [41], in which *Pa. chlamydospora* is considered a prevalent pathogen on GTDs symptomatic vines.

Cryptovalsa ampelina was previously reported in several regions of Spain [42,43]. This pathogen is mainly found on pruning debris and rarely on standing vines showing symptoms of trunk diseases [42].

The basidiomycete species *Ph. mori* was only isolated in one vine of the 'Monastrell' cultivar together with the esca pathogen *Pa. chlamydospora*. This species has not been reported so far as a pathogen associated with GTDs worldwide. The genus *Phellinus* appears to be associated with a secondary stage of the esca disease, colonizing grapevines initially infected by *Pa. chlamydospora* and *Pm. minimum*, which are more prevalent and virulent species [5].

The *Diaporthe* species have been associated with several major diseases of grapevines, such as Phomopsis cane and leaf spot and Diaporthe dieback [44–47]. The most frequent species isolated of this genus in Europe are *D. eres* and *D. ampelina* (syn. *Phomopsis viticola*) [48]. Both species are shown to be pathogenic on grapevine [48]. Recently, two new *Diaporthe* species were isolated from symptomatic vines collected in Spain, namely *Diaporthe hispaniae* and *Diaporthe hungariae*. These species were closely related, but clearly separated based on morphological and molecular characteristics from *D. ampelina*, historically known as the most virulent *Diaporthe* species of grapevine [47].

The use of tolerant cultivars would be an interesting and sustainable alternative strategy to control GTD infections. This study allowed for classifying several grapevine cultivars according to external symptoms associated with natural infections caused by fungal grapevine trunk pathogens in La Mancha DO, as well as to characterize the pathogenic microflora associated with symptomatic vines in this area. Knowledge of tolerant cultivars to fungal trunk pathogens may help growers to reduce their reliance on fungicides and costs for controlling GTDs. Further research is needed to evaluate the correlation between foliar symptoms and wood deterioration, and to explore the mycoflora associated with asymptomatic vines.

4. Material and Methods

4.1. Study Area

A plot of the Instituto Regional de Investigación y Desarrollo Agroalimentario y Forestal (IRIAF) was planted with a grapevine germplasm collection and located at 663 m.a.s.l. (latitude: 39.176753N, longitude: −3.000247W) was used. This plot includes a broad range of *V. vinifera* cultivars authorized in the Castilla–La Mancha wine region, which can be considered representative of the vineyards in La Mancha DO. The soil is classified as Calcixerept petric. Its main features are: shallow (<40 cm depth), well-drained, with about 40% of coarse elements and loam to sandy-clay-loam textures (46.6% sand; 32.2% silt, 2.12% clay) and 3.2% of organic matter content [49]. The region has a temperate climate with high differences in temperature between winter and summer. According to the Winkler index, this region is classified as Region IV, and it records scarce rainfall during the year (about 350 mm), with less than 50% occurring in the vine growing season (between vine sprouting and harvesting). The reference evapotranspiration value (ET_0) is about 1300 mm/year, exceeding 1000 mm during the active vegetation period.

4.2. Plant Material

The germplasm collection consists of 22 white ('Airén', 'Alarije', 'Albilla Dorada', 'Albillo Real', 'Chardonnay', 'Chelva', 'Gewürztraminer', 'Jaén Blanco', 'Macabeo', 'Malvar', 'Malvasía Aromática', 'Merseguera', 'Moscatel de Alejandría', 'Moscatel de Grano Menudo', 'Pardillo', 'Parellada', 'Pedro Ximénez', 'Riesling', 'Sauvignon Blanc', 'Verdejo', 'Vermentino', and 'Viognier'), and 25 red ('Bobal', 'Cabernet Franc', 'Cabernet Sauvignon', 'Coloraillo', 'Forcallat Tinta', 'Garnacha Peluda', 'Garnacha Tinta', 'Garnacha Tintorera', 'Graciano', 'Malbec', 'Mazuela', 'Mencía', 'Merlot', 'Monastrell', 'Moravia Agria', 'Moribel', 'Petit Verdot', 'Pinot Noir', 'Prieto Picudo', 'Rojal', 'Syrah', 'Tempranillo', 'Tinto de la Pámpana Blanca', 'Tinto Velasco', and 'Touriga Nacional') grapevine cultivars grafted onto 110 Richter rootstock and planted in 2002 with one panel of 139 vines per cultivar.

The information used in this research referred to 47 cultivars with 139 grapevines each (6533 vines in total). The planting pattern was 3 m between rows and 1.5 m between plants (density of 2222 grapevines/ha). All vines were double cordon trained and spur pruned, with no specific strategies to control GTDs. Vines were cultivated under irrigated conditions by a drip system with two drippers per grapevine. Irrigation was applied considering about 25% of crop evapotranspiration, and it represented 120–150 mm per year, on average. The rows were positioned 120°E-SE/300°W-NW.

4.3. GTD Assessment and Fungal Isolations

All cultivars were inspected four to five times per season during the vegetative period during the growing seasons 2016, 2017, and 2018, between flowering and maturity, which is the time when GTDs symptoms are most evident. The cultivar susceptibility was assessed for esca foliar symptoms according to a scale ranging from 0 to 5, depending on the affectation level and GTD-associated cordon dieback (partially or totally dry) [50] (Table 4).

Table 4. Scale of foliar/external symptoms incidence according to percentage of affected vegetation.

Level	Foliar/External Symptom Incidence (%)
0	0
1	0–10
2	11–25
3	26–50
4	51–80
5	>80

Vines showing either esca foliar or cordon dieback symptoms were marked and recorded. GTD fungal isolations were performed from these vines at the end of the study (2018) to correlate foliar symptoms and fungal incidence. In total, 37 samples were collected from 18 different cultivars.

Fungal isolations were carried out according to the methodology described by [51]. Pieces of wood from symptomatic canes and arms were debarked and cut into transverse slices approximately 1 mm thick. These slices were then surface disinfected by immersion in 70% alcohol for 1 or 2 min, depending on thickness, and air dried on sterile filter paper. Later, they were placed in plates of malt extract agar supplemented with 0.5 g/L of streptomycin sulfate (MEAS) and incubated at 25 °C in darkness for 10 days. The plates were observed daily to check the growth of the mycelium. Fungal colonies were transferred to new Petri dishes with Potato Dextrose Agar (PDA) and incubated at 25 °C in darkness, to obtain pure cultures. From these primary isolations, single spore or hyphal tipped isolates were obtained previously to their identification.

4.4. Morphological and Molecular Identification of Fungal Cultures

Preliminary morphological identification of the isolates at different taxonomical levels was carried out by observing the cultural and microscope characters of the colonies under stereoscope and microscope, respectively [52]. Colonies were then tentatively grouped as Basidiomycetes, or fungi belonging to the families Botryosphaeriaceae, and Diatrypaceae, the genus *Phaeoacremonium*, and the species *Pa. chlamydospora*.

For species identity confirmation, total fungal DNA was extracted from fungal cultures grown on PDA medium, using the E.Z.N.A. Plant DNA Kit (Omega Bio-tek, Norcross, GA, USA), following the manufacturer's instructions. Diatrypaceae, Basidiomycetes, and *Pa. chlamydospora* isolates were identified based on the sequence of the ITS region, Botryosphaeriaceae isolates were identified based on the sequences of ITS and a portion of translation elongation factor 1-alpha (tef-1α) region, while, for *Phaeoacremonium* isolates, part of the β-tubulin gene (tub) was used. The primer pairs used for amplification and sequencing of each region were as follows: ITS1-F [53] and ITS4 [54] for ITS, EF1-688F and EF1-1251R for tef-1α [55] and BtCadF and BtCadR for tub [56]. Amplification by polymerase chain reaction (PCR) was performed in a total volume of 25 μL using HotBegan™ Taq DNA Polymerase (Canvax Biotech SL, Córdoba, Spain), according to the manufacturer's instructions on a Peltier Thermal Cycler-200 (MJ Research). One reaction was composed of 1× PCR Buffer B, 2.5 mM of MgCl2, 0.8 mM of dNTPs, 0.4 μM of each primer, 1 U of HotBegan Taq DNA Polymerase, and 1 μL of purified template DNA. The PCR cycling conditions consisted of an initial step of 3 min at 94 °C, followed by 35 cycles of denaturation at 94 °C for 30 s, annealing at 55 °C for 30 s and elongation at 72 °C for 45 s. A final extension was performed at 72 °C for 5 min. PCR products were confirmed by 1.2% agarose gel electrophoresis and were purified and sequenced by Macrogen Inc. (Madrid, Spain) using both PCR primers. Sequences were assembled and edited using Sequencher software 5.0 (Gene Codes Corp., Ann Arbor, MI, USA). The isolate identities were based mainly on BLASTn searches in NCBI, but, for Botryosphaeriaceae isolates, the multiple sequence alignments and maximum likelihood phylogenetic analyses were conducted in MEGA X [57], using closely related ex-type or representative species as phylogenetic reference.

4.5. Data Analysis

Mean, standard deviation, and sum were calculated using the descriptive process of the software Statistix 10 (Analytical Software; Tallahassee, FL, USA). The effect of the evaluation season and the grapevine cultivar on the disease severity were examined using Friedman's test. For that, the back-transformation of the rating scale was used. Friedman's test was used because the dependent variable does not satisfy the requirements of parametric tests. The means were compared using Dunn's test with a Bonferroni adjustment after a Kruskall–Wallis test at $p = 0.05$ [58]. A Zar's test of multiple comparisons of proportions was performed to study the effect of the cultivar on disease presence (1) or absence (0) in the last studied season (2018) [59]. Data were analyzed using the software Statistix 10 (Analytical Software; Tallahassee, FL, USA) and SPSS (version 19; SPSS Inc., Chicago, IL, USA).

Supplementary Materials: The following are available online at https://www.mdpi.com/article/10.3390/plants10061171/s1, Figure S1: Maximum likelihood phylogeny inferred from the alignment of combined sequences of a portion of translation elongation factor 1-alpha (tef-1α) region and internal transcribed spacers (ITS). Support values higher than 70% are given at the nodes. The tree was rooted using *Neofusicoccum luteum* (CBS110299 and CBS140738) as outgroup sequences. Scale bar shows expected changes per site. Species isolated in this study are indicated in bold. This analysis was conducted in MEGAX, and involved 26 nucleotide sequences. There were a total of 825 positions in the final dataset (EF:1-313 and ITS:314-825).

Author Contributions: Conceptualization, D.G.; methodology, D.G., M.L., J.A., J.M., and J.L.C.-V.; investigation, J.L.C.-V., J.A., P.M.I.-C., J.M., and J.M.-G.; resources, D.G., J.A., and J.L.C.-V.; writing—original draft preparation, J.L.C.-V.; writing—review and editing, J.L.C.-V., D.G., J.A., and J.M.; project administration, D.G., and J.L.C.-V.; funding acquisition, D.G. All authors have read and agreed to the published version of the manuscript.

Funding: This research was founded by FEDER funding through a State Program of I+D+i oriented to the Challenges of Society (RTA2015-00015-C02-00), supported by The National Institute for Agricultural and Food Research and Technology (INIA). D.G. and J.M. were supported by the Ramón y Cajal program, Spanish Government (RyC-2017-23098 and RyC-028404).

Institutional Review Board Statement: Not applicable.

Informed Consent Statement: Not applicable.

Conflicts of Interest: The authors declare no conflict of interest. The funders had no role in the design of the study; in the collection, analyses, or interpretation of data; in the writing of the manuscript, or in the decision to publish the results.

References

1. Gramaje, D.; Úrbez-Torres, J.R.; Sosnowski, M.R. Management trunk diseases with respect to etiology and epidemiology: Current strategies and future prospects. *Plant Dis.* **2018**, *102*, 12–39. [CrossRef]
2. Bertsch, C.; Ramírez-Suero, M.; Magnin-Robert, M.; Larignon, P.; Chong, J.; Abou-Mansour, E.; Fontaine, F. Grapevine trunk diseases: Complex and still poorly understood. *Plant Pathol.* **2013**, *62*, 243–265. [CrossRef]
3. Mugnai, L.; Graniti, A.; Surico, G. Esca (black measles) and brown wood-streaking: Two old and elusive diseases of grapevines. *Plant Dis.* **1999**, *83*, 404–416. [CrossRef]
4. Chiarappa, L. Esca (black measles) of grapevine. An overview. *Phytopathol. Mediterr.* **2000**, *39*, 11–15.
5. Mondello, V.; Larignon, P.; Armengol, J.; Kortekamp, A.; Váczy, K.Z.; Prezman, F.; Serrano, E.; Rego, C.; Mugnai, L.; Fontaine, F. Management of grapevine trunk diseases: Knowledge transfer, current strategies and innovative strategies adopted in Europe. *Phytopathol. Mediterr.* **2018**, *57*, 369–383. [CrossRef]
6. Yan, J.Y.; Xie, Y.; Zhang, W.; Wang, Y.; Liu, J.K.; Hyde, K.D.; Li, X.H. Species of Botryosphaeriaceae involved in grapevine dieback in China. *Fungal Divers.* **2013**, *61*, 221–236. [CrossRef]
7. Úrbez-Torres, J.R.; Leavitt, G.M.; Voegel, T.M.; Gubler, W.D. Identification and distribution of *Botryosphaeria* spp. associated with grapevine cankers in California. *Plant Dis.* **2006**, *90*, 1490–1503. [CrossRef] [PubMed]
8. Úrbez-Torres, J.R. The status of Botryosphaeriaceae species infecting grapevines. *Phytopathol. Mediterr.* **2011**, *50*, 5–45.
9. Eskalen, A.; Feliciano, A.J.; Gubler, W.D. Susceptibility of grapevine pruning wounds and symptom development in response to infection by *Phaeoacremonium aleophilum* and *Phaeomoniella chlamydospora*. *Plant Dis.* **2007**, *91*, 1100–1104. [CrossRef]
10. Úrbez-Torres, J.R.; Gubler, W.D. Double pruning, a potential method to control Bot canker disease of grapes, and susceptibility of grapevine pruning wounds to infection by Botryosphaeriaceae. *Phytopathol. Mediterr.* **2008**, *48*, 185.
11. Moyo, P.; Allsopp, E.; Roets, F.; Mostert, L.; Halleen, F. Arthropods vector grapevine trunk disease pathogens. *Phytopathology* **2014**, *104*, 1063–1069. [CrossRef]
12. Feliciano, A.J.; Eskalen, A.; Gubler, W.D. Differential susceptibility of three grapevine cultivars to *Phaeoacremoium aleophilum* and *Phaeomoniella chlamydospora* in California. *Phytopathol. Mediterr.* **2004**, *43*, 66–69.
13. Sosnowski, M.R.; Wicks, T.J.; Lardner, R.; Scott, E.S. The influence of grapevine cultivar and isolate of *Eutypa lata* on wood and foliar symptoms. *Plant Dis.* **2007**, *91*, 924–931. [CrossRef]
14. Savocchia, S.; Steel, C.C.; Stodart, B.J.; Somers, A. Pathogenicity of *Botryosphaeria* species isolated from declining grapevines in sub-tropical regions of eastern Australia. *Vitis* **2007**, *46*, 27–32.
15. Zanzotto, A.; Gardiman, M.; Lovat, L. Effect of *Phaeomoniella chlamydospora* and *Phaeoacremonium* sp. on in vitro grapevine plants. *Sci. Hortic.* **2008**, *116*, 404–408. [CrossRef]
16. Landi, L.; Murolo, S.; Romanazzi, G. Colonization of *Vitis* spp. wood by sGFP-transformed *Phaeomoniella chlamydospora*, a tracheomycotic fungus involved in esca disease. *Phytopathology* **2012**, *102*, 290–297. [CrossRef]
17. Travadon, R.; Rolshausen, P.E.; Gubler, W.D.; Cadle-Davidson, L.; Baumgartner, K. Susceptibility of cultivated and wild *Vitis* spp. to wood infection by fungal trunk pathogens. *Plant Dis.* **2013**, *97*, 1529–1536. [CrossRef] [PubMed]
18. Billones-Baaijens, R.; Jones, E.E.; Ridgway, H.J.; Jaspers, M.V. Susceptibility of common rootstock and scion varieties of grapevines to Bottryosphaeriaceae species. *Australas. Plant Pathol.* **2014**, *43*, 25–31. [CrossRef]
19. Guan, X.; Essakhi, S.; Laloue, H.; Nick, P.; Bertsch, C.; Chong, J. Mining new resources for grape resistance against Botryosphaeriaceae: A focus on *Vitis vinifera* subsp. *sylvestris*. *Plant Pathol.* **2016**, *65*, 273–284. [CrossRef]
20. Sosnowski, M.; Ayres, M.; Wicks, T.; McCarthy, M.; Scott, E. Investigating potential for resistance to grapevine trunk diseases. *Wine Vitic. J.* **2016**, *31*, 41–45.
21. Markakis, E.A.; Koubouris, G.C.; Sergentani, C.K.; Ligoxigakis, E.K. Evaluation of Greek grapevine cultivars for resistance to *Phaeomoniella chlamydospora*. *Eur. J. Plant Pathol.* **2017**, *149*, 277–283. [CrossRef]
22. Sofia, J.; Mota, M.; Gonçalves, M.T.; Rego, C. Response of four Portuguese grapevine cultivars to infection by *Phaeomoniella chlamydospora*. *Phytopathol. Mediterr.* **2018**, *57*, 506–518.
23. Martínez-Diz, M.P.; Díaz-Losada, E.; Barajas, E.; Ruano-Rosa, D.; Andrés-Sodupe, M.; Gramaje, D. Screening of Spanish *Vitis vinifera* germplasm for resistance to *Phaeomoniella chlamydospora*. *Sci. Hortic.* **2019**, *246*, 104–109. [CrossRef]
24. Chacón, J.L.; Gramaje, D.; Izquierdo, P.M.; Martínez, J.; Mena, A. Evaluation of six red grapevine cultivars inoculated with *Neofusicoccum parvum*. *Eur. J. Plant Pathol.* **2020**, *158*, 811–815. [CrossRef]
25. Marchi, G. Susceptibility to esca of various grapevine (*Vitis vinifera*) cultivars grafted on different rootstocks in a vineyard in the province of Siena (Italy). *Phytopathol. Mediterr.* **2001**, *40*, 27–36.
26. Fussler, L.; Kobes, N.; Bertrand, F.; Maumy, M.; Grosman, J.; Savary, S. Characterization of grapevine trunk diseases in France from data generated by National Grapevine Wood Diseases Survey. *Phytopathology* **2008**, *98*, 571–579. [CrossRef]
27. Murolo, S.; Romanazzi, G. Effects of grapevine cultivar, rootstock and clone on esca disease. *Australas. Plant Pathol.* **2014**, *43*, 215–221. [CrossRef]
28. Berlanas, C.; Songy, A.; Clément, C.; Fontaine, F.; Gramaje, D. Variation amongst 'Tempranillo' clones in susceptibility to *Neofusicoccum parvum*. *Phytopathol. Mediterr.* **2017**, *56*, 545.
29. Eskalen, A.; Gubler, W.D.; Khan, A. Rootstock susceptibility to *Phaeomoniella chlamydospora* and *Phaeoacremonium* spp. *Phytopathol. Mediterr.* **2001**, *40*, S433–S438.

30. Gubler, W.D.; Baumgartner, K.; Browne, G.T.; Eskalen, A.; Rooney-Latham, S.; Petit, E.; Bayramian, L.A. Root diseases of grapevines in California and their control. *Australas. Plant Pathol.* **2004**, *33*, 157–165. [CrossRef]
31. Jaspers, M.V.; Bleach, C.M.; Harvey, I.C. Susceptibility of grapevine rootstocks to *Cylindrocarpon* disease. *Phytopathol. Mediterr.* **2007**, *46*, 114.
32. Alaniz, S.; García-Jiménez, J.; Abad-Campos, P.; Armengol, J. Susceptibility of grapevine rootstocks to *Cylindrocarpon liriodendri* and *C. macrodidymum*. *Sci. Hortic.* **2010**, *125*, 305–308. [CrossRef]
33. Gramaje, D.; García-Jiménez, J.; Armengol, J. Field evaluation of grapevine rootstocks inoculated with fungi associated with Petri disease and esca. *Am. J. Enol. Vitic.* **2010**, *61*, 512–520. [CrossRef]
34. Rolshausen, P.E.; Greve, L.C.; Labavitch, J.M.; Mahoney, N.E.; Molyneux, R.J.; Gubler, W.D. Pathogenesis of *Eutypa lata* in grapevine: Identification of virulence factors and biochemical characterization of cordon dieback. *Phytopathology* **2008**, *98*, 222–229. [CrossRef]
35. Pouzoulet, J.; Pivovaroff, A.L.; Santiago, L.S.; Rolshausen, P.E. Can vessel dimension explain tolerance toward fungal vascular wilt diseases in woody plants? Lessons from Dutch elm disease and esca disease in grapevine. *Front. Plant Sci.* **2014**, *5*, 1–11. [CrossRef]
36. Andreini, L.; Caruso, G.; Bertolla, C.; Scalabrelli, G.; Viti, R.; Gucci, R. Gas exchange, stem water potential and xylem flux on some grapevine cultivars affected by esca disease. *S. Afr. J. Enol. Vitic.* **2009**, *30*, 142–147. [CrossRef]
37. Andreini, L.; Guarino, L.; Viti, R.; Scalabrelli, G. Evaluation of the effect of esca disease on bud break in *Vitis vinifera* L.: Possible relationship between cultivars and rootstocks. *Vitis* **2013**, *52*, 33–40.
38. Serra, S.; Ligios, V.; Schianchi, N.; Prota, V.A.; Deidda, A.; Scanu, B. Incidence of grapevine trunk diseases on four cultivars in Sardinia, Southern Italy. *Vitis* **2021**, *60*, 35–42.
39. Sosnowski, M.; Ayres, M.; Wicks, T.; McCarthy, M. In search of resistance to grapevine trunk diseases. *Wine Vitic. J.* **2013**, *28*, 55–58.
40. Bruez, E.; Lecomte, P.; Grosman, J.; Doublet, B.; Bertsch, C.; Guerin-Dubrana, L. Overview of grapevine trunk diseases in France in the 2000s. *Phytopathol. Mediterr.* **2013**, *52*, 262–275.
41. Luque, J.; Martos, S.; Aroca, A.; Raposo, R.; García-Figueres, F. Symptoms and fungi associated with declining mature grapevine plants in northeast Spain. *J. Plant Pathol.* **2009**, *91*, 381–390.
42. Luque, J.; Sierra, D.; Torres, E.; García, F. *Cryptovalsa ampelina* on grapevines in N.E. Spain: Identification and pathogenicity. *Phytopathol. Mediterr.* **2006**, *45*, S101–S109.
43. Martín, M.T.; Martín, L.; de-Francisco, M.T.; Cobos, R. First report of *Lasidiplodia theobromae* and *Cryptovalsa ampelina* associated with grapevine decline from Castilla y Leon, Spain. *Plant Dis.* **2009**, *93*, 545. [CrossRef]
44. Mostert, L.; Crous, P.W.; Kang, J.C.; Phillips, A.J. Species of *Phomopsis* and a *Libertella* sp. occurring on grapevines with specific reference to South Africa: Morphological, cultural, molecular and pathological characterization. *Mycologia* **2001**, *93*, 146–167. [CrossRef]
45. van Niekerk, J.M.; Groenewald, J.Z.; Farr, D.F.; Fourie, P.H.; Halleer, F.; Crous, P.W. Reassessment of Phomopsis species on grapevines. *Australas. Plant Pathol.* **2005**, *34*, 27–39. [CrossRef]
46. Rawnsley, B.; Wicks, T.J.; Scott, E.S.; Stummer, B.E. *Diaporthe perjuncta* does not cause Phomopsis cane and leaf spot disease of grapevine in Australia. *Plant Dis.* **2004**, *88*, 1005–1010. [CrossRef]
47. Úrbez-Torres, J.R.; Peduto, F.; Smith, R.J.; Gubler, W.D. Phomopsis dieback: A grapevine trunk disease caused by *Phomopsis viticola* in California. *Plant Dis.* **2013**, *97*, 1571–1579. [CrossRef]
48. Guarnaccia, V.; Groenewald, J.Z.; Woodhall, J.; Armengol, J.; Cinelli, T.; Eichmeier, A.; Crous, P.W. Diaporthe diversity and pathogenicity revealed from a broad survey of grapevine diseases in Europe. *Pers. Mol. Phylogenet. Evol.* **2018**, *40*, 135–153. [CrossRef]
49. Amorós-Ortíz-Villajos, J.A.; Bravo-Martín-Consuegra, S.; García-Navarro, F.J.; Pérez-de-los-Reyes, C.; Chacón-Vozmediano, J.L.; Martínez-Gascueña, J.; Jiménez-Ballesta, R. *Atlas de Suelos Vitícolas de Castilla-La Mancha*, 1st ed.; Martínez, F.J., Ed.; Arte y Comunicación Calatrava: Almagro, Spain, 2015; pp. 152–153. (In Spanish)
50. Lecomte, P.; Darrieutort, G.; Liminana, J.M.; Comont, G.; Muruamendiaraz, A.; Fermaud, M. New insights into esca of grapevine: The development of foliar symptoms and their association with xylem discoloration. *Plant Dis.* **2012**, *96*, 924–934. [CrossRef]
51. Larignon, P.; Dubos, B. Fungi associated with esca disease in grapevine. *Eur. J. Plant Pathol.* **1997**, *103*, 147–157. [CrossRef]
52. Armengol, J.; Vicent, A.; Torné, L.; García-Figueres, F.; García-Jiménez, J. Fungi associated with esca and grapevine declines in Spain: A three-year survey. *Phytopathol. Mediterr.* **2001**, *40*, S325–S329.
53. Gardes, M.; Bruns, T.D. ITS primers with enhanced specificity for basiodiomycetes-applications to the identification of mycorrhizae and rusts. *Mol. Ecol.* **1993**, *2*, 113–118. [CrossRef]
54. White, T.J.; Bruns, T.D.; Lee, S.B.; Taylor, J.W. Amplification and direct sequencing of fungal ribosomal RNA genes for phylogenetics. In *PCR Protocols: A Guide to Methods and Applications*; Innis, M.A., Gelfand, D.H., Sninsky, J.J., White, T.J., Eds.; Academic Press: New York, NY, USA, 1990; Volume 18, pp. 315–322.
55. Alves, A.; Crous, P.W.; Correia, A.; Phillips, A.J.L. Morphological and molecular data reveal cryptic speciation in *Lasiodiplodia theobromae*. *Fungal Divers.* **2008**, *28*, 1–13.
56. Travadon, R.; Lawrence, D.P.; Rooney-Latham, S.; Gubler, W.D.; Wilcox, W.F.; Rolshausen, P.E.; Baumgartner, K. *Cadophora* species associated with wood-decay of grapevine in North America. *Fungal Biol.* **2015**, *119*, 53–66. [CrossRef] [PubMed]

57. Kumar, S.; Stecher, G.; Li, M.; Knyaz, C.; Tamura, K. MEGA X: Molecular evolutionary genetics analysis across computing platforms. *Mol. Biol. Evol.* **2018**, *35*, 1547–1549. [CrossRef]
58. Demšar, J. Statistical comparisons of classifiers over multiple data sets. *J. Mach. Learn. Res.* **2006**, *7*, 1–30.
59. Zar, J.H. *Biostatistical Analysis*, 5th ed.; Pearson Education India: New Delhi, India, 2010; pp. 724–725.

plants

Article

Silverleaf (*Chondrostereum purpureum*) Effects on Japanese Plum (*Prunus salicina*)

Daina Grinbergs [1,*], Javier Chilian [1,*], Carla Hahn [1], Marisol Reyes [1], Mariana Isla [1], Andrés France [1] and Jorunn Børve [2]

[1] Laboratorio de Fitopatología de Frutales INIA Quilamapu, Instituto de Investigaciones Agropecuarias, INIA, Av. Vicente Méndez 515, Chillán 3780000, Chile; carlahahn@gmail.com (C.H.); mreyes@inia.cl (M.R.); marisla@udec.cl (M.I.); afrance@inia.cl (A.F.)

[2] Norwegian Institute of Bioeconomy Research, NIBIO, P.O. Box 115, 1431 Ås, Norway; Jorunn.Borve@nibio.no

* Correspondence: dgrinbergs@inia.cl (D.G.); jchilian@inia.cl (J.C.); Tel.: +56-42-2206-832 (D.G. & J.C.)

Abstract: Silverleaf is an important fungal trunk disease of fruit crops, such as Japanese plum (*Prunus salicina*). It is known that infection by *Chondrostereum purpureum* results in discolored wood, "silvered" foliage, and tree decline. However, effects on fruit yield and quality have not been assessed. Therefore, the objectives of this study were to determine *C. purpureum* pathogenicity on *P. salicina* and the effects on physiology, fruit yield, and quality, in Chile, in 2019 and 2020. Wood samples from affected plum trees were collected in the Chilean plum productive area. Fungi were isolated by plating wood sections from the necrosis margin on culture media. Morphological and molecular characteristics of the isolates corresponded to *C. purpureum* (98%). Representative isolates were inoculated from healthy plum plants and after 65-d incubation, wood necrotic lesions and silver leaves were visible. Fungi were reisolated, fulfilling Koch's postulates. To determine Silverleaf effects, xylem water potential and fruit yield and quality were measured in healthy and Silverleaf-diseased plum trees 'Angeleno'. Water potential was altered in diseased trees, and fruit yield was reduced by 51% (2019) and by 41% (2020) compared to fruit from healthy trees. Moreover, cover-colour, equatorial-diameter, and weight were reduced, and fruit were softer, failing to meet the criteria to be properly commercialized and exported to demanding markets.

Keywords: Silverleaf disease; *Chondrostereum purpureum*; *Prunus salicina*; Japanese plum

Citation: Grinbergs, D.; Chilian, J.; Hahn, C.; Reyes, M.; Isla, M.; France, A.; Børve, J. Silverleaf (*Chondrostereum purpureum*) Effects on Japanese Plum (*Prunus salicina*). *Plants* **2021**, *10*, 2777. https://doi.org/10.3390/plants10122777

Academic Editor: Carlos Agustí-Brisach

Received: 5 August 2021
Accepted: 14 September 2021
Published: 16 December 2021

1. Introduction

The Japanese plum (*P. salicina* L.) is a deciduous stone fruit tree native to China. It is grown globally, and Chile cultivates 4520 ha, mainly 'Angeleno', 'Black-Amber', and 'Friar' cultivars. Chilean production volume is ~400,000 t (2019–2020), and Chile is the major fresh plum exporter worldwide [1].

Fungal trunk diseases have increased in recent years, in fruit, ornamental, forest, and woody species, worldwide [2]. Some of the reasons of this increase in trunk diseases in fruit crops are changes in climate, the incorporation of intensive plant production management measures such as high density plantations, the use of dwarfing rootstocks, and severe pruning, in order to accelerate plant and fruit production, which stress the plants and alter their behaviour [2–4]. Fungal trunk diseases are one of the main pathological problems for *Prunus*, which are affected by several fungal taxonomic groups [5]. *Prunus salicina* has been reported as a host of different wood fungal pathogens, such as the ones from the genera *Armillaria* [6], *Botryosphaeria* [7], *Diplodia* [8], *Calosphaeria*, *Jattaea* [9], *Lasiodiplodia* [10], *Neofusicoccum* [11], *Phaeoacremonium*, *Tonignia* [12], and *Chondrostereum* [13].

The basidiomycete *C. purpureum* (Pers.) Pouzar is an important wood pathogen that causes Silverleaf disease. While this fungal species has attracted considerable interest in several countries as a biological agent for woody weed control in coniferous forest plantations [13–17], it can cause severe and destructive disease in woody plants, including ornamental, wild, forest, and fruit crop species in some areas of the world [4,17–23]. More than 230 species are recorded as hosts of *C. purpureum* [15,18,19,24,25], i.e., the Rosaceae family [3,4,18,19,26–28] and genus *Prunus*, one of the most susceptible to this pathogen [18], such as *P. armeniaca* [29], *P. avium* [30], *P. persica* 49 [31], *P. serotina* [19], *P. domestica* [32], and *P. salicina* [13].

This fungus reproduces through basidiospores, which are airborne disseminated from basidiocarps, developing in dying trunks and branches of infected trees, to fresh wood wounds [33]. Mycelia grow through the xylem tissue, producing central wood discolouration [34] and subsequently occluding the vessels. In addition, the fungus produces a specific endopolygalacturonase (endoPG) enzyme, which moves to the foliage, inducing silver-greyish colour symptoms on leaves. This silver colour is usually visible several seasons after the infection [27,35]. Subsequently, the fungus kills the plant and develops purple resupinated carpophores that release spores under high humidity and mild temperatures [33].

Silverleaf is an important disease in countries such as New Zealand, the United States (Washington and Oregon) [32], Chile [3], Australia [36], France [18], Poland [37], Latvia [38], Norway [39], and other European countries [32]. It can produce severe losses in orchards, and it is the main cause of mortality in peach and nectarine in New Zealand, with losses of 8% per year [40]. In Chile, it has caused disease in different fruit crop species, such as almond, apples, blueberries, cherry, Chilean guava, nectarine, quince, pear tree, and plum [3,4,23], showing silver leaves and brown central discoloration in branches and the main trunks. In blueberry, the effects of Silverleaf on plants and fruit were measured, with water potential and stomatal conductance the most affected physiological parameters, as well as a yield reduction of 40% and problems in fruit quality, such as reduction of maturation, colour, and weight [4]. Nevertheless, there is a lack of information about the effects of Silverleaf in other fruit crops.

New wood pathogens are being reported every day, and in a wider range of hosts and geographical locations [5,9,12,17,28,41,42]. However, little is known about their impact on fruit yield and quality. The most exhaustive research has been performed on grapevines, revealing that wood diseases are the main biotic factor limiting vineyard productivity and longevity, causing major economic losses [43].

Similar to that for other other trunk diseases, the information about Silverleaf effects on fruit crops and in Japanese plum is scarce. Therefore, the objectives of the present study were to investigate the etiology of the fungal pathogen causing foliar silvering in *P. salicina* in Chile, to determine its pathogenicity in plum, and to assess the effects of *C. purpureum* on plum physiology, fruit yield, and quality.

2. Results

2.1. *Chondrostereum purpureum* Isolates

Ninety-seven wood samples from Japanese plum trees showing foliar silvering (Figure 1A) and internal brown circular discolorations in transverse section (Figure 1B) (ragged, tapered cylinders in axial section), in affected branches and main trunks, were collected from 2018 to 2020. Some samples also showed other wood symptoms such as wedge-shaped and dark brown irregular discolorations, suggesting the presence of a diversity of pathogens. In 95 isolates, obtained from 98% of the wood samples, macroscopic and microscopic morphological characteristics of the colonies were consistent with those described for *C. purpureum* [44]. White-cottony mycelia (Figure 1C) with clamp connections grew from wood sections from the necrosis progress area, which were surface disinfected and plated on Petri plates containing acidified potato dextrose agar (APDA) (25% PDA, acidified with 0.2% v/v 85% lactic acid) and water agar (WA). After 14–21 days of incubation at 25 °C, 52% of the isolates developed beige to light pink pseudo-basidiocarps on the edges of the plates. The fruiting bodies of most of the isolates developed hyaline, apiculate, and ovoid basidiospores, 5.3 (4.7–6.8) μm × 3.6 (3.0–4.5) μm (n = 50).

Figure 1. Silverleaf disease caused by *Chondrostereum purpureum* on Japanese plum in Chile; (**A**) foliar symptoms and (**B**) wood discoloration symptoms on plum trees 'Angeleno', (**C**) isolate HMCi7 Genbank: MW938164, and (**D**) Polimerase Chain Reaction (PCR) products using APN1 species-specific primers for 22 *C. purpureum* isolates (Lanes 2 to 11 and 15 to 26). Lane 12: positive control isolate RGM 122 GenBank: MK22253.1. Lane 27: negative control. Lanes 1, 13, 14, and 28: molecular weight standards. The white arrow indicates the fragment of interest, whose size is 500 base pairs.

Moreover, the identity of 57 isolates representing different localities and host cultivars was confirmed though DNA amplification using APN1 primers, developing intensely discrete 500-bp bands on agarose gels (Figure 1D) [14]. The internal transcribed spacer of representative isolates, selected from different host cultivars and collection localities (HMCi7; HMCi121; HMCi147; and HMCi148), was also amplified using Internal Transcribed Spacer sequence (ITS) ITS1 and ITS4 [45]. Sequences were deposited in GenBank (MW938164, MW938165; MW938166; and MW938167, respectively) (Table 1).

In 24% of the wood samples, it was possible to isolate other fungal pathogens, most of them associated with wedge-shaped and irregular dark-brown discolorations. Fungi were identified as *Cytospora* (12.6%), *Phomopsis* (9.2%), *Schizophyllum* (17.2%), and *Stereum* (7.9%) species in the Botryosphaeriaceae family (42.8%), in addition to other fungi (10.3%).

Table 1. Isolate, plant host, geographic origin, DNA amplification using APN1 *Chondrostereum purpureum*-specific primers, and GenBank Accession number of 17 representative *C. purpureum* isolates, morphologically and molecularly identified in this study.

Isolate	Species	Host		Geographic Origin		APN1	ITS GenBank Accession Number
HMCi 314	*Chondrostereum purpureum*	*Prunus domestica* subsp. *domestica*	D'Agen	Colbún	35°45′01.1592″ S, 71°25′46.4889″ W	Positive	-
HMCi 325	*Chondrostereum purpureum*	*Prunus domestica* subsp. *domestica*	D'Agen	Sagrada Familia	34°59′50.0820″ S, 71°21′48.2976″ W	Positive	-
HMCi 331	*Chondrostereum purpureum*	*Prunus domestica* subsp. *domestica*	D'Agen	San Javier	35°38′54.7944″ S, 71°36′47.9340″ W	Positive	-
HMCi 341	*Chondrostereum purpureum*	*Prunus domestica* subsp. *italica*	Reina Claudia	Chillán	34°58′10.7536″ S, 71°21′14.8464″ W	Positive	-
HMCi 308	*Chondrostereum purpureum*	*Prunus domestica* subsp. *italica*	Reina Claudia	San Rafael, Maule	35°18′37.0044″ S, 71°29′06.5832″ W	Positive	-
HMCi 290	*Chondrostereum purpureum*	*Prunus domestica* subsp. *italica*	Reina Claudia	Yungay	37°07′21.6127″ S, 72°00′02.1028″ W	Positive	-
HMCi 249	*Chondrostereum purpureum*	*Prunus salicina*	Angeleno	Codegua	34°01′12.3420″ S, 70°41′50.0352″ W	Positive	-
HMCi 7	*Chondrostereum purpureum*	*Prunus salicina*	Angeleno	Curicó	34°58′58.21″ S, 71°16′37.01″ W	Positive	MW938164
HMCi 340	*Chondrostereum purpureum*	*Prunus salicina*	Angeleno	Portezuelo	36°34′43.9356″ S, 72°33′19.7424″ W	Positive	-
HMCi 121	*Chondrostereum purpureum*	*Prunus salicina*	Black amber	Curicó	36°37′27.1128″ S, 72°00′27.8532″ W	Positive	MW938165
HMCi 272	*Chondrostereum purpureum*	*Prunus salicina*	Black amber	Romeral	34°57′17.2836″ S, 71°08′11.2560″ W	Positive	-
HMCi 276	*Chondrostereum purpureum*	*Prunus salicina*	Black amber	Teno	34°52′39.4149″ S, 71°05′19.0032″ W	Positive	-
HMCi 168	*Chondrostereum purpureum*	*Prunus salicina*	Fortune	Melipilla	33°41′09.3696″ S, 71°06′23.7960″ W	Positive	-
HMCi 253	*Chondrostereum purpureum*	*Prunus salicina*	Friar	Paine	33°52′11.7156″ S, 70°44′21.5700″ W	Positive	-
HMCi 148	*Chondrostereum purpureum*	*Prunus salicina*	Larry Ann	Curicó	34°58′56.0352″ S, 71°16′34.0896″ W	Positive	MW938167
HMCi 147	*Chondrostereum purpureum*	*Prunus salicina*	Larry Ann	Rio Claro	35°12′0.18936″ S, 71°14′36.1140″ W	Positive	MW938166
HMCi 157	*Chondrostereum purpureum*	*Prunus salicina*	Larry Ann	Yungay	37°08′50.452″ S, 71°52′20.673″ W	Positive	-

2.2. Pathogenicity Tests

Healthy nursery plants were successfully inoculated with *C. purpureum* isolates and reproduced silver foliar symptoms (Grade 3–8) (Figures 2 and 3A,B). After the incubation period, brown central staining was visible when the inoculated branches were transversally cut (Figure 3A,C). Moreover, one of the inoculated plants, with the HMCi121 isolate, developed resupinated purple carpophores in the main trunk, above the soil level, after 20 months of incubation (Figure 3E).

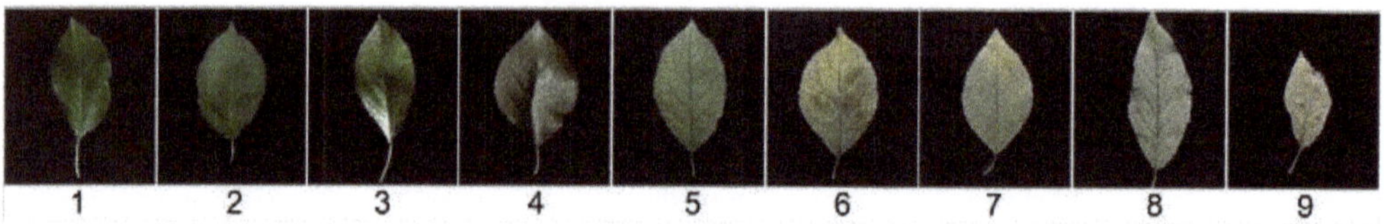

Figure 2. Silverleaf disease foliar symptom visual severity scale (one to nine) for Japanese plum. 1 = healthy or apparently healthy leaf, 3 = epidermis starts detaching from the mesophyll and the color is lighter than in healthy leaves, 5 = more than 75% of the leaf epidermis is detached from the mesophyll and the leaf is beginning to look gray–silver, 7 = 100% of the epidermis is detached from the mesophyll and the leaf is completely white–silver, 9 = the epidermis peels off from the leaf and the mesophyll begins to oxidize. Grades 2, 4, 6, and 8 indicate intermediate symptom severity between the previous and the following grade.

Figure 3. Pathogenicity tests of four *Chondrostereum purpureum* isolates on 2-year old Japanese plum plants 'Angeleno'; (**A**) transversal cut on an inoculated branch 20 cm from the inoculation point, (**B**) foliar symptoms, and (**C**) necrotic symptoms developed by inoculated plants after a 65-day incubation. (**D**) *Chondrostereum purpureum* DNA banding patterns on agarose gel, after amplification with APM22 fingerprinting markers. Lanes 2 and 3: inoculated and reisolated *C. purpureum* HMCi7, Lanes 5 and 6: isolated and reisolated HMCi121, Lanes 8 and 9: isolated and reisolated HMCi147, and Lanes 11 and 12: isolated and reisolated HMCi148. Lanes 4, 7, 10, and 13 are negative controls. Lanes 1 and 14: molecular weight standards. (**E**) Fruiting bodies of isolate HMCi121 after 20 months of incubation.

It was possible to reisolate *C. purpureum* from inoculated plants (100%) on APDA, while the fungus was not reisolated from controls. The banding patterns in agarose gels, produced by DNA amplification using Sequence Characterized Amplified Region (SCAR) fingerprinting primers APM22D13 [14], were identical between inoculated and reisolated *C. purpureum* isolates (Figure 3D), fulfilling Koch's postulates.

2.3. *Silverleaf Effects on Plum*

To confirm the presence or absence of *C. purpureum* in symptomatic and asymptomatic plants, respectively, the fungus was detected by amplifying fungal DNA, directly from the trunk sawdust, using APN1 species-specific primers, in 2019 and 2020. A 500-bp band was reproduced in agarose gels from diseased trees, while DNA from asymptomatic ones did not reproduce the band (Figure 4). In 2020, symptoms were less severe than in 2019 (Grade 3, average), and two of the previously selected trees did not show any foliar symptoms, but they were positive for APN1 amplification.

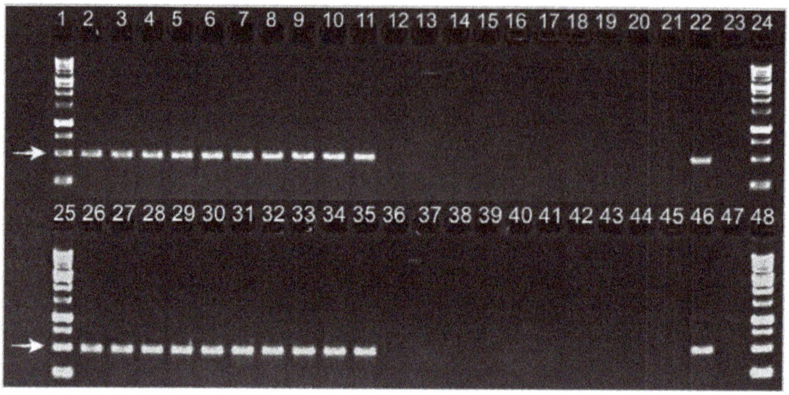

Figure 4. *Chondrostereum purpureum* DNA amplification using APN1 species-specific primers in Silverleaf-diseased plum plants, showing foliar symptoms (Lanes 2–11 and 26–35), and healthy ones (Lanes 12–21 and 36–45) (2019). Lanes 22 and 46: positive controls (isolate HMCi147 GenBank: MW938166). Lanes 23 and 47: negative controls. Lanes 1, 24, 25, and 48: molecular weight standards. White arrows indicate the fragment of interest, whose size is 500 base pairs.

2.3.1. Water Potential

Xylem potential was 19% lower in diseased plants (−11 bar) than in healthy ones (−9.2 bar) in 2019 (Student's *t*-test, $p < 0.0001$). Similarly, in 2020, the water potential of diseased plants (−12 bar) was 22.5% lower than that in healthy ones (−9.4 bar) (Student's *t*-test, $p < 0.0001$) (Figure 5). It was measured a few days after harvest in both years.

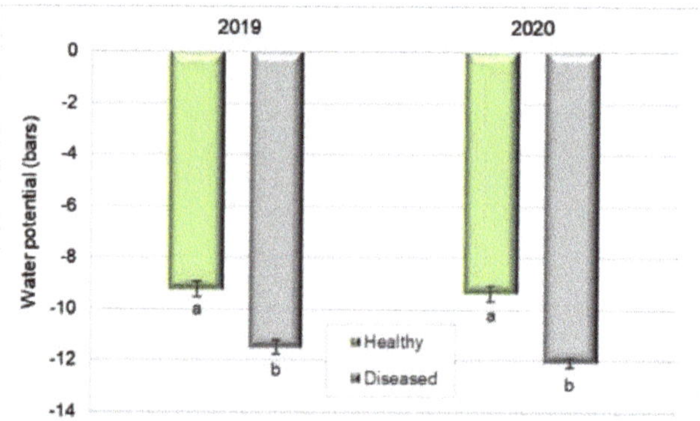

Figure 5. Water potential of leaves of healthy and Silverleaf-diseased Japanese plums 'Angeleno', based on ten replicates of two trees and five leaf samples per tree (2019) and 9 × 2 × 5 (2020). Bars represent the standard error of the means, and columns with different letters are statistically different. Student's *t*-test: $p < 0.0001$ in 2019 and 2020.

2.3.2. Yield Assessment

The total yield of the Silverleaf-diseased trees was 51% lower (12.1 kg per tree) than that of healthy ones (24.8 kg per tree) (Student's *t*-test, $p = 0.0001$) in 2019 (Figure 6). In 2020, the total yield was 41% lower (13.7 kg per tree) than that of healthy plants (23.5 kg per tree) (Student's *t*-test, $p = 0.0003$) (Figure 6).

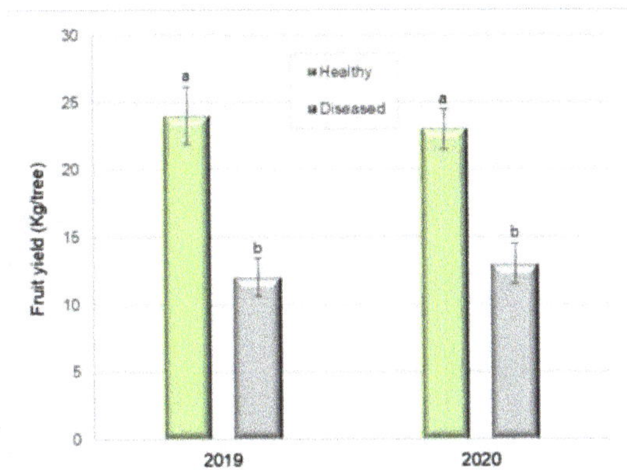

Figure 6. Fruit yield (kg) of healthy and Silverleaf-diseased Japanese plums 'Angeleno'. Ten replicates of two trees (2019) and 9 × 2, in 2020. Bars represent the standard error of the means, and columns with different letters are statistically different. Student's *t*-test: $p = 0.0001$ in 2019 and $p = 0.0003$ in 2020.

2.3.3. Fruit Quality

In the 2019 period, fruit harvested from healthy trees had 89% cover colour, compared to 73% in diseased ones (Pearson Chi-squared test = 16.37, $p = 0.0001$). In the next season, the cover colour of fruit from healthy trees was higher than that in the previous year (91%), while the colour of fruit from diseased trees remained similar to that in the previous year (Pearson's Chi-squared test = 18, $p < 0.0001$) (Figure 7A).

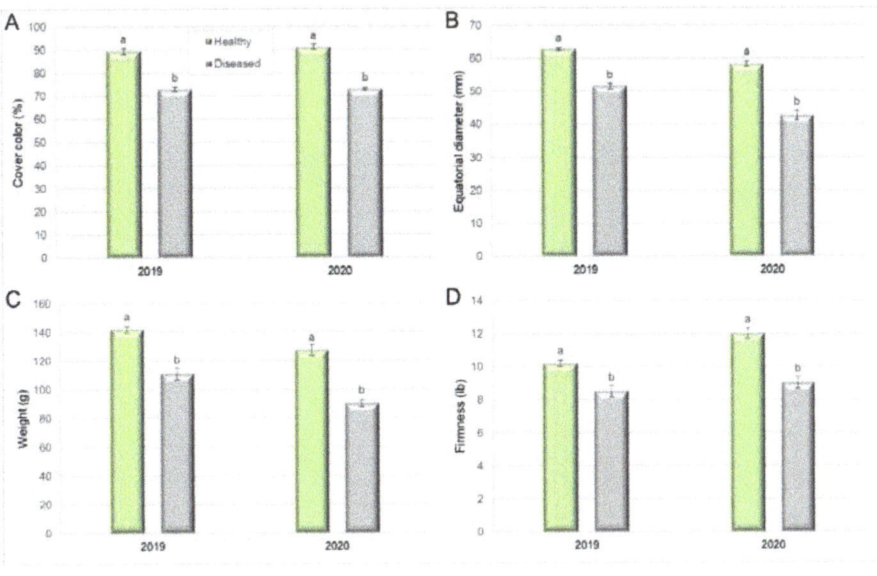

Figure 7. Quality of fruit from healthy and Silverleaf-diseased trees in a Japanese plum orchard ('Angeleno'): (**A**) cover color (Pearson Chi-squared test, $p = 0.0001$ in 2019 and $p < 0.0001$ in 2020), (**B**) equatorial diameter (Kruskal–Wallis, $p = 0.0002$ in 2019; $p < 0.0001$ in 2020), (**C**) individual weight (Student's *t*-test, $p < 0.0001$ in 2019 and 2020) and (**D**) firmness (Kruskal–Wallis, $p = 0.0005$ in 2019 and $p < 0.0001$ in 2020). Mean of 20 trees × 10 fruit samples (2019) and 18 × 10 (2020). Bars represent the standard error of the means, and columns with different letters are statistically different.

The fruit equatorial diameter was also negatively affected by the disease. In 2019, it was 62.8 mm in fruit from healthy trees, compared to 51.5 mm in fruit from diseased ones (Kruskal–Wallis, $p = 0.0002$). In 2020, the equatorial diameter was lower than in the previous season, both for fruit from healthy trees (58.5 mm) and diseased ones (42.4 mm) (Kruskal–Wallis, $p < 0.0001$) (Figure 7B).

Fruit individual weight was also a parameter affected by Silverleaf. In 2019, fruit harvested from healthy trees weighed 141.7 g compared with 111.3 g for diseased trees (Student's t-test, $p < 0.0001$). In the next season, fruit from healthy trees weighed 127.5 g compared with 90 g for diseased trees (Figure 7C) (Student's t-test, $p < 0.0001$). Finally, fruit from diseased trees (4.54 kgf) was softer than fruit from healthy ones (3.75 kgf), in 2019 (Kruskal–Wallis, $p = 0.0005$), as well as in 2020 (5.44 kgf for fruit from diseased and 3.88 kgf for fruit from healthy trees (Kruskal–Wallis, $p < 0.0001$)) (Figure 7D), indicating a shorter postharvest life for fruit from diseased trees.

The other measured quality parameters (pH, titratable acidity, soluble-solids, and background colour) were not different between fruit harvested from diseased and healthy trees (data not shown).

3. Discussion

In the present study, *C. purpureum* was successfully isolated from the necrotic margin of stained wood of Japanese plums trees showing Silverleaf foliar symptoms. The symptoms were similar to those described for other fruit hosts [3,23,46]. *Chondrostereum purpureum* is a primary invader of woody angiosperms and enters its host through a fresh wound, followed by the infection of aggressive saprobic fungi such as *Trametes versicolor* and *Schizophyllum commune* [15,16]. However, *C. purpureum* is still present in these trees and is also able to produce foliar symptoms, as was demonstrated in this study, with the symptom observation and *C. purpureum* molecular detection and isolation.

The isolated fungus developed white-cottony mycelia, and it was also possible to detect it directly in wood, through clear staining bands on agarose gels, when fungal DNA was amplified with APN1 species-specific primers [14]. Moreover, the Japanese plum plants inoculated with *C. purpureum* isolates clearly showed foliar symptoms, developing silver-greyish leaves in the inoculated branches, as well as internal wood necrosis and fruiting bodies. Reisolated fungal specimens showed the same cultural features as the inoculated ones, as well as identical banding patterns on agarose gels when SCAR fingerprinting markers [14] were used to amplify their DNA, thus fulfilling Koch's postulates.

It was demonstrated that wood diseases alter physiological parameters in different hosts [47,48]. Likewise, in this study, the water potential was lower in diseased plants compared to healthy ones. Similarly, in 2020, the water potential of diseased plants was lower than that in healthy ones. These differences can be explained by *C. purpureum* growth through the xylem tissues of the host [33], as well as from the systematic silvering of leaves [26]. During the infection process, *C. purpureum* produces the occlusion of tree vessels [31]. Moreover, transpiration, stomatal conductance, and leaf area decrease significantly while leaf-silvering intensity increases [26]. The resulting physiological disruption and dehydration, combined with fungal toxins, finally cause the death of the host [21,33,49].

On the other hand, intensity of symptoms may change between seasons, such as in some trees in the present study. Intensity was lower in some plants, despite the presence and viability of the fungus, which was confirmed by PCR and microbiological isolation, as occurred in apple [50], where the reversion of foliar symptoms was first described. Damaged wood on the branches and trunk is the battleground of microorganisms, pathogens, and endophytes. While vascular tissue is being destroyed by pathogens [51], some endophytic microorganisms could be acting as their antagonists and/or inducing plant resistance [50] and thus modulating disease expression [52].

Foliar symptoms remained similar in their intensity (Grades 3–6 on the severity scale) (Figure 2) in most of the diseased analysed trees (95%) during 2019 and 2020. In 2019, a reduction of 51% in total fruit weight was recorded in Silverleaf-diseased plants compared to healthy ones. Moreover, in 2020, the reduction was 41%. The fungal pathogen inhabiting the vessels and necrotic tissue affected the number and weight of fruits per tree. Similarly, fruit yield decreases due to several wood pathogens [41]. The results were consistent with yield assessments performed in other fruit crops in Chile, such as apples and blueberries [4,53]. Although there are authors who point out yield losses due to the damaging effects of *C. purpureum* infections in stone and pome fruits [18,26,36,40,44], as well as blueberries [4], as far as we are concerned, this is the first record about yield losses in Japanese plum related to the detrimental effects caused by *C. purpureum*.

Regarding fruit quality, our results showed that important quality components such as fruit weight, cover colour, equatorial diameter, and firmness were influenced by Silverleaf disease. Similar results were reported by [25,26] in apple orchards, with smaller and lower numbers of fruit, reduced colour, and an increased incidence of physiological damage such as a water core, short post-harvest storage, and softening of fruit. Furthermore, [27] stated that the spread of *C. purpureum* in woody tissues and the loss of photosynthetic capability eventually leads to tree death.

In conclusion, our work has demonstrated that *C. purpureum* alters physiological parameters such as the water potential of Japanese plum trees and negatively impacts fruit yield and quality. The latter can decrease the orchard productivity and, moreover, the fruit harvested from diseased trees does not meet the requirements to be properly commercialized or exported to demanding markets.

4. Materials and Methods

4.1. Collection of Samples

Collections of wood samples (97) from plum trees showing foliar silvering and internal wood discoloration symptoms were conducted in the Chilean Japanese plum productive area, from the Metropolitana Region (33°42′16.11″ S, 70°59′11.82″ W) to the Ñuble Region (36°37′24.98″ S, 72°0′23.39″ W). Two nurseries and 27 orchards of different Japanese plum cultivars, mainly 'Angeleno', 'Black Amber', 'Friar', 'Fortune', and 'Larry Anne', were examined. Samples of *Prunus domestica* subsp. *domestica* 'D'Agen' and *Prunus domestica* subsp. *italica* 'Reina Claudia' were included (Table 1). Symptoms were recorded and photographed.

4.2. Isolation and Purification

In the laboratory, bark was removed from the samples, and 0.5 cm wooden pieces were cut from the margin of the discoloration area. These pieces were superficially disinfected using 10% *v*/*v* sodium hypochlorite (4.9% chlorine) for 4 min and aseptically plated on Petri plates containing acidified potato dextrose agar (APDA) (25% potato dextrose agar (PDA), acidified with 0.2% *v*/*v* 85% lactic acid) (PDA Difco, Baltimore, MD, USA) and water agar (WA) (Winkler, Santiago, Chile). Plates were incubated at 25 °C in darkness until mycelial development. Pure cultures were obtained by transferring hyphal tips to fresh PDA plates and incubating them at 25 °C.

4.3. Identification and Characterization

The identification was focused on isolates resembling *C. purpureum*. Fungal colonies showing white to beige cottony mycelia were preliminary selected (n = 95). Subsequently, the isolates were identified by their cultural characteristics after 7 and 14 days of incubation on PDA at 25 °C. The morphology of the mycelia, presence of clamp connections, and morphometry of spores from pseudo-basidiocarps were determined using an optical microscope (Eclipse 80i, Nikon, Tokyo, Japan) and the software Nikon NIS-elements D2.30 (Tokyo, Japan).

Furthermore, representative isolates were identified by molecular means (n = 57). Pure cultures were incubated on PDA at 25 °C for 7 days. Total nucleic acids were isolated from fresh mycelium using the CTAB method, and genomic DNA (20 ng) was amplified using APN 1 *C. purpureum*-specific primers (Table 2), following the protocols described by [3]. The PCR products (20 μL) were analysed on 1.5% agarose gels, using a 1 kb DNA ladder (Maestrogen Inc., Xiangshan Dist., Hsinchu, Taiwan) as a molecular size standard. Electrophoresis was performed at 7.5 V/cm for 1 h, and gels were stained with ethidium bromide (1 μg/mL). Gels were visualized under UV light (λ = 365 nm) transillumination (Clear View standard UV transilluminator, UK) and digitally recorded (PC2010, Cannon, Nagasaki, Japan).

Table 2. Primer sequences used in the molecular analysis. APN1 and APM22 are specific primers for *Chondrostereum purpureum*.

Primer	Target	Sense	Sequence (5′-3′)	TM (°C)	Reference
ITS1	ITS	Forward	CTTGGTCATTTAGAGGAAGTAA	51	[45]
ITS4	ITS	Reverse	TCCTCCGCTTATTGATATGC	52	[45]
APN1-F	IGS	Forward	GCACGGAGAAGGAGAAGATTGGCT	61.6	[14]
APN1-R	IGS	Reverse	TTTCGGACTTTTGGGGCTCATTTCG	64.7	[14]
APM22D13F	SCAR	Forward	GGGGTGACGAGGACGACGGTG	63.2	[14]
APM22D13R	SCAR	Reverse	GGGGTGACGACATTATACTGCAGGTAGTAG	60	[14]

The internal transcribed spacer of four representative isolates selected from different host cultivars and collection localities was amplified using ITS1 and ITS4 primers (Table 1) and GoTaq® Green Master Mix 2X (Promega, Madison, WI, USA). The PCR conditions were an initial denaturation at 94 °C for 5 min, followed by 30 cycles of denaturation at 94 °C for 1 min, annealing at 52 °C for 1.5 min, and extension at 72 °C for 2 min. Fungal DNA was quantified using a fluorometer (Qubit 2.0, Invitrogen, Carlsbad, CA, USA). Electrophoresis, staining, and visualization were conducted as described for APN1 gels. The PCR products were purified and sequenced by Macrogen (Macrogen Inc., Seoul, Korea), and the nucleotide sequences were assembled and edited using the Sequencher software version 5.4.6 (Gene Codes Corporation, Ann Arbor, MI, USA). Subsequently, sequences were compared with the GenBank database using the BLAST software (Basic Local Alignment Search Tool program), (National Center for Biotechnology Information (NCBI), Rockville Pike, Bethesda, MD, USA).

Furthermore, colonies showing different characteristics such as a dark colour were preliminary identified by the morphometry of their reproductive structures. Basidiomycete fungi, of which the colonies are similar to those of *C. purpureum*, were discriminated by the colony shape and density and also by their DNA amplification using APN1 species-specific primers [23].

4.4. Pathogenicity

Four representative isolates were inoculated on healthy 2-year-old nursery plants 'Angeleno'. Fresh cuts were aseptically performed on the main 1-year-old lateral branches (1–1.5 cm diameter). Cuts were inoculated with 0.7-cm-diameter mycelial plugs collected from growing colonies of *C. purpureum* representative isolates HMCi7, HMCi121, HMCi147, and HMCi148 and incubated on PDA for 7 days at 25 °C. Sterile agar was used for controls. The inoculum was covered with petroleum jelly and plastic film to prevent dehydration. Five plants were inoculated per treatment based on a completely randomized block design at the beginning of September 2019. Plants were incubated in a screenhouse for 65 days at an environmental temperature of 5–18 °C and were periodically monitored to observe the occurrence of foliar symptoms, of which the severity was recorded using a visual scale (Figure 2). After the incubation period, branches were cut and analysed in the laboratory. Branches were cut longitudinally, and necrotic symptoms were recorded. Small wood pieces 0.5 cm were cut from the necrotic margin and aseptically plated on APDA. The reisolated fungi were purified, following the protocol described above.

Genomic DNA was extracted from the reisolated fungi and amplified using *C. purpureum* SCAR species-specific primers [14] (Table 1). PCR conditions were those indicated by the authors. Electrophoresis, staining, and visualization were conducted as described above. Subsequently, banding patterns from the inoculated isolates were compared with the reisolated ones to confirm Koch's postulates.

4.5. Silverleaf Effects on Plum

To determine the Silverleaf effects on plum trees, water potential, fruit yield, and quality were measured in a 22-year-old orchard ('Angeleno') on Mariana 2624 rootstock, with planting distances of 4.5 × 3 m, located in the Maule Region (34°58′58.21″ S, 71°16′37.01″ W), in the 2019 and 2020 harvest periods. Twenty healthy and 20 diseased trees were selected, homogeneous in their height and architecture, and located on four adjacent rows, to avoid topography and other differences among them. Each experimental unit consisted of two adjacent trees of the same treatment, with 10 replicates.

Trees were selected based on the absence of foliar silver symptoms for healthy ones and on the presence of them in diseased ones (Grade 3 or higher on the symptom severity scale) and were widespread throughout the canopy for diseased plants, meaning more than the 80% of the leaves had Silverleaf symptoms, from slightly to severely affected.

The Silverleaf foliar symptom visual severity scale (Grades 1–9) was devised using leaves from 'Angeleno' plum trees, naturally infected with *C. purpureum*, showing different levels of Silverleaf symptoms. Grades were assigned to the different foliar symptom visual intensities: 1 = healthy or apparently healthy leaf, 3 = epidermis starts detaching from the mesophyll and the colour is lighter than that in healthy leaves, 5 = more than 75% of the leaf adaxial epidermis is detached from the mesophyll, and the leaf is beginning to look grey–silver, 7 = 100% of the epidermis is detached from the mesophyll, and the leaf is completely white–silver, 9 = the epidermis peels off from the leaf, and the mesophyll begins to oxidize. Grades 2, 4, 6, and 8 indicate intermediate symptom severity between the previous and the following grade. Epidermis detachment on the leaf adaxial surface was confirmed using a Stereo Microscope at 15 × (Olympus SZ61, Tokyo, Japan).

Additionally, the presence or absence of *C. purpureum* was confirmed through the amplification of DNA isolated from sawdust collected from the main trunk [3], using APN1-specific primers [14], following the previously described protocol.

4.5.1. Water Potential

Xylem potential was measured on diseased and healthy trees on 1 March 2019, and 10 March 2020 (about 3–6 days after harvest). For both periods, it was measured on light-exposed leaves (five leaves per plant), which were previously covered for 2 h to avoid sunlight. Xylem potential was measured using a Scholander-type pressure chamber (Model 615, PMS Instruments, Albany, NY, USA).

4.5.2. Yield Assessment

Fruit was harvested on 27 February 2019 and 4 March 2020, the same dates that producers harvested the orchard. Before harvesting the fruit from each selected healthy and diseased tree, 10 random fruits from one lateral representative branch of each tree were set aside for quality parameter analysis (see Section 4.5.3) in the laboratory. Subsequently, the remaining fruit of each tree was individually collected and weighed.

4.5.3. Fruit Quality

Quality parameters were measured on the 10 previously reserved fruit. The measured parameters were individual weight (g), cover colour (%), and background colour (%), based on a plum colour chart [54], calibre (polar and equatorial diameter measured using a digital calliper), titratable acidity measured by potentiometric titration with NaOH 0.1 N (Hanna, pH 211, Nușfalău, Romania) expressed as citric acid proportion (%) (AOAC, 2000), pH, and soluble solids, measured with a digital refractometer (Brix degrees) (Atago, Pocket PAL-refractometer, Japan). For firmness measurement, two cheeks per fruit were cut, and the firmness was measured using a manual firmness penetrometer (FT 327, Facchini, Alfonsine, Italy) inserted in an aluminium holder for penetrometers (Dimeri, Santiago, Chile). The probe diameter was 8 mm. Media from both cheeks were subsequently analysed. The average of measurements (n = 10) was calculated for each parameter for further analysis.

5. Experimental Design and Statistical Analyses

The experimental unit consisted of two adjacent trees from the same treatment. The treatments were Silverleaf-diseased and healthy conditions, with 10 replicates from four rows in the orchard based on a randomized complete block design.

Fruit yield (total weight) was directly analysed, while for water potential, 10 samples (leaves) per plot were measured and averaged. Subsequently, averages of each plot were statistically analysed. For each fruit quality parameter, 20 samples (fruits) per plot were measured, and the averages were analysed.

Before testing for statistical significance, to detect differences between treatment means, normality and homoscedasticity were evaluated by the Shapiro–Wilk test, residual independence analysis, and graphical methods. When assumptions of normality and homoscedasticity were fulfilled, Student's *t*-test was used to compare healthy and diseased means of water potential, yield, and individual fruit weight.

Cover colour was analysed by Pearson's Chi-squared test, and fruit firmness and calibre by the Kruskal–Wallis test. The statistical analyses used depended on the nature of the data and distribution. Analyses were performed using InfoStat 2021 statistical software (Universidad Nacional de Córdoba, Córdoba, Argentina).

Author Contributions: Conceived the idea of the research, D.G.; scientific experiments design and methodology, D.G. and J.C.; Field surveys and collections, D.G., J.C. and M.I.; isolation, morphological characterization and pathogenicity tests, D.G. and M.I.; fungal sequencing, molecular detection and genetic characterization, J.C., physiological measurements, M.R.; yield assessment, D.G., J.C. and M.I.; fruit quality evaluations, C.H. and M.I.; writing and editing, D.G., J.C. and M.I.; critical review and editing, A.F. and J.B.; project administration and funding acquisition, J.B. All authors have read and agreed to the published version of the manuscript.

Funding: This research was supported by funding from The Research Council of Norway (project nr. 282257).

Data Availability Statement: Data is contained within the article.

Conflicts of Interest: The authors declare no conflict of interest.

References

1. Odepa, Estudios y Políticas Agrarias. Ministerio de Agricultura. Evolución de la Fruticultura Chilena en los Últimos 20 Años. Oficina de estudios y políticas Agrarias. 2021. Available online: https://www.odepa.gob.cl/rubros/frutas-frescas (accessed on 27 April 2021).
2. Gramaje, D.; Baumgartner, K.; Halleen, F.; Mostert, L.; Sosnowski, M.R.; Úrbez-Torres, J.R.; Armengol, J. Fungal trunk diseases: A problem beyond grapevines? *Plant Pathol.* **2016**, *65*, 355–356. [CrossRef]
3. Grinbergs, D.; Chilian, J.; Carrasco-Fernandez, J.; France, A.; Moya-Elizondo, E.; Gerding, M. A PCR-based method for the rapid detection of *Chondrostereum purpureum* in apple. *Plant Dis.* **2020**, *104*, 702–707. [CrossRef]
4. France, A.; Grinbergs, D.; Carrasco, J. First detection of Silverleaf (*Chondrostereum purpureum*) on rabbiteye blueberry (*Vaccinium ashei*) and disease damages. *Acta Hortic.* **2017**, *1180*, 277. [CrossRef]
5. Gramaje, D.; Agustí-Brisach, C.; Pérez-Sierra, A.; Moralejo, E.; Olmo, D.; Mostert, L.; Damm, U.; Armengol, J. Fungal trunk pathogens associated with wood decay of almond trees on Mallorca (Spain). *Persoonia* **2012**, *28*, 1–13. [CrossRef] [PubMed]
6. Cha, J.; Lee, S.; Chun, K.; Lee, S.; Ohga, S. Armillaria root rot caused by *Armillaria Tabescens* on *Prunus Salicina* in a Korean garden. *J. Fac. Agric. Kyushu Univ.* **2009**, *54*, 1273–1277.
7. Ko, Y.; Yao, K.; Chen, C.; Liu, C.; Maruthasalam, S.; Lin, C. First report of gummosis disease of plum (*Prunus salicina*) caused by a *Botryosphaeria* sp. in Taiwan. *Plant Dis.* **2008**, *92*, 483. [CrossRef] [PubMed]
8. Mojeremane, K.; Lebenya, P.; du Plessis, I.; van der Rijst, M.; Mostert, L.; Armengol, J.; Halleen, F. Cross pathogenicity of *Neofusicoccum australe* and *Neofusicoccum stellenboschiana* on grapevine and selected fruit and ornamental trees. *Phytopathol. Mediterr.* **2020**, *59*, 581–593. [CrossRef]
9. Damm, U.; Crous, P.; Fourie, P. Afissitunicate ascus mechanism in the Calosphaeriaceae, and novel species of *Jattaea* and *Calosphaeria* on *Prunus* wood. *Persoonia* **2008**, *20*, 39–52. [CrossRef]
10. Damm, U.; Crous, P.; Fourie, P. Botryosphaeriaceae as potential pathogens of *Prunus* species in South Africa, with descriptions of *Diplodia africana* and *Lasiodiplodia plurivora* sp. nov. *Mycologia* **2007**, *99*, 664–680. [CrossRef]

11. Li, Q.; Tang, L.; Sun, W.; Huang, S.; Guo, T.; Mo, J.; Fan, M.; Zhang, A.; Hsiang, T. First report of stem canker and dieback caused by *Neofusicoccum parvum* on plum in Guangxi, Southern China. *Plant Dis.* **2019**, *103*, 2952. [CrossRef]
12. Damm, U.; Mostert, L.; Crous, P.; Fourie, P. Novel *Phaeoacremonium* species associated with necrotic wood of *Prunus* trees. *Persoonia* **2008**, *20*, 87–102. [CrossRef]
13. Ramsfield, R.D. Risk assessment of inundative biological control with *Chondrostereum purpureum* in New Zealand. *N. Z. J. For. Sci.* **2006**, *36*, 11–20.
14. Becker, E.; Ball, L.; Hintz, W. PCR Based genetic markers for detection and infection frequency analysis of the biocontrol fungus *Chondrostereum purpureum* on Stika Alder and Trembling Aspen. *Biol. Control.* **1999**, *15*, 71–80. [CrossRef]
15. Becker, E.; Shamoun, S.; Hinz, W. Efficacy and environmental fate of *Chondrostereum purpureum* used as a biological control for red alder (*Alnus rubra*). *Biol. Control.* **2005**, *33*, 269–277. [CrossRef]
16. Shamoun, S. Application of biological control to vegetation management in forestry. In Proceedings of the X International Symposium on Biological Control of Weeds, Bozeman, MT, USA, 4–14 July 1999; pp. 87–96.
17. Hamberg, L.; Saksa, T.; Hantula, J. Role of *Chondrostereum purpureum* in biocontrol of trees. *Appl. Microbiol. Biotechnol.* **2021**, *105*, 431–440. [CrossRef] [PubMed]
18. Bus, V.; Spiers, A.; Brewster, D.; Hofstee, M. Preliminary screening of apple germplasm for resistance to Silverleaf infection. *N. Z. J. Crop Hortic. Sci.* **1996**, *24*, 1–6. [CrossRef]
19. Setliff, E.C. The wound pathogen *Chondrostereum purpureum*, its history and incidence on trees in North America. *Austral. J. Bot.* **2002**, *50*, 645–651. [CrossRef]
20. Willoughby, I.H.; Seier, M.K.; Stokes, V.J.; Thomas, S.E.; Varia, S. Synthetic herbicides were more effective than a bioherbicide based on *Chondrostereum purpureum* in reducing resprouting of *Rhododendron ponticum*, a host of *Phytophthora ramorum* in the UK. *Forestry* **2015**, *88*, 336–344. [CrossRef]
21. Hamberg, L.; Hantula, J. The efficacy of six elite isolates of the fungus *Chondrostereum purpureum* against the sprouting of European aspen. *J. Environ. Manag.* **2016**, *171*, 217–224. [CrossRef]
22. Merlet, L.; Wiseman, M.; Serdani, M.; Putnam, M.L. First report of Silver Leaf caused by *Chondrostereum purpureum* on *Vaccinium corymbosum* in Oregon. *Plant Dis.* **2018**, *102*, 2041. [CrossRef]
23. Grinbergs, D.; Chilian, J.; Lisboa, K.; France, A. First report of Silverleaf disease caused by *Chondrostereum purpureum* on murta (*Ugni molinae* Turcz.) in Chile. *Plant Dis.* **2019**, *103*, 2140. [CrossRef]
24. Spiers, A.G.; Hopcroft, D.H. Factors affecting *Chondrostereum purpureum* infection of Salix. *Eur. J. For. Pathol.* **1988**, *18*, 257–278. [CrossRef]
25. Spiers, A.G.; Brewster, D. Evaluation of chemical and biological treatments for control of *Chondrostereum purpureum* infection of pruning wounds in willows, apples, and peaches. *N. Zeal. J. Crop Hort.* **1997**, *25*, 19–31. [CrossRef]
26. Spiers, A.; Edwards, W.; Hopcroft, D. Effects of Silverleaf infection on ultrastructure of foliage of *Prunus*, *Rosa*, and *Populus*. *N. Z. J. Bot.* **1987**, *25*, 411–423. [CrossRef]
27. Simpson, R.M.; Van Hekezen, R.; Van Lune, F.; Brewster, D.; Spiers, A.G. Extracellular enzymes of *Chondrostereum purpureum*, causal fungus of Silverleaf disease. *N. Z. Plant Prot.* **2001**, *54*, 202–208. [CrossRef]
28. Cloete, M.; Fourie, P.; Damm, U.; Crous, P.; Mostert, L. Fungi associated with die-back symptoms of apple and pear trees, a possible inoculum source of grapevine trunk disease pathogens. *Phytopathol. Mediterr.* **2011**, *50*, 176–190. [CrossRef]
29. Ozgonen, H.; Erkilic, A.; Baloglu, S. First report of *Chondrostereum purpureum* causing Silverleaf disease of apricot (*Prunus armeniaca*) in Malatya, Turkey. *N. Z. J. Crop Hortic. Sci.* **2006**, *34*, 357. [CrossRef]
30. Børve, J.; Ippolito, A.; Tanovic, B.; Michalecka, M.; Sanzani, S.M.; Ponitowska, A.; Mari, M.; Hrustic, J. Fungal diseases. In *Cherries. Botany, Production and Uses*; Quero-García, J., Iezzoni, A., Pulawska, J., Lang, G.A., Eds.; CABI: Wallingford, CT, UK, 2017; pp. 338–364.
31. Spiers, A.; Brewster, D.; Bus, V.; Hopcroft, H. Seasonal variation in susceptibility of xylem tissue of *Malus*, *Pyrus*, *Prunus*, and *Salix* species to *Chondrostereum purpureum* in New Zealand. *Mycol. Res.* **1998**, *102*, 881–890. [CrossRef]
32. Ogawa, J.; English, H. *Diseases of Temperate Zone Tree Fruit and Nut Crops*; University of California: Oakland, CA, USA, 1991; Volume 3345, 461p.
33. Pearce, R.B.; Sumer, S.; Doran, S.J.; Carpenter, T.A.; Hali, L.D. Non-invasive imaging of fungal colonization and host response in the living sapwood of sycamore (*Acer pseudoplatanus* L.) using nuclear magnetic resonance. *Physiol. Mol. Plant Path.* **1994**, *45*, 359–384. [CrossRef]
34. Vartiamäki, H.; Hantula, J.; Uotila, A. Susceptibility of silver birch pruning wounds to infection by white-rot fungus (*Chondrostereum purpureum*), a potential bioherbicide. *Silva Fenn.* **2009**, *43*, 537–547. [CrossRef]
35. Senda, M.; Narita, T.; Akada, S.; Okuno, T.; Miyairi, K. Characterization of an endopolygalacturonase gene cppgl from phytopathogenic fungus *Chondrostereum purpureum*. *J. Gen. Plant Pathol.* **2001**, *67*, 41–44. [CrossRef]
36. Bishop, G.C. Studies on Silver Leaf Disease of Stone and Pome Fruit Trees. Ph.D. Thesis, University of Adelaide, Adelaide, Australia, 1978; 128p.
37. Głowacka, A.; Rozpara, E. Evaluation of several dessert cultivars of plum, new under climatic conditions of Poland. *Hort. Sci. (Prague)* **2017**, *44*, 126–132. [CrossRef]
38. Kaufmane, E.; Grâvîte, I.; Ikase, L. Plum research and growing in Latvia. *Proc. Latv. Acad. Sci. Sect. B* **2019**, *73*, 195–206. [CrossRef]

39. Børve, J.; Talgø, V.; Stensvand, A. Silver Leaf in Norwegian Prunus domestica Plum Orchards. In Proceedings of the IOBC/WPRS Working Groups: Pheromones and Other Semiochemicals in IP & Integrated Protection of Fruit Crops, PheroFIP 2019, Lisbon, Portugal, 20–25 January 2019; p. 219.
40. Atkinson, J.D. *Diseases of Tree Fruits in New Zealand*; Bulletin 81; Department of Scientific and Industrial Research: Auckland, New Zealand, 1971; 406p.
41. Bien, S.; Damm, U. *Prunus* trees in Germany—A hideout of unknown fungi? *Mycol. Prog.* **2020**, *19*, 667–690. [CrossRef]
42. Espargham, N.; Mohammadi, H.; Gramaje, D. A survey of trunk disease pathogens within *Citrus* trees in Iran. *Plants* **2020**, *9*, 754. [CrossRef]
43. Bertsch, C.; Ramirez-Suero, M.; Magnin-Robert, M.; Larignon, P.; Chong, J.; Abou-Mansour, E.; Spagnolo, A.; Clément, C.; Fontaine, F. Grapevine trunk diseases: Complex and still poorly understood. *Plant Pathol.* **2013**, *62*, 243–265. [CrossRef]
44. Spiers, A.; Brewster, D.; Slade, A.; Gardiner, S. Characterization of New Zealand isolates of *Chondrostereum purpureum* with regard to morphology, growth, pathogenicity and RAPD bandings patterns. *Mycol. Res.* **2000**, *104*, 395–402. [CrossRef]
45. White, T.J.; Bruns, T.; Lee, S.; Taylor, J. Amplification and direct sequencing of fungi ribosomal RNA genes for phylogenetics. In *PCR Protocols. A Guide to Methods and Applications: 315–322*; Academic Press: San Diego, CA, USA, 1990.
46. France, A.; Santelices, C.; Buddie, A.; Kirk, P. Silverleaf: First worlwide report of a new and harmful disease on blueberry. *Acta Hortic. (ISHS)* **2009**, *810*, 341–344. [CrossRef]
47. Petit, A.N.; Vaillant, N.; Boulay, M.; Clément, C.; Fontaine, F. Alteration of photosynthesis in grapevines affected by esca. *Phytopathology* **2006**, *96*, 1060–1066. [CrossRef] [PubMed]
48. Fontaine, F.; Pinto, C.; Vallet, J.; Clément, C.; Gomes, A.; Spagnolo, A. The effects of grapevine trunk diseases (GTDs) on vine physiology. *Eur. J. Plant Pathol.* **2015**. [CrossRef]
49. Hamberg, L.; Vartiamäki, H.; Hantula, J. Breeding Increases the Efficacy of *Chondrostereum purpureum* in the Sprout Control of Birch. *PLoS ONE* **2015**, *10*, e0117381. [CrossRef] [PubMed]
50. Grinbergs, D.; Chilian, J.; Padilla, N.; Reyes, M.; France, A.; Gerding, M.; Ernesto, A. Moya-Elizondo.. Endophytic microorganisms associated with reversion of Silverleaf disease symptoms in apple. *Phytopathology* **2021**. [CrossRef] [PubMed]
51. Xing, J.; Li, P.; Zhang, Y.; Li, J.; Liu, Y.; Lachenbruch, B.; Su, X.; Zhao, J. Fungal pathogens of canker disease trigger canopy dieback in poplar saplings by inducing functional failure of the phloem and cambium and carbon starvation in the xylem. *Physiol. Mol. Plant Path.* **2020**, *112*, 101523. [CrossRef]
52. Bailey, B.; Bae, H.; Strem, M.; Crozier, J.; Thomas, S.; Samuels, G.; Vinyard, B.; Holmes, K. Antibiosis, mycoparasitism, and colonization success for endophytic *Trichoderma* isolates with biological control potential in *Theobroma cacao*. *Biol. Control* **2008**, *46*, 24–35. [CrossRef]
53. Grinbergs, D.; France, A.; Chilian, J. Silverleaf disease in apple orchards: Understanding and preventing an increasing problem. In Proceedings of the 11th International IOBC-WPRS Workshop on Pome Fruit Diseases, Jurmala, Latvia, 26–30 June 2017; Stensvand, A., Lennox, C., Wenneker, M., Morocko-Bicevska, I., Rancane, R., Eds.; International Organization for Biological and Integrated Control of Noxious Animals and Plants, West Palearctic Regional Section (IOBC-WPRS): Darmstadt, Germany, 2018; Volume 138, p. 100.
54. Rivero, M.; Quiroga, M.; Gonzalez, O.; Moraga, L. Plum postharvest manual. INTA, Instituto de Investigaciones Agropecuarias. 2013. Available online: https://inta.gob.ar/sites/default/files/script-tmp-ficha_n_1_-_cosecha.pdf (accessed on 27 April 2021).

plants

Article

Investigations on Fungi Isolated from Apple Trees with Die-Back Symptoms from Basilicata Region (Southern Italy)

Stefania Mirela Mang [1,*], Carmine Marcone [2], Aurel Maxim [3] and Ippolito Camele [1,*]

1 School of Agricultural, Forestry, Food and Environmental Sciences (SAFE), University of Basilicata, Viale dell'Ateneo Lucano 10, 85100 Potenza, Italy
2 Department of Pharmacy, University of Salerno, Via Giovanni Paolo II 132, 84084 Salerno, Italy; cmarcone@unisa.it
3 Department of Engineering and Environmental Protection, Faculty of Agriculture, University of Agricultural Sciences and Veterinary Medicine, No. 3-5, Calea Manastur Street, 400372 Cluj-Napoca, Romania; aurel.maxim@usamvcluj.ro
* Correspondence: stefania.mang@unibas.it (S.M.M.); ippolito.camele@unibas.it (I.C.); Tel.: +39-0971205519 (S.M.M.); +39-0971205544 (I.C.)

Abstract: Val d'Agri is an important orchard area located in the Basilicata Region (Southern Italy). A phenomenon affecting cv. "Golden Delicious" apples which lead to tree death has been observed in the past several years in this area. This phenomenon has already been detected in about 20 hectares and is rapidly expanding. The symptoms observed were "scaly bark" and extensive cankers, mainly located in the lower part of the trunk, associated with wood decay. Dead plants ranged from 20% to 80% and, in many cases, trees were removed by farmers. In order to identify the causes of this phenomenon, investigations were started in autumn/winter 2019. In order to determine the possible causal agents, fungal and bacterial isolations, from symptomatic tissues, were performed in laboratory. Bacterial isolations gave negative results, whereas pure fungal cultures (PFCs) were obtained after 3–4 passages on potato dextrose agar (PDA) media. Genetic material was extracted from each PFC and amplified by PCR using three pairs of primers: ITS5/4, Bt2a/Bt2b and ACT-512F/ACT-783R. The amplicons were directly sequenced, and nucleotide sequences were compared with those already present in the NCBI GenBank nucleotide database. All isolated fungi were identified based on morphological features and multilocus molecular analyses. *Neofusicoccum parvum*, *Diaporthe eres* and *Trametes versicolor* were most frequently isolated, while *Pestalotiopsis funerea*, *Phomopsis* spp. and *Diaporthe foeniculina* were less frequently isolated. All nucleotide sequences obtained in this study have been deposited into the EMBL database. Pathogenicity tests showed that *N. parvum* was the most pathogenic and aggressive fungus, while *Phomopsis* sp. was demonstrated to be the less virulent one. All the investigated fungi were repeatedly reisolated from artificially inoculated twigs of 2-year-old apple trees, cv. "Golden Delicious", and subsequently morphologically and molecularly identified. The role played by the above-mentioned fungi in the alterations observed in field is also discussed.

Keywords: apple die-back; canker; fungi; multi-loci phylogeny; wood decay

Citation: Mang, S.M.; Marcone, C.; Maxim, A.; Camele, I. Investigations on Fungi Isolated from Apple Trees with Die-Back Symptoms from Basilicata Region (Southern Italy). *Plants* **2022**, *11*, 1374. https://doi.org/10.3390/plants11101374

Academic Editors: Paula Baptista and Carlos Agustí-Brisach

Received: 2 March 2022
Accepted: 18 May 2022
Published: 21 May 2022

1. Introduction

Apple die-back syndrome is a complex disease initially characterized by a stunted appearance of the plants and the presence of chlorosis symptoms on leaves. As the disease develops, cracks and necrotic lesions of the cortex appear mostly at the base of the stem and at the grafting point. Subsequently, "scaly bark" and extensive cankers, generally located in the lower part of the trunk develop, which are also associated with wood decay, and the progressive death of the trees is registered. Many authors worldwide investigated the apple die-back syndrome, attributing it to different causal agents, such as various phytopathogens or other possible physiological causes [1–5]. Furthermore, in 2011, Cloete et al. [6] reported the presence of the die-back syndrome and cankers on apples and pears in South Africa, from which they isolated several fungal pathogens belonging to the *Diplodia*, *Phaeoacremonium*, *Phomopsis*, *Neofusicoccum* and *Eutypa* genera. A very recent study by Di Francesco et al. [5], which characterized, in Brazil, apple cultivars' susceptibility to *Neofusicoccum parvum* (Pennycook & Samuels) P.W. Crous, Slippers & A.J.L. Phillips, stated that due to climate change, this fungus is emerging as a new pathogen on species of the *Rosaceae* plant family. An apple tree die-back syndrome causing severe tree losses was also observed in the main apple producing regions in Tunisia, as described by Souli et al. [7]. The authors identified, both morphologically and molecularly, *Phytophthora* and *Pythium* species as being the causal agents and the factors that promoted apple tree die-back. They also showed that soil salinity contributed to increase the disease severity [7].

In Italy, apple cultivation is mainly concentrated in the northeast part of the country, specifically in Trentino Alto Adige/Südtirol region. This area comprises about half of the harvested hectares and has an intensive growing system, producing almost 70% of the Italian apples. However, some other Italian regions such as Veneto, Emilia-Romagna, Piedmont, Campania and Basilicata also successfully cultivate apple trees with quite relevant productions. Overall, in 2019 in Italy, the surface cultivated with apple was about 58,000 hectares and the apple production reported in the same year was about 23 million quintals [8].

Due to the economic importance of this crop in Italy, many studies regarding its cultivation and phytosanitary status have been performed. The phytosanitary surveys on apples cultivated in Bolzano area, was investigated by Lindner [9]. The author reported the cortical damage of apple cvs. "Gala", "Golden Delicious" and "Red Delicious", in spring 2007, and alterations very much resembling the "blister bark" and "paper bark" symptoms often associated the with withering and drying of the branches were also described. Nevertheless, the author reported that they did not isolate any bacterial or fungal pathogen from the plants, concluding that the cause of the observed symptoms was probably to be found among particularly unfavorable climatic conditions registered during the winter and spring periods. The same author also observed damages at the grafting point level and identified two fungi, known as canker agents, namely *Phomopsis mali* and *Diplodia malorum*.

The apple tree die-back syndrome was reported during 2008–2009 in many apple orchards in the north of Italy and since that period many other trees have become infected, especially young plants.

The phytosanitary status of the apples from Trentino region (Northern Italy) was investigated by Prodorutti et al. [10]. The authors reported an increase in the die-back symptoms on apple trees, showing that plants were usually stunted with cracking and necrosis in the lower part of the trunk and on the graft union site. The trees died during the growing season. Furthermore, the same authors reported that the incidence of the disease was, in some cases, very high (reaching almost 80%), and that the most affected trees were the youngest ones of about 2–5 years old that had been subjected to various types of stress. One bacterium, *Pseudomonas syringae* pv. *syringae*, and a few fungi, such as *Phomopsis* spp., *Neonectria* spp. And some *Botryosphaeriaceae*, were isolated from trunk tissues, taken from symptomatic trees. Their role in the syndrome expression was also demonstrated, even if they appeared not to cause the death of the artificially inoculated branches [10].

The Val d'Agri area, located in the Basilicata Region (Southern Italy), has geographical and climatic conditions that favor apple cultivation. In particular, "Fuji" and "Golden Delicious" apple cultivars are mostly grown in the area. Apple orchards belong to private farmers and are kept as small-scale cultivation systems. During the autumn/winter 2019, in the Val d'Agri area, on about 20 hectares cultivated with apples, symptoms characterized by "scaly bark" and extensive cankers, mainly located in the lower part of the trunk and associated with wood decay, were observed. Additionally, dead plants ranged from 20 to 80%, and in many cases, trees were removed by farmers. In order to identify the causes of this phenomenon, investigations were started in 2019. It is to be mentioned that despite good prevention and control measures employed so far against the key diseases on fruit trees in the south of Italy, including the Basilicata region, knowledge about the distribution and the pathogens involved in apple die-back syndrome is still missing. These data are very important since the presence of the die-back syndrome could economically affect the growers in the region. The identification and characterization of various fungal and bacterial pathogens attacking fruit trees, including apples, were initially based on only the morphological features of the pure cultures obtained in vitro [11,12]. However, over time, despite the ease of application, the morphological features proved to be inefficient to further classify fungal and bacterial pathogens. Therefore, other solutions, including molecular approaches, were investigated in order to identify and characterize the phytopathogens associated with the die-back symptoms [6,7,13,14]. Nowadays, several gene regions or genes, such as the Internal Transcribed Spacer (ITS) of the ribosomal DNA (rDNA), β-tubulin (*TUB-2*) and actin (*ACT*) protein-coding genes, are extensively utilized to identify and characterize phytopathogens [15–23].

The aim of the present study was to investigate the die-back syndrome on apple orchards from the Val d'Agri area (Basilicata region, Southern Italy). More precisely, the main objectives of the present study were to: (1) identify fungi or bacteria eventually associated with the die-back symptoms observed on apple trees; and (2) perform pathogenicity tests on apple trees in order to verify the involvement of the identified pathogens in the apple die-back disease observed in the Val d'Agri area.

2. Results

2.1. Pathogens Isolations

Pure culture fungal isolates on PDA media from die-back symptomatic material obtained in this study were selected for further characterization through morphological and cultural characteristics, DNA sequencing and phylogenetic analysis (Table 1).

From apple die-back symptomatic samples, the above-described fungi were isolated with different frequencies. Among the most frequently isolated fungi were the *N. parvum*, with a 55% isolation frequency (IF%), followed by *D. eres*, with a 15% IF, and *T. versicolor*, with a 14% IF. All the other fungi were less frequently isolated with an IF ranging from 10–12%, except for *Phomopssis* spp., which was very rarely isolated (<5% IF).

Despite repeated trials to isolate bacteria from symptomatic apple wood, no bacterial colonies were ever obtained. During the investigation for the identification of the apple die-back disease cause no symptom or damage of the root system were noticed. Furthermore, all isolation attempts, performed from roots taken from the symptomatic apple trees, gave negative results.

2.2. Morphological Identification

Based on their cultural and morphological features, pure fungal isolates were classified in five distinct genera: *Neofusicoccum* [24], *Diaporthe* [25–28], *Trametes* [29–33], *Pestalotiopsis* [34–36] and *Phomopsis* [6,37–40] (Table 1 and Figure 1). In particular, in the case of *Diaporthe* grayish or white colonies on PDA and alpha and beta conidia were observed; in the case of *Neofusicoccum*, grey-black colonies and fusiform conidia, nonseptate when young and biseptate ellipsoidal (partially light brown with a darker middle center) when old, were detected. In the case of *Phomopsis*, white colonies and alpha and beta conidia were observed. For *Pestalotiopsis*, reddish colonies and 4-septate conidia, fusiform to ellipsoid and straight to slightly curved, were noticed. *Trametes* genus was identified based on white colonies and the presence of clavate basidia with an inflated epibasidial segment, 4-spored, clamped at the base and basidiospore cylindrical in large spores with slightly inflated top, ellipsoid to ovoid.

Table 1. Fungal isolates obtained during this study with their respective GenBank accession numbers and percentage of identity when compared to reference nucleotide sequences for the same species from the NCBI nucleotide database.

Isolate Name	Species	GenBank Accession Number *			Identity (%) *		
		ITS **	TUB-2 **	ACT **	ITS **	TUB-2 **	ACT **
Noefp1	*Neofusicoccum parvum*	LR757960	OU022063	OU023206	>99–100	>99	100
Neofp2	*N. parvum*	LR757961	OU022064	OU023207	-//-	-//-	-//-
Neofp3	*N. parvum*	LR757962	OU022065	OU023208	-//-	-//-	-//-
Neofp4	*N. parvum*	LR757963	OU022066	OU023209	-//-	-//-	-//-
Neofp5	*N. parvum*	LR757964	OU022067	OU023210	-//-	-//-	-//-
Neofp6	*N. parvum*	LR757965	OU022068	OU023211	-//-	-//-	-//-
Neofp7	*N. parvum*	LR757966	OU022069	OU023212	-//-	-//-	-//-
Tramtv1	*Trametes versicolor*	LR759930	-	-	>99–100	-	-
Tramtv2	*T. versicolor*	LR759931	-	-	-//-	-	-
Tramtv3	*T. versicolor*	LR759932	-	-	-//-	-	-
Tramtv4	*T. versicolor*	LR759933	-	-	-//-	-	-
Tramtv5	*T. versicolor*	LR759934	-	-	-//-	-	-
Tramtv6	*T. versicolor*	LR759935	-	-	-//-	-	-
Diapore1	*Diaporthe eres*	OU020696	OU022056	OU023199	>99	>99–100	>99
Diapore2	*D. eres*	OU020697	OU022057	OU023200	-//-	-//-	-//-
Diapore3	*D. eres*	OU020698	OU022058	OU023201	-//-	-//-	-//-
Diapore4	*D. eres*	OU020699	OU022059	OU023202	-//-	-//-	-//-
Diaporf1	*Diaporthe foeniculina*	OU020700	OU022060	OU023203	>99–100	100	>99
Diaporf2	*D. foeniculina*	OU020701	OU022061	OU023204	-//-	-//-	-//-
Diaporf3	*D. foeniculina*	OU020702	OU022062	OU023205	-//-	-//-	-//-
Pestf1	*Pestalotiopsis funerea*	OU020703	OU022070	-	>99	>99	-
Pestf2	*P. funerea*	OU020704	OU022071	-	-//-	-//-	-
Pestf3	*P. funerea*	OU020705	OU022072	-	-//-	-//-	-
Phomp1	*Phomopsis* sp.	OU026160	-	-	>99	-	-
Phomp2	*Phomopsis* sp.	OU026161	-	-	-//-	-	-
Phomp3	*Phomopsis* sp.	OU026162	-	-	-//-	-	-

Note: * The percentage of identity was established after comparing the nucleotide sequences from this study with at least two of the reference species existent in the database for each fungal species. ** ITS = Nuclear ribosomal internal transcribed spacer regions; TUB-2 = β-tubulin 2 gene; ACT = actin gene. "-" = no data were obtained/or exist in the GenBank nucleotide database. "-//-" = identical values as those in the previous row are reported.

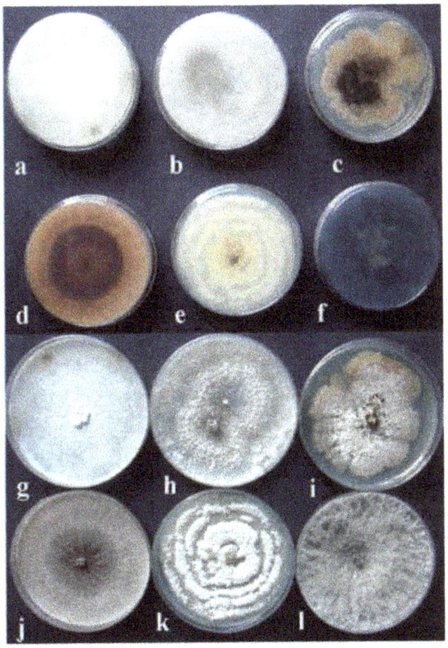

Figure 1. Pure fungal cultures on PDA obtained from samples of apples with dieback symptoms. (**a**,**g**) = *Trametes versicolor;* (**b**,**h**) = *Diaporthe eres;* (**c**,**i**) = *Diaporthe feoniculina;* (**d**,**j**) = *Pestalotiopsis funerea;* (**e**,**k**) = *Phomopsis* spp.; (**f**,**l**) = *Neofusicoccum parvum.*

2.3. Molecular Characterization

The PCR amplifications for each gene investigated yielded amplicons of expected sizes: ITS5/ITS4 (~700 bp), *tub-2* (~500 bp) and *ACT* (~300 bp), which, after direct sequencing in both directions, using the same primers as for the amplification, led to 26 nucleotide sequences (Table 1). A megablast search, excluding "uncultured/environmental sample sequences", performed in the NCBI's nucleotide database (www.ncbi.org, accessed on 12 January 2022) for all nucleotide sequences obtained in this study, identified at Genus level all fungal isolates (Table 1).

2.4. Phylogenetic Analysis

Single locus analysis gave consistent results for all three loci (ITS, *tub-2* and *ACT*), and the topology of trees was congruent in terms of species grouping. All sequences obtained in this study have been deposited in the European Molecular Biology Laboratory (EMBL-EBI) nucleotide database (www.ebi.ac.uk, accessed on 12 January 2022) and their GenBank accession numbers are presented in Table 1.

The final alignment dataset, for the ITS region, was composed of a total number of 658 characters. It contained 58 nucleotide sequences, including five outgroup species, namely: *Diaporthella corylina* (acc. no. KC343004) utilized for *Diaporthe* fungi; *Valsa japonica* (acc. no AF191185) for *Phomopsis* sp. fungi; *Sordaria alcina* (acc. no. AY681198) for the *Pestaliopsis* sp. Fungal group; *Grifola frondosa* (acc. No. AY049140) utilized for *Trametes* sp. Fungi; and for the *Neofusicoccum parvum* group, *Diplodia seriata* (acc. no. MH221102). Phylogenetic analysis based on the ITS region variation showed that fungal isolates from the same species clustered together into the same clade and, as expected, outgroup species were placed separately from the other groups (Figure 2).

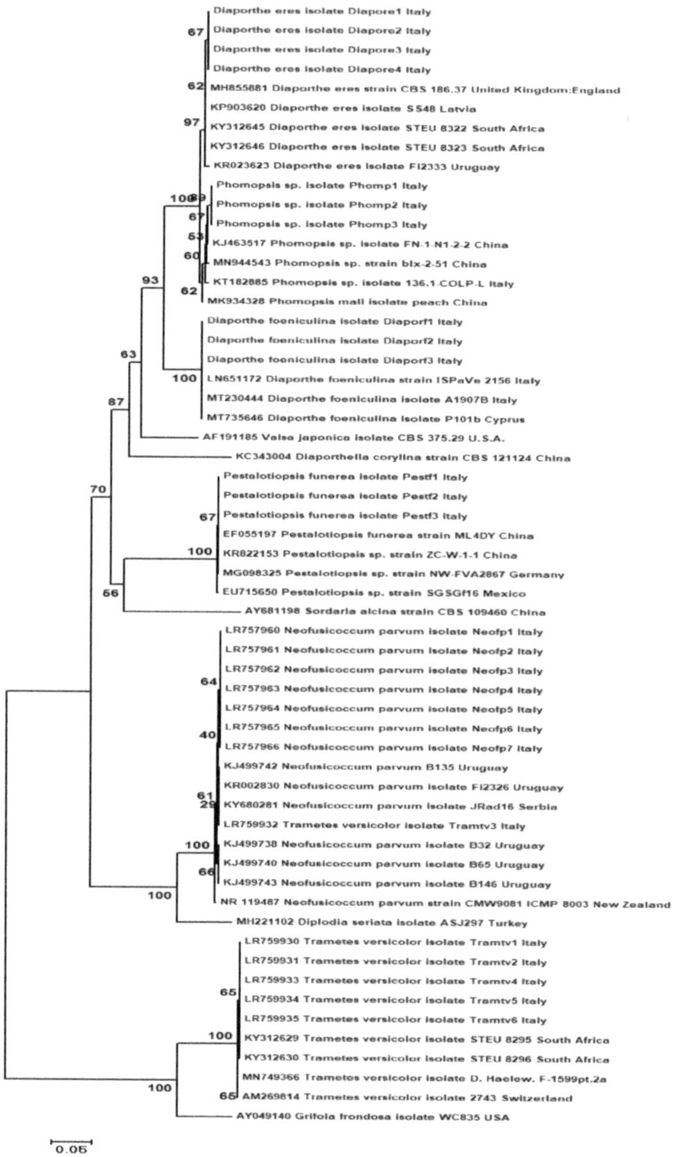

Figure 2. Molecular phylogenetic tree obtained through the neighbor-joining (NJ) method, based on the 58 ITS region sequences data (658 bp) from fungal isolates in the present study and published sequences. Five fungal species (*Diaporthella corylina, Valsa japonica, Sordaria alcina, Grifola fondosa* and *Diplodia seriata*) were used as outgroups in the analysis. The optimal tree with the sum of branch length = 1.38532669 is shown. The confidence probability estimated using the bootstrap test (1000 replicates) is shown next to the branches. The tree is drawn to scale, with branch lengths in the same units as those of the evolutionary distances used to infer the phylogenetic tree. The evolutionary distances were computed using the Tajima–Nei method and are in the units of the number of base substitutions/site. Scientific names of the fungi along with collection place, isolate abbreviation and GenBank AC number are shown in the trees.

Overall, the 53 nucleotide sequences obtained in this study, based on the ITS region sequence analysis, clustered in two clades which contained all six fungal species. Two of these species belonged to *Diaporthe* genus, namely, *Diaporthe eres* Nitschke and *D. foeniculina* (Sacc.) Udayanga & Castl., and others have been identified as *N. parvum*, *Pestalotiopsis funerea* (Desm.) Steyaert, *Phomopsis* sp. Sacc. & Roum. and *T. versicolor* (L.) Lloyd (Figure 2). Within the phylogenetic tree, the first clade grouped together five of the species previously mentioned, while the second clade contained only one species, *Trametes versicolor*, clearly separated from the others. Additionally, the fungal species isolated and identified in this study were positioned close to similar reference species downloaded from the GenBank for each fungus, and their location was well supported by very high (97–100%) bootstrap values (Figure 2). The ITS data confirmed the previous preliminary fungal identification based on morphological features.

Since ITS alone does not provide sufficient resolution to exactly classify fungi at species level, other loci were considered for the phylogeny-based identification of the taxa investigated in this study. In particular, the β-tubulin (*tub-2*) gene, a very well-known molecular locus extensively used in phylogenetic studies of phytopathogenic fungi [35,38–40], was examined. A total number of 39 nucleotide sequences of the *tub-2* partial gene were obtained and employed along with the species *D. corylina*, *B. dothidea* and *S. alcina*, used as outgroups in phylogenetic analysis, which was carried out using the NJ method, as performed for the ITS (Figure 3).

Results from the phylogenetic analysis of the fungal species from this study, based on the *tub-2* gene, have shown that they grouped together with similar species from the GenBank database. A better separation within the clades and subclades compared to what obtained from the ITS was also observed (Figure 3). Two separate clades were obtained from the nucleotide sequences investigated in this study. In one, *Neofusicoccum* isolates grouped together into the same subgroup, which was very well supported (98% bootstrap support value), and were separated from the *Pestalotiopsis* subclade, which was also highly supported (99% bootstrap support value), whereas the *Diaporthe* isolates were all placed in a separate clade and were clearly distinguished in two subclades as species, e.g., *D. eres* and *D. foeniculina*, both sharing their vicinity with the same outgroup, *D. corylina* (Figure 3). All fungal species based the *tub-2* gene variation were grouped together with the same species from the GenBank with an elevated bootstrap support (98–99%). The separation of the fungal species within each clade or subclade, strongly supported by high bootstrap values, was 99% for *D. eres*, *D. feoniculina* and *N. parvum* and 98% in the case of *P. funerea* (Figure 3).

Regarding the third gene, namely actin, despite our repeated PCR trials, amplicons could not be obtained for all fungal species (Figure 4). Therefore, the alignment for the *ACT* gene used in phylogenetic analysis contained 298 characters and involved only 24 nucleotide sequences. Moreover, *ACT* gene analysis showed that the two *Diaporthe* species identified in this study along with their reference species from the GenBank were well separated from the *Neofusicoccum* sp. isolates and, thus, clustered in two different clades also supported by very high bootstrap values of 94% and 99%, respectively. In addition, *Neofusicoccum* isolates were grouped together with their reference species with a 99% bootstrap support value (Figure 4). The phylogenetic reconstruction based on the *ACT* gene reconfirmed the molecular identification based on other loci at species level for each fungal species analyzed and also was in concordance with the preliminary morphological characterization.

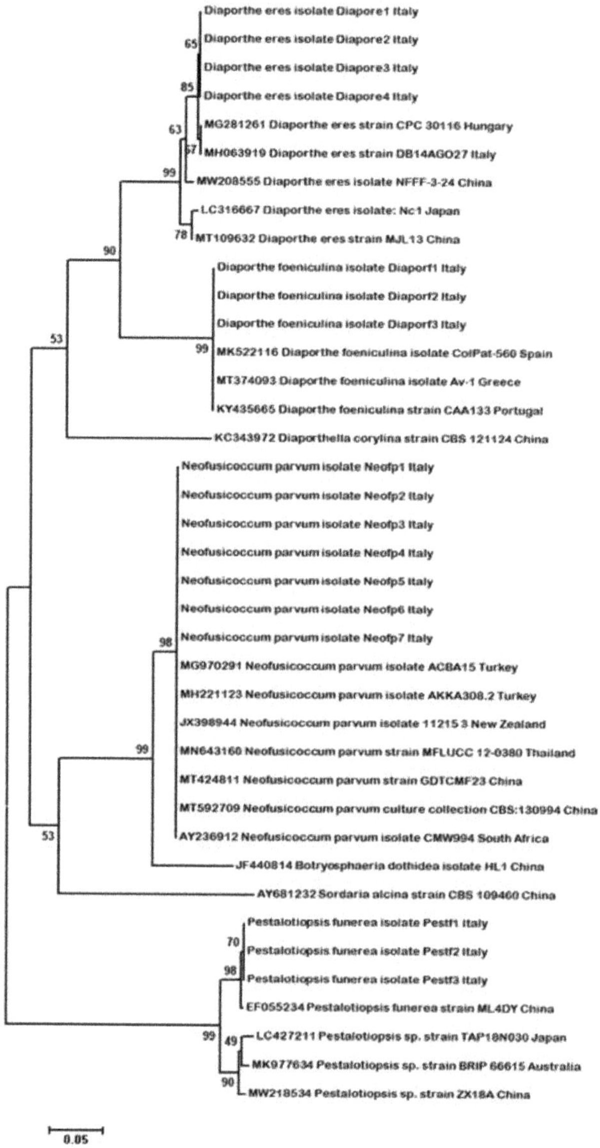

Figure 3. Molecular phylogenetic tree obtained through the neighbor-joining (NJ) method, based on the 39 *tub-2* gene sequences data (500 bp) from fungal isolates in the present study and published sequences. The fungal species (*D. corylina*, *B. dothidea* and *S. alcina*) were used as outgroups in the analysis. The optimal tree with the sum of branch length = 1.08594872 is shown. The confidence probability estimated using the bootstrap test (1000 replicates) is shown next to the branches. The tree is drawn to scale, with branch lengths in the same units as those of the evolutionary distances used to infer the phylogenetic tree. The evolutionary distances were computed using the Tajima–Nei method and are in the units of the number of base substitutions/site. Scientific names of the fungi along with collection place, isolate abbreviation and GenBank AC number are shown in the trees.

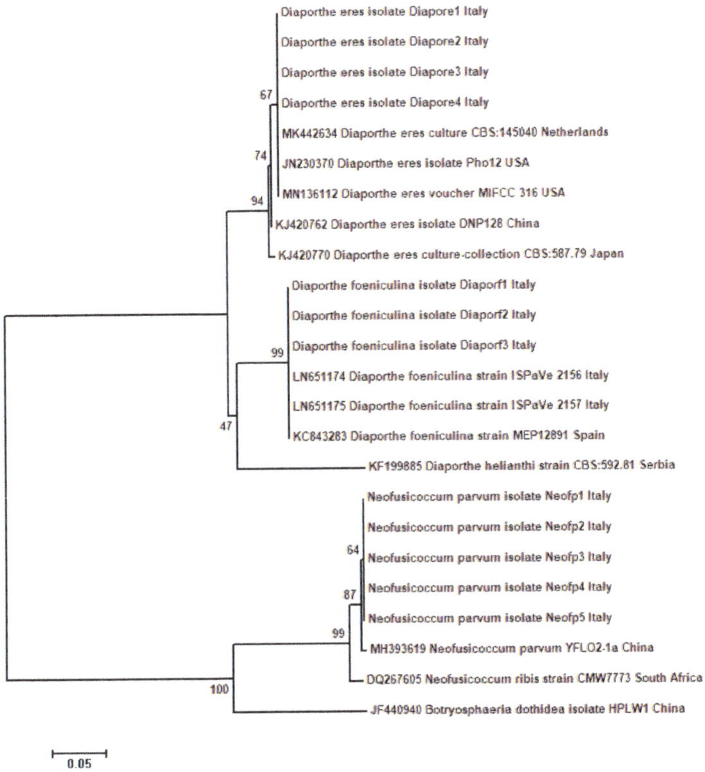

Figure 4. Molecular phylogenetic tree obtained through neighbor-joining (NJ) method based on the 24 *ACT* gene sequences data (298 bp) from fungal isolates in the present study and published sequences. The fungal species (*D. helianthi* and *B. dothidea*) were used as outgroups in the analysis. The optimal tree with the sum of branch length = 0.90526841is shown. The confidence probability estimated using the bootstrap test (1000 replicates) is shown next to the branches. The tree is drawn to scale, with branch lengths in the same units as those of the evolutionary distances used to infer the phylogenetic tree. The evolutionary distances were computed using the Tajima–Nei method and are in the units of the number of base substitutions/site. Scientific names of the fungi along with collection place, isolate abbreviation and GenBank AC number are shown in the trees.

Multilocus phylogenetic analyses for *Diaporthe, Neofusicoccum* and *Pestalotiopsis* spp. isolates showed that the topology of the trees was congruent in terms of grouping for all fungal species investigated, supporting the single locus phylogenetic outcomes (Figures S1–S3 in Supplementary Material).

2.5. Pathogenicity Trial

In the artificial inoculations test, on twigs of 2-year-old apple trees (cv. "Golden Delicious"), using the six fungi investigated in this study, the size of the observed lesions greatly varied among isolates ($F_{6,77} = 390$, $p < 0.001$). Furthermore, the tested fungal isolates produced lesions in the host that were always larger than those observed from the control (Tukey *t*-tests, $p < 0.001$ in all cases). Most of the fungal isolates (*D. eres, P. funerea, Phomopsis* spp. and *T. versicolor*) developed smaller lesions than *D. foeniculina* and *N. parvum*. No lesion developed after control treatment (Figure 5).

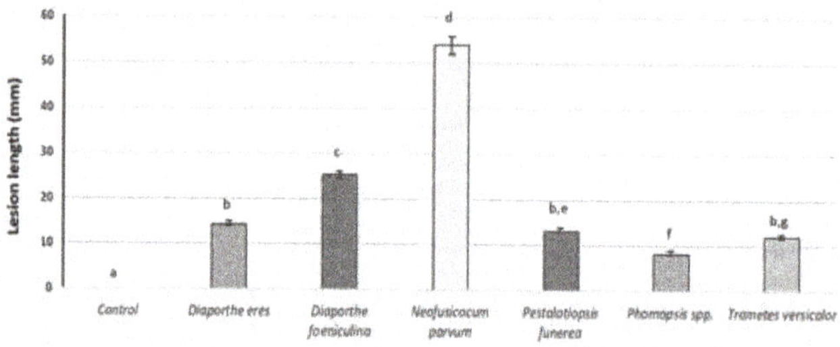

Figure 5. Results of artificial inoculation of the apple twigs with the six fungal isolates investigated. The length of the lesions developed was measured 1 month after inoculation. The experiment was carried out only once with three replications (twigs) and four wounds per replicate. Columns indicate the average length of the lesions with standard errors. Means followed by different letters are significantly different according to Tukey's test ($p = 0.01$).

Among all fungi investigated, *N. parvum* produced the longest lesions (53.98 mm). It proved to be also the most pathogenic since the inoculated tree showed very strong die-back symptoms, such as reddish-brown cankers on the twigs, associated with internally brown necrosis. Finally, the death of all twigs and whole branches was observed at one month after artificial infection. *D. feoniculina* produced 25 mm length cankers on twigs with internally brown necrosis and death of some twigs and branches was also noticed. *D. eres*, *P. funerea* and *Phomopsis* formed similar lesions as described above but of a shorter length, which ranged between 8–14 mm. In the case of *T. versicolor*, symptoms of wood caries were also seen (Figure 5).

All the inoculated fungi were always reisolated from the lesions and based on molecular methods were identical to the cultures used for inoculation.

3. Discussion

This study is the first to address the presence of the die-back syndrome on apple orchards in the Val d'Agri region and to isolate and further characterize fungal species which could be involved in the observed disease through morphology, DNA sequencing and phylogenetic analysis. The search for the causal agent of the apple die-back syndrome contributed to in vitro isolation of six fungi, already known to be involved in different diseases in apple and other plant species. Both cultural and morphological features of the five phytopathogenic genera, namely *Neofusicoccum, Diaporthe, Trametes, Pestalotiposis* and *Phomopsis*, identified in the present study were consistent and resembled the above-mentioned ones.

The morphological classification of fungi is an inexpensive and rapid tool but has also many limitations. As a consequence, current mycotaxonomy has changed a lot, now employing other methodological approaches, such as phylogeny, chemotaxonomy, genetics, ecology or molecular biology [35,41–46]. The preliminary identification of the fungi isolated from the apple plant, in the present study based on morphological features, was confirmed by molecular outcomes gained from the sequencing of the ITS region, (*tub-2* and actin (*ACT*) genes. A lesser variation was noticed over the ITS region, for all fungal species investigated, and only in the case of the reference sequences downloaded from the database, which was probably due to sequencing errors. It is widely accepted that sometimes the sequences deposited in the GenBank are of poor quality and around 30% of the ITS sequences deposited may be associated with the wrong taxon [47].

ITS locus alone, despite its advantages and official recognition as a DNA barcoding marker [48–52], can be limited in providing enough resolution in the case of closely-related fungal species [53]. Considering all these limitations, for an accurate fungal species identification other loci like *tub*-2 and *ACT* were explored, showing that nucleotide variation was higher in *ACT* gene, followed by the *tub*-2 gene for all fungal species investigated.

Typically, different wood-rotting fungi have been associated with the die-back syndrome in apple over the years as *Coriolus* spp., *Stereum* spp., *Schizophyllum commune* [1]. Apart from these, *Sphaeropsis pyriputrescens* Xiao & J.D Dogers fungus was reported to cause cankers and twig die-back on apple and crabapple trees in the USA [54]. Cloete et al. [6] found that apple and pear trees in South Africa are the hosts of many fungi associated with the die-back symptoms, such as *Diplodia* spp., *Neofusicoccum* spp., *Phaeoacremonium* spp. and *Phomopsis* sp. Very recent studies of Jabiri et al. [55] reported symptoms of dieback disease, such as root rot, yellow leaves and wilting, caused by *Phytopytium vexans* on young apple trees (6–10 years old) of cv. "Golden Delicious" in Morocco.

The *Diaporthe* genus has also been associated to the shoot canker or fruit rot in pear [56,57]. Dissanayake et al. [28,58], based on molecular phylogenetic analysis, revealed seven new species, within the above-mentioned genus, in Italy. Among the *Diaporthe* species, *D. eres* has recently been reported to be linked to necrosis and stem cankers and caused the death of young apple rootstocks in Canada [59]. Moreover, *D. eres*, is among the most serious phytopathogenic fungi affecting many plant species all over the world [26,28,56–63]. The outcomes from this study, showing the frequent isolation from apple with die-back symptoms of *D. eres*, agree with the previous studies by Sessa et al. [64] who reported the *D. eres* isolation from peach and apple with wood disease symptoms, such as wedge-shaped necrosis and canker. Additionally, the identification of more than one species of *Diaporthe*, namely *D. eres* and *D. neotheicola*, on the same plant species in our study is in agreement with the earlier works of van Niekerk et al. [40] and of Thompson et al. [65], who reported that the same host plant may be colonized by different *Diaporthe* spp. at the same time.

Neofusicoccum parvum, which recently emerged as a phytopathogen, was also among the most frequently isolated fungi in the present study. The *Neofusicoccum* Crous, Slippers and A.J. L. Phillips Genus was described by Crous et al. [24], aiming to include *Botryosphaeriaceae* with *Fusicoccum*-like anamorphs [66]. This fungus has been already reported to cause cankers on many cultivable plant species [67], including apple and pear [68], but it seems that it is expanding its host range, as demonstrated by the very recent study by Choi et al., in Korea [69]. Moreover, *N. parvum* was the most aggressive fungus in our study, as demonstrated by both natural and artificial inoculations on apple from Val d'Agri, also causing the death of the entire tree.

Trametes versicolor was among the wood fungi frequently isolated in the present investigation. Fungi from the Genus *Trametes* are white rot polypores. Nearly 60 species are known worldwide on many hosts, and some are used for medicinal purposes [70–72], The taxonomy situation within *T. versicolor* is still complex, since unresolved phylogenies and unclear species boundaries exist [31,33]. The study of Kile [73], who examined host-pathogen relationships between the apple tree, *T. versicolor* and factors affecting host susceptibility, showed that the fungus was a facultative parasite which caused the white rot of the sapwood and the susceptibility of living wood to fungal decay increased with the age of the tree, due to a natural decline in host plant resistance. The frequent isolation of *T. versicolor* from apple trees showing die-back symptoms agree with the study of Darbyshire et al. [74], which associated the die-back of apple trees in Australia to the wood-rotting fungus *T.* versicolor and also showed that it is a low-sugar disease. An association between *Coriolus versicolor* (syn. *T. versicolor*) and the die-back disease of apples in Washington state, described by Dilley and Covey, supported the present study outcomes [75]. We can assume that frequent in vitro isolation of *T. versicolor* from apple trees obtained in our study and the die-back symptoms observed after natural and/or artificial inoculation to this host can be associated with this disease.

Many studies reported that the *Diaporthe* (*Phomopsis*) fungi has been associated with shoot blight and canker, decay, wilting, necrosis of bark and fruit rot in several fruit tree species worldwide [27,37,76–78]. The isolation of *Phomopsis* sp. in this study agrees with the earlier studies by Pertot and Vindimian [79], who reported the diffusion of *P. mali*, causing the dieback of young apple trees in Trentino (Northern Italy). Cloete et al. [6] also found three *Phomopsis* sp. isolates from pear and apple exhibiting die-back symptoms and considered them as a possible inoculum source for grapevine trunk disease pathogens. Our outcomes concord with the results of Bai et al. [56], who identified the presence of these fungi on pear in China, and of Kanematsu et al. [37], who, in Japan, showed that they were responsible for shoot cankers.

Regarding *Diaporthe* species, our results from pathogenicity tests are similar to those reported by Sessa et al. [64], who investigated the diversity and the virulence of the *Diaporthe* species associated with wood disease symptoms in deciduous fruit trees in Uruguay. The same authors recognized them to be the causal agents of twig and branch cankers, showing that *D. eres* and *D. foeniculina* produced necrosis. Furthermore, another study by Abramczyk et al. [62] characterized isolates of *D. eres* based on morphological and pathological characteristics, which were isolated from fruit plants and genetically identified as *D. eres* species complex [78]. Additionally, they demonstrated that in pathogenicity tests *D. eres* produced small necrosis of about 12–17 mm in diameter, occurring at the site of inoculation. *D. eres* colonies were obtained from the artificially inoculated tissue, again confirming the results obtained in our study on this fungus, showing its pathogenic abilities towards apple trees.

Pathogenicity tests results showing that *N. parvum* was the most virulent among all fungal species isolated and identified on apple trees from the Val d'Agri region match those by Cloete et al. [6]. The authors analyzed fungi associated with die-back symptoms of apple and pear trees cultivated in proximity of grapevine in Western Cape, South Africa, and found that a species of *Neofusicoccum* (*N. australe*) was among the most virulent species towards apple, with mean necrotic lesions of about 40.2 mm length. In the same study, *Phomospsis* sp. was observed to be less virulent (necrotic lesions of about 11.8 mm in length), and this was similar to the results obtained in the present study (necrotic lesions of about 8 mm in length). Another study by Espinoza et al. [80] found that *Neofusicoccum* spp. was associated with the stem canker and dieback of blueberry in Chile and reported, for the first time, *N. parvum* as a canker-causing agent on blueberry. In their study, the same authors performed pathogenicity tests on kiwi, blueberry and apple and found that *N. parvum* was the most aggressive fungus, in all hosts, and this is also in accordance with our results.

4. Materials and Methods

4.1. Biologic Material

Pieces of symptomatic trunks from the apple cv. "Golden Delicious", showing die-back symptoms, were collected in autumn/winter 2019. During this period, apple orchards located in the Val d'Agri area were surveyed for the presence of apple die-back symptoms. A total number of 50 samples, made of pieces of living material (bark and cankered trunks) showing die-back symptoms, more specifically, "scaly bark" and extensive cankers, mainly located in the lower part of the trunk, and wood decay were obtained from trees 3–12 years old (Figure 6). They were brought to the Plant Pathology Laboratory at the University of Basilicata and stored in fridges, at 4 °C, until used.

Figure 6. Apple trees located in the Val d'Agri area showing die-back symptoms (**a**–**c**) and death of tree caused by die-back (**d**). Figure a—courtesy of Dr. Camilla Nigro, ALSIA, Basilicata Region.

4.2. Pathogen Isolation

Symptomatic wood pieces were cut under laminar flow sterile conditions into small parts, surface-sterilized by soaking in a 70% ethanol solution for 1 min, in a 1% NaOCl solution for 1 min, in 70% ethanol solution for another 30 sec and finally rinsed in sterile water for 2 min. After sterilization, the trunk pieces were dried on a sterile paper and cut into small parts. Small parts of about 2 × 2 mm taken from the margins between necrotic and healthy tissue were placed on petri plates containing potato dextrose agar (PDA, Oxoid Ltd., Hants, UK), amended with streptomycin sulphate (40 mg L^{-1}, MerckKGaA, Darmstadt, Germany) and were incubated at 25 °C in the dark until growth could be detected. Subcultures were performed from the growing hyphae onto PDA and incubated under the same conditions. Pure cultures were created for all obtained PDA plates.

To isolate and identify bacterial pathogens probably linked to the die-back syndrome symptomatic wood trunk, samples were first surface sterilized and prepared, as reported by Schaad et al. [81].

4.3. Morphological Identification

All fungal isolates obtained in this study were stored, as pure cultures (PFC), in the culture collection of the Plant Pathology Laboratory of the School of Agriculture, Forestry, Food and Environmental Sciences (SAFE) at the University of Basilicata on PDA slants and maintained at 4 °C in fridge.

Fungal isolates were examined using a Axioscope microscope (Zeiss, Jena, Germany) and preliminary identified by morphological characteristics.

4.4. Molecular Characterization

For molecular characterization, genomic DNA was extracted from fresh PFC mycelia of each isolate, 7–10 days old, through an extraction protocol described by Mang et al. [43]. Genomic DNA quality and quantity were checked using a Nanodrop ND-1000 spectrophotometer (Thermo Scientific Inc., Willmington, DE, USA) and the material was stored at −20 °C in 1.5 mL Eppendorf tubes until further use. In order to determine the fungal species, three different genes/regions were amplified. Namely, the internal transcribed spacers (ITS1 and ITS2) of the ribosomal RNA (ITS); β-tubulin (*tub*-2) and actin (*ACT*). The oligonucleotides used for PCR amplifications were: ITS5/ITS4 [82], Bt2a/Bt2b [83] and ACT512F/ACT783R [84] (Table 2).

Table 2. Details of primers pairs used in this study for the amplification and sequencing of fungal DNA.

Locus *	Primer	Sequences 5′→3′	Reference
ITS	ITS5 ITS4	5′-GGA AGT AAA AGT CGT AAC AAG G-3′ 5′-TCC TCC GCT TAT TGA TAT GC-3′	White et al., 1990
TUB-2	Bt2a Bt2b	5′-GGT AAC CAA ATC GGT GCT GCT TTC-3′ 5′-ACC CTC AGT GTA GTG ACC CTT GGC-3′	Glass and Donaldson, 1995
ACT	ACT-512F ACT-783R	5′-ATG TGC AAG GCC GGT TTC GC-3′ 5′-TAC GAG TCC TTC TGG CCC AT-3′	Carbone and Kohn, 1999

* ITS: internal transcribed spacer regions and intervening 5.8S rRNA gene; TUB-2: partial beta tubulin gene; ACT: actin gene.

PCR amplifications were performed under the conditions explained in Mang et al. [43,85] for ITS only. For the other two genes the Phire Direct PCR Master mix (Thermo Scientific Inc., USA) was used, following manufacturer's instructions with some modifications. PCR mixtures were composed of 10 μL of 2X Phire Plant PCR Buffer (including 1.5 mM $MgCl_2$ and 20 μM of dNTPs), Primers 0.5 μM each; 0.4 μL of Phire Hot Start II DNA polymerase enzyme, 5 μL of template DNA (20 ng/μL) and double distilled water up to 20 μL. The PCR cycling protocol consisted of: an initial denaturation at 98 °C for 5 min for 1 cycle; then 40 cycles of denaturation at 98 °C for 5 s; annealing at 60 °C for ITS and at 62 °C for *tub-2* and *ACT* genes for 5 s; extension at 72 °C for 20 s, followed by a final extension at 72 °C for 1 min for 1 cycle. All PCR products were separated in 1.5% agarose gels in Tris-Acetic acid-EDTA (TAE) buffer and visualized under the UV after staining with SYBR Safe DNA Gel Stain (ThermoFisher Scientific™, Carlsbad, CA, USA). A 100-bp GeneRuler Express DNA Ladder (ThermoFisher Scientific™ Baltics UAB, Vilnius, Lithuania) was used as a molecular weight marker. Direct sequencing of all PCR products was performed by BMR Genomics [Padua, Italy], using a 3130xl automatic sequencer in both directions and using the same primers as for the PCR. Subsequently, the sequence information was analyzed by the local alignment search tool using BLASTn [86,87] in the National Center for Biotechnology Information (NCBI) database (http://www.ncbi.nlm.nih.gov/BLAST, accessed on 12 January 2022). Annotations were based on BLAST searches with a minimum of 99–100% identity over at least 80% of the length of the nucleotide sequence, which are the commonly used thresholds for reliable sequence annotation [88]. Nucleotide sequences primary identification was carried out through the BLASTn search tool program [86,87] of the NCBI by comparing all sequences obtained in this study with those already present in the database.

4.5. Sequences Alignments and Phylogenetic Analysis

All nucleotide sequences produced by this study and identified based on high sequence identity (>99–100%) to similar species already present in nucleotide databases, along with few additional reference sequences downloaded from GenBank (http://www.ncbi.nlm.nih.gov/GenBank, accessed on 12 January, 2022), were used for the phylogenetic analysis. Subsequently, they were manually edited and aligned with the ClustalX version 2.0 [88] program, using the MEGA X (Molecular Evolutionary Genetic Analysis) [89] phylogeny package to build representative alignments (Table 3). As reported in previous studies by Slippers et al. [66] and by Crous et al. [24], *N. parvum* and *N. ribis* are closely related cryptic species within the recently described Genus *Neofusicoccum* (*Botryosphaeriaceae*, *Ascomycetes*). Therefore, in case of the *ACT* gene, when no other reference species were available in the GenBank nucleotide database, this fungal species was also used, allowing us to perform the phylogenetic investigation (Table 3).

Table 3. List of taxa, fungal isolates and GenBank accession numbers of the genes analyzed in this study and used for phylogenetic analysis.

Taxa	Isolate Name	Culture No.	Gene	GenBank Accession Number *			Reference
	ITS	TUB-2	ACT	ITS	TUB-2	ACT	
Diaporthe eres	CBS 186.37	CPC 30116	CBS:145040	MH855881	MG281261	MK442634	Wu et al., 2019
D. eres	SS48	NEFF 3-23-4	Pho12	KP903620	MW208555	JN230370	GenBank
D. eres	STEU 8322	DB14AGO27	MIFCC 316	KY312645	MH063919	MN136112	GenBank
D. eres	STEU 8323	Nc1	CBS:587.79	KY312646	LC316667	KJ420770	GenBank
D. eres	Fi2333	MJL13	DNP128	KR023623	MT109632	KJ420762	GenBank
Diaporthe foeniculina	ISPaVe 2156	ColPat-560	ISPaVe 2156	LN651172	MK522116	LN651174	GenBank [a] Lopez-Moral et al., 2020
D. foeniculina	A1907B	Av-1	ISPaVe 2157	MT230444	MT374093	LN651175	GenBank [c] Udayanga et al., 2014
D. foeniculina	P101b	CAA133	MEP12891	MT735646	KY435665	KC843283	GenBank [a] Mathioudakis et al., 2020 [c] Udayanga et al., 2014
** *Diaporthella corylina*	CBS121124	CBS121124	CBS:592.81	KC343004	KC343972	N/A	Gomes et al., 2013
** *Diaporthe helianthi*	N/A	N/A	AR4131	N/A	N/A	KF199885	[c] GenBank
Neofusicoccum parvum	B32	ACBA15	YELO-21a	KJ499738	MG970291	MH393619	GenBank
N. parvum	B65	AKKA308.2	CMW 7773	KJ499740	MH221123	- [e]	GenBank [c] Hunter et al.,2006
[d] *N. ribis*	- [e]	- [e]	CMW 7773	- [e]	- [e]	DQ267605	GenBank
N. parvum	B135	11215_3	N/A	KJ499742	JX398944	N/A	GenBank
N. parvum	B146	MFLUCC_12-0380	N/A	KJ499743	MN643160	N/A	GenBank
N. parvum	Fi2326	GDTCMF23	N/A	KR002830	MT424811	N/A	GenBank
N. parvum	JRad16	CBS:130994	N/A	KY680281	[a] MT529709	N/A	GenBank [a] Zhang et al., 2021
N. parvum	CMW9081; ICMP 8003	CMW994	N/A	NR119487	[a] AY236912	N/A	GenBank [a] Slippers et al., 2004
** *Diplodia seriata*	ASJ297	HL1	N/A	MH221102	JF4040814	N/A	GenBank
[b] *Botryosphaeria dothidea*	N/A	N/A	HPLW1	N/A	N/A	JF440940	[a] Tang et al., 2012
Pestalotiopsis funerea	ML4DY	ML4DY	N/A	EF055197	EF055234	N/A	GenBank
Pestalotiopsis sp.	SGSGf16	TAP18N030	N/A	EU715650	LC427211	N/A	GenBank
Pestalotiopsis sp.	ZC-W-1-1	BRIP 66615	N/A	KR822153	MK977634	N/A	GenBank
Pestalotiopsis sp.	NW-FWA2867	ZX18A	N/A	MG098325	MW218534	N/A	GenBank
** *Sordaria alcina*	CBS 109460	CBS 109460	N/A	AY681198	[a] AY681232	N/A	Liu et al., 2010 [a] Cai et al., 2006

Table 3. *Cont.*

Taxa	Isolate Name	Culture No.	Gene	GenBank Accession Number *			Reference
	ITS	TUB-2	ACT	ITS	TUB-2	ACT	
Phomopsis sp.	FN-1-N1-2-2	N/A	N/A	KJ465317	N/A	N/A	GenBank
Phomopsis sp.	blx-2-51	N/A	N/A	MN944543	N/A	N/A	GenBank
Phomopsis sp.	Peach	N/A	N/A	MK934328	N/A	N/A	GenBank
Phomopsis sp.	136.1-COLP-L	N/A	N/A	KT182885	N/A	N/A	GenBank
** *Valsa japonica*	CBS375.29	N/A	N/A	AF191185	N/A	N/A	Adams et al., 2002
Trametes versicolor	D. HaelewF-1599pt.2a	N/A	N/A	MN749366	N/A	N/A	GenBank
T. versicolor	STEU 8295	N/A	N/A	KY312629	N/A	N/A	GenBank
T. versicolor	STEU 8296	N/A	N/A	KY312630	N/A	N/A	GenBank
T. versicolor	2473	N/A	N/A	AM269814	N/A	N/A	GenBank
** *Grifola frondosa*	WC835	N/A	N/A	AY049140	N/A	N/A	Shen et al., 2002

Notes: * ITS-Nuclear ribosomal internal transcribed spacer regions; TUB- (*tub*-2) gene; *ACT*-actin gene. ** This fungal species has been used as outgroup for the phylogenetic analysis. The sign "-" in the column table indicates either that the nucleotide sequences do not exist in the GenBank or were not obtained in this study for that particular fungal isolate. [a] This study was considered only for the *tub*-2 nucleotide gene sequence. [b] This fungal species was used only for the phylogenetic analysis involving the *tub*-2 gene nucleotide sequences. [c] This GenBank origin of the accession regards only *ACT* gene nucleotide sequences used for the phylogenetic analysis. [d] This fungal species was used only for the phylogenetic analysis involving the *ACT* gene. N/A: not available.[e] Data not considered.

Phylogeny reconstructions were performed with MEGAX [89] for each gene using the neighbor-joining (NJ) statistical method [90] with an interior branch test and 1000 bootstrap replications [91,92], nucleotide substitution type and the Tajima–Nei substitution model [93] with uniform rates among sites. A deletion was used as treatment for gaps and missing data and the codon positions included were 1st, 2nd, 3rd and noncoding sites. The evolutionary distances computed using the Tajima–Nei method [93] are in the units of the number of base substitutions/site. The same procedure described previously was used for the *tub-2* and *ACT* genes. An unequal number of nucleotide sequences were involved in the phylogenetic analyses for each gene investigated, which was caused by the lack of positive PCR and sequencing results for some of the genes and also the nonexistence of nucleotide sequences in the GenBank database. Therefore, only single gene phylogenies could be performed, each one containing all nucleotide sequences obtained in this study for the examined gene plus reference species downloaded from the GenBank.

In particular, the reference sequences representing the relevant species used to build alignments for species identification were: for *D. eres* and *D. foeniculina* (*D. corylina* strain CBS121124 and only for *ACT* gene *D. helianthi* strain AR4131), *N. parvum* (*Diplodia seriata* strain ASJ297; or *Botryosphaeria dothidea* strains HL1 and HPLW1 for the *tub-2* and *ACT* genes, respectively), *Pestalotiopsis* sp. (*Sordaria alcina* strain CBS 109460), *Phomopsis* sp. (*Valsa japonica* isolate CBS375.29) and *Trametes versicolor* (*Grifola frondosa* isolate WC835) (Table 3).

A different number of nucleotide sequences were obtained for each gene and fungal species in this study; therefore, a multilocus phylogeny with three genes (*ITS* + *tub-2* + *ACT*) was possible only for *Diaporthe* and *Neofusicoccum* spp., while for *Pestalotiopsis* spp. a two-gene phylogeny (*ITS* + *tub-2*) was performed using the Seaview5 program, as presented in Table S1 and in Figures S1–S3. (Supplementary material).

4.6. Pathogenicity Trials

A trial was conducted under field conditions to examine the formation of lesions on twigs of 2-year-old apple trees (cv. "Golden Delicious"), using a common protocol. In particular, the pathogenicity tests were performed using 4 mm diameter mycelial plugs taken from the margins of 7-day-old cultures on PDA amended with antibiotic streptomycin sulphate (40 mg L^{-1}, MerckKGaA, Darmstadt, Germany). An equal number of young apple shoots were equally treated but using only sterile agar plugs, which were left as controls. One fungal isolate was used for each apple tree, according to the fungal species identified and characterized in this study, and each treatment was replicated four times. A wound of the about the size of the agar plug was made on each woody shoot, in the phloem and cortex tissue, with a sterile scalpel. Immediately after wounding, the plug was positioned in the center of the wound and covered by a sterile water wetted cotton piece. In order to avoid a rapid dehydration, lesion sites were wrapped with parafilm (Pechiney Plastic Packaging, Menasha, WI, USA). For each fungal pathogen the trial layout was a randomized block design with four repetitions using twigs as experimental units. The whole pathogenicity trial consisted of six fungal pathogens, isolated in this study from apple trees in Val d'Agri, and an agar plug only. Following inoculation, all young apple trees were placed in a greenhouse, where they were kept under natural light conditions at 22 °C and at about 70% relative humidity. After 30 days of inoculation, apple twigs were inspected for lesion development and after 45 days post-inoculation, when their necrosis was evident, the twigs were removed and brought to the laboratory for immediate analysis. The number of twigs with necrosis was recorded and, after the removal of the bark, the length of the developed canker lesions was measured. In order to reisolate the causal agent, small pieces (approx. 5 mm length) of diseased wooden tissue were cut from the edge of the necrotic lesions from the inoculated twigs and, after surface disinfection, were placed in petri dishes containing PDA and antibiotic streptomycin sulphate. Plates were incubated for 7 days at 20 °C in an incubator, under dark conditions, until growth was detected. Subsequently, the identification of the reisolated fungi was carried out by both morphological features and molecular analysis, using the protocols for morphological identification, DNA extraction and PCR conditions described above.

4.7. Statistical Analysis

Since the data obtained from the lesion measurements were normally distributed (Shapiro–Wilk tests [94], followed by a Holm–Bonfferoni [95] correction), a one-way ANOVA was used to test for mean differences among the investigated fungal isolates. Tukey *post hoc* tests for multiple comparisons of means were also performed to detect significant differences among the treatments. The statistical analyses performed in this study were performed using the R version 3.6.2 software (R Core Team, Vienna, Austria) [96].

5. Conclusions

Fungal species investigated in this study are well known to be involved in the fruit tree trunk diseases. The present study demonstrated that among all fungi investigated, *N. parvum* was the most aggressive and may be involved in the heavy decline of apple trees in the Val D'Agri area. In addition, other fungi, such as *D. eres*, *D. foeniculina*, *P. funerea*, *T. versicolor* and *Phomopsis* spp., could have contributed to the aggravation of the existing symptoms. Our field observations allowed us to assume that fungi, and in particular *N. parvum*, could penetrate the trees through wounds created by cuttings. Therefore, to avoid this, it is necessary to protect the wounds, in particular after cuttings.

Given the economic importance of apples worldwide, more investigations related to the role played by the phytopathogens discovered in this study, which are involved in die-back disease on apple trees, seem necessary. Future outcomes will be expected to add beneficial knowledge to better understand this complex disease in order to establish appropriate strategies to protect this regionally relevant and worldwide nutritionally important crop.

Supplementary Materials: The following supporting information can be downloaded at: https://www.mdpi.com/article/10.3390/plants11101374/s1, Figure S1: PhyML tree (ITS + *tub-2* + *ACT)* for *Diaporthe* spp.; Figure S2: PhyML tree (ITS + *tub-2* + *ACT*) for *Neofusicoccum* spp.; Figure S3: PhyML tree (ITS + *tub-2*) for *Pestalotiopsis* spp.; Table S1: Multilocus phylogeny parameters used in the study.

Author Contributions: Conceptualization, S.M.M. and I.C.; methodology, S.M.M.; software, S.M.M. and A.M.; validation, S.M.M., C.M., A.M. and I.C.; formal analysis, S.M.M.; investigation, S.M.M.; resources, I.C.; data curation, S.M.M., I.C., C.M. and A.M.; writing—original draft preparation, S.M.M.; writing—review and editing, S.M.M., I.C., C.M. and A.M.; visualization, S.M.M., I.C., C.M. and A.M.; supervision, I.C. and S.M.M.; project administration, I.C. and S.M.M.; funding acquisition, I.C. All authors have read and agreed to the published version of the manuscript.

Funding: This research was funded by the Basilicata Region-Phytosanitary Office (Matera, Italy), under the project entitled: "Epidemiological studies regarding the presence and spread in Basilicata of pathogens of agricultural and forestry plants with particular concern on those of quarantine. Molecular characterization of the pathogens and possibilities of fighting against them".

Institutional Review Board Statement: Not applicable.

Informed Consent Statement: Not applicable.

Data Availability Statement: The authors declare that the data supporting the findings of this study are available within the article and in the supplementary materials (Figures S1–S3 and Table S1).

Acknowledgments: The authors thank Giuseppe Malvasi and Vincenzo Pucciariello (Phytosanitary Office of Matera, Department of Agricultural and Forestry Politics, Basilicata Region, Italy) for furnishing the information regarding the presence of the disease investigated in this study and for their valuable help with the field investigations.

Conflicts of Interest: The authors declare no conflict of interest.

References

1. Doepel, R.F. Wood-rotting fungi associated with dieback in apple trees: Progress report: Orchard surveys and laboratory studies. *J. Dep. Agric. West. Aust.* **1962**, *3*, 668–678. Available online: https://researchlibrary.agric.wa.gov.au/journal_agriculture4/vol3/iss9/2 (accessed on 15 January 2022).
2. Jeffers, S.N.; Aldwinckle, H.S.; Burr, T.J.; Arnerson, P.A. *Phytophthora* and *Phytium* species associated with crown rot in New York apple orchards. *Phytopathology* **1982**, *72*, 533–538. [CrossRef]
3. Latorre, B.A.; Rioja, M.E.; Wilcox, W.F. *Phytophthora* species associated with crown rot of apple in Chile. *Plant Dis.* **2001**, *85*, 603–605. [CrossRef] [PubMed]
4. Mazzola, M.; Andrews, P.K.; Reganold, J.P.; Lévesque, A. Frequency, virulence, and metalaxyl sensitivity of *Phytium* spp. isolated from apple roots under conventional and organic production systems. *Plant Dis.* **2002**, *86*, 669–674. [CrossRef] [PubMed]
5. Di Francesco, A.; Rusin, C.; Di Foggia, M.; Marceddu, S.; Rombolà, A.; Botelho, R.V.; Baraldi, E. Characterization of apples cultivars susceptibility to *Neofusicoccum parvum* Brazilian strains. *Eur. J. Plant Pathol.* **2020**, *156*, 939–951. [CrossRef]
6. Cloete, M.; Fourie, P.H.; Damm, U.; Crous, P.W.; Moster, L. Fungi associated with die-back symptoms of apple and pear trees, a possible inoculum source of grapevine trunk disease pathogens. *Phytopathol. Mediterr.* **2011**, *50*, S176–S190. [CrossRef]
7. Souli, M.; Paloma, A.-C.; Prez-Sierra, A.; Fattouch, S.; Armengol, J.; Boughalleb, N. Etiology of apple tree dieback in Tunisia and abiotic factors associated with the disease. *Afric. J. Microbiol. Res.* **2014**, *8*, 2272–2281. [CrossRef]
8. ISTAT 2019. Coltivazioni Legnose Fruttifere. Available online: http://dati.istat.it (accessed on 15 January 2022).
9. Lindner, L. Analisi diagnostiche. In *Diagnostica Fitopatologica. Relazioni D'attività anni 2014–2006. Centro di Sperimentazione Laimburg, Bolzano (BZ), 77 Pag. (Diagnostic Analyses. Phytopathology Diagnostic. Activity Reports Years 2012–2006*; Experimental Center Laimburg: Bolzano, Italy, 2014; 77p.
10. Prodorutti, D.; Cainelli, C.; Gualandri, V.; Profaizer, D.; Dallago, G.; Branz, A.; Deliati, I.; Pertot, I.; Angeli, G. Moria e deperimento del melo in Trentino. *Atti Giornate Fitopatol.* **2012**, *2*, 619–621.
11. Kienholz, J.R. Comparative study of the apple anthracnose and perennial canker fungi. *J. Agric. Res.* **1939**, *59*, 635–665.
12. Pennycook, S.R.; Samuels, G.J. *Botryosphaeria* and *Fusicoccum* species associated with ripe fruit rot of *Actinidia deliciosa* (Kiwifruit) in New Zealand. *Mycotaxon* **1985**, *24*, 445–458.
13. Hantula, J.; Lilja, A.; Nuorteva, H.; Parikka, P.; Werres, S. Pathogenicity, morphology and genetic variation of *Phytophthora cactorum* from strawberry, apple, rhododendron, and silver birch. *Mycol. Res.* **2000**, *104*, 1062–1068. [CrossRef]
14. Petróczy, M.; Palkovics, L. First report of *Monilia polystroma* on apple in Hungary. *Eur. J. Plant Pathol.* **2009**, *125*, 343–347. [CrossRef]
15. Hibett, D.S. Ribosomal RNA and fungal systematics. *Trans. Mycol. Soc. Jpn.* **1992**, *33*, 533–556.
16. Arzanlou, M.; Bakhshi, M. ITS-rDNA sequences differentiate a new lineage of *Diplodia* associated with canker disease of apple in Iran. *Plant Pathol. Quar.* **2012**, *2*, 132–141. [CrossRef]
17. Diaz, G.A.; Mostert, L.; Halleen, F.; Lolas, M.; Gutierrez, M.; Ferrada, E.; Latorre, B.A. *Diplodia seriata* Associated with *Botrypshaeria* Canker and Dieback in Apples Trees in Chile. *Plant Dis.* **2019**, *103*, 1025. [CrossRef]
18. McKean, P.G.; Vaughan, S.; Gull, K. The extended tubulin superfamily. *J. Cell Sci.* **2001**, *114*, 2723–2733. [CrossRef] [PubMed]
19. Santos, L.; Phillips, A.J.L.; Crous, P.; Alves, A. *Diaporthe* species on Rosaceae with descriptions of *D. pyracanthae* sp. nov. and *D. malorum* sp. nov. *Mycosphere* **2017**, *8*, 485–511. [CrossRef]
20. Pan, T.-T.; Xue, D.-S.; Li, B.-H.; Lian, S.; Wang, C.-X. Characterization of the pathogen causing shoot dieback on apple. *Acta Phytopathol. Sinica* **2020**, *50*, 107–111. [CrossRef]
21. Li, H.; Zhang, Z.; Qin, G.; He, C.; Li, B.; Tian, S. Actin is required for cellular development and virulence of *Botrytis cinerea* via the mediation of secretory proteins. *Msystems* **2020**, *5*, e00732-19. [CrossRef]
22. Tang, W.; Ding, Z.; Zhou, Z.Q.; Wang, Y.Z.; Guo, L.Y. Phylogenetic and pathogenic analyses show that the causal agent of apple ring rot in China is *Botryosphaeria dothidea*. *Plant Dis.* **2012**, *96*, 486–496. [CrossRef]
23. Wennecker, M. Fungal Pathogens in Pome Fruit Orchards and Causal Agents of Postharvest Decay. Ph.D. Thesis, Wageningen University, Wageningen, The Netherlands, 25 February 2019; p. 200.
24. Crous, P.W.; Slippers, B.; Wingfield, M.J.; Rheeder, J.; Marasas, W.F.O.; Phillips, A.; Alves, A.; Burgess, T.I.; Barber, P.A.; Groenewald, J.Z. Resolving phylogenetic lineages in the *Botryosphaeriaceae*. *Stud. Mycol.* **2006**, *55*, 235–253. [CrossRef]
25. Udayanga, D.; Liu, X.Z.; Crous, P.W.; McKenzie, E.H.C.; Chukeatirote, E.; Hyde, K.D. A multi-locus phylogenetic evaluation of *Diaporthe* (*Phomopsis*). *Fungal Divers.* **2012**, *56*, 157–171. [CrossRef]
26. Udayanga, D.; Castlebury, L.A.; Rossman, A.Y.; Chukeatirote, E.; Hyde, K.D. Insights into the genus *Diaporthe*: Phylogenetic species delimitation in the *D. eres* species complex. *Fungal Divers.* **2014**, *67*, 203–229. [CrossRef]
27. Gomes, R.R.; Glienke, C.; Videira, S.I.R.; Lombard, L.; Groenewald, J.Z.; Crous, W. *Diaporthe*: A genus of endophitic, saprofitic and plant pathogenic fungi. *Persoonia* **2013**, *31*, 1–41. [CrossRef] [PubMed]
28. Dissanayake, A.J.; Phillips, A.J.L.; Hyde, K.D.; Yan, J.Y.; Li, X.H. The current status of species in *Diaporthe*. *Mycosphere* **2017**, *8*, 1106–1156. [CrossRef]
29. Gilbertson, R.L.; Ryvarden, L. *North American Polypores Vol. 2: Megasporoporia–Wrightoporia*; Fungiflora: Oslo, Norway, 1987; pp. 437–885.
30. Ryvarden, L.; Gilbertson, R.L. *European Polypores: Part 2: Meripilus–Tyromyces*; Fungiflora: Oslo, Norway, 1994; pp. 394–743.

31. Justo, A.; Hibbett, D.S. Phylogenetic classification of *Trametes* (Basidiomycota, Polyporales) based on a five-marker dataset. *Taxon* **2011**, *60*, 1567–1583. [CrossRef]
32. Welti, S.; Moreau, P.-A.; Favel, A.; Courtecuisse, R.; Haon, M.; Navarro, D.; Taussac, S.; Lesage-Meessen, L. Molecular phylogeny of *Trametes* and related genera, and description of a new genus *Leiotrametes*. *Fungal Divers.* **2012**, *55*, 47–64. [CrossRef]
33. Carlson, A.; Justo, A.; Hibbett, D.S. Species delimitation in *Trametes*: A comparison of ITS, RPB1, RPB2 and TEF1 Gene Phylogenies. *Mycologia* **2014**, *106*, 735–745. [CrossRef]
34. Crous, P.W.; Summerell, B.A.; Swart, L.; Denman, S.; Taylor, J.E.; Beziudenhout, C.M.; Palm, M.E.; Mariconowitz, S.; Groenewald, J.Z. Fungal pathogens of Proteaceae. *Persoonia* **2011**, *27*, 20–45. [CrossRef]
35. Maharachchikumbura, S.S.N.; Guo, L.-D.; Chukeatirote, E.; Bahkali, A.H.; Hyde, K.D. Pestalotiopsis—Morphology, phylogeny, biochemistry and diversity. *Fungal Divers.* **2011**, *50*, 167–187. [CrossRef]
36. Maharachchikumbura, S.S.N.; Hyde, K.D.; Groenewald, J.Z.; Xu, J.; Crous, P.W. Pestalotiopsis revisited. *Stud. Mycol.* **2014**, *79*, 121–186. [CrossRef]
37. Kanematsu, S.; Kobayashi, T.; Kudo, A.; Ohtsu, Y. Conidial morphology, pathogenicity and culture characteristics of *Phomopsis* isolates from peach, Japanese pear and apple in Japan. *Jpn. J. Phytopathol.* **1999**, *65*, 264–273. [CrossRef]
38. Santos, J.M.; Phillips, A.J.L. Resolving the complex of *Diaporthe* (*Phomopsis*) species occurring on *Feoniculum vulgare* in Portugal. *Fungal Divers.* **2009**, *34*, 111–125.
39. Van Niekerk, J.M.; Crous, P.W.; Groenewald, J.Z.; Fourie, P.H.; Halleen, F. DNA phylogeny, morphology and pathogenicity of Botryosphaeria species on grapevines. *Mycologia* **2004**, *96*, 781–798. [CrossRef] [PubMed]
40. Van Niekerk, J.M.; Groenewald, J.Z.; Farr, D.F.; Fourie, P.H.; Halleen, F.; Crous, P.W. Reassessment of *Phomopsis* species on grapevine. *Australas. Plant Pathol.* **2005**, *34*, 27–39. [CrossRef]
41. Koenraadt, S.; Somerville, S.C.; Jones, A.L. Characterization of mutations in the beta-tubulin gene of benomyl-resistant field strains of *Venturia inequalis* and other plant pathogenic fungi. *Phytopathology* **1992**, *82*, 1348–1354. [CrossRef]
42. Begerow, D.; Beate, J.; Oberwinkler, F. Evolutionary relationships among β-tubulin gene sequences of basidiomycetous fungi. *Mycol. Res.* **2004**, *108*, 1257–1263. [CrossRef]
43. Mang, S.M.; Scrano, L.; Camele, I. Preliminary studies on fungal contamination of two rupestrian churches from Matera (Southern Italy). *Sustainability* **2020**, *12*, 6988. [CrossRef]
44. Hyde, K.D.; Dong, Y.; Phookamsak, R.; Jeewon, R.; Bhat, D.J.; Jones, E.B.; Liu, N.G.; Abeywickrama, P.D.; Mapook, A.; Wei, D.; et al. *Fungal diversity notes* 1151–1276: Taxonomic and phylogenetic contributions on genera and species of fungal taxa. *Fungal Diver.* **2020**, *100*, 5–277. [CrossRef]
45. Mang, S.M.; Raciopi, R.; Camele, I.; Rana, G.L.; D'Auria, M. Use of metabolic profiles to distinguish three *Monilinia* species. *Plant Ptahol.* **2015**, *97*, 55–59. [CrossRef]
46. Sofo, A.; Elshafie, H.S.; Scopa, A.; Mang, S.M.; Camele, I. Impact of airborne zinc polution on the antimicrobial activity of olive oil and teh microbial metabolic profiles of Z-contaminated soils in an Italian olive orchard. *J. Trace Elem. Med. Biol.* **2018**, *49*, 276–284. [CrossRef]
47. Hofstetter, V.; Buyck, B.; Eyssartier, G.; Schnee, S.; Gindro, K. The unabearable lightness of sequence-based identification. *Fungal Diver.* **2019**, *96*, 243–284. [CrossRef]
48. Senanayake, I.C.; Jeewon, R.; Chomnunti, P.; Wanasinghe, D.N.; Norphanphoun, C.; Karunarathna, A.; Pem, D.; Perera, R.H.; Camporesi, E.; McKenzie, E.H.; et al. Taxonomic circumscription of *Diaporthales* based on multigene phylogeny and morphology. *Fungal Divers.* **2018**, *93*, 241–443. [CrossRef]
49. Senanayake, I.C.; Rathnayaka, A.R.; Marasinghe, D.S.; Calabon, M.S.; Gentekaki, E.; Lee, H.B.; Hurdeal, V.G.; Pem, D.; Dissanayake, L.S.; Wijesinghe, S.N.; et al. Morphological approaches in studying fungi: Collection, examination, isolation, sporulation and preservation. *Mycosphere* **2020**, *11*, 2678–2754. [CrossRef]
50. Wibberg, D.; Stadler, M.; Lambert, C.; Bunk, B.; Spröer, C.; Rückert, C.; Kalinowski, J.; Cox, R.J.; Kuhnert, E. High quality genome sequences of thirteen *Hypoxylaceae* (Ascomycota) strengthen the phylogenetic family backbone and enable the discovery of new taxa. *Fungal Divers.* **2021**, *106*, 7–28. [CrossRef]
51. Vu, D.; Groenewald, M.; de Vries, M.; Gehrmann, T.; Stielow, B.; Eberhardt, U.; Al-Hatmi, A.; Groenewald, J.Z.; Cardinali, G.; Houbraken, J.; et al. Large-scale generation and analysis of filamentous fungal DNA barcodes boosts coverage for kingdom fungi and reveals thresholds for fungal species and higher taxon delimitation *Stud. Mycol.* **2019**, *92*, 135–154. [CrossRef]
52. Schoch, C.L.; Seifert, K.A.; Huhndorf, S.; Robert, V.; Spouge, J.L.; Levesque, C.A.; Chen, W.; Fungal Barcoding Consortium; Fungal Barcoding Consortium Author List; Bolchacova, E. Nuclear ribosomal internal transcribed spacer (ITS) region as a universal DNA barcode marker for Fungi. *Proc. Natl. Acad. Sci. USA* **2012**, *109*, 6241–6246. [CrossRef]
53. Kiss, L. Limits of nuclear ribosomal DNA internal transcribed spacer (ITS) sequences as species barcode for fungi. *Proc. Natl. Acad. Sci. USA* **2012**, *109*, E1811. [CrossRef]
54. Xiao, C.L.; Boal, R.J. A New Canker and Twig Dieback Disease of Apple and Crabapple Trees Caused by *Sphaeropsis pyriputrescens* in Washington State. *Plant Dis.* **2005**, *89*, 10. [CrossRef]
55. Jabiri, S.; Lahlali, R.; Bahra, C.; Amraoui, M.B.; Tahiri, A.; Amiri, S. First report of *Phytopythium vexans* associated with dieback disease of apple trees in Morocco. *J. Plant Pathol.* **2020**, *102*, 1319. [CrossRef]

56. Bai, Q.; Zhai, L.F.; Chen, X.R.; Hong, N.X.; Xu, W.X.; Wang, G.P. Biological and molecular characterization of five *Phomopsis* species associated with pear shoot canker in China. *Plant Dis.* **2015**, *99*, 1704–1712. Available online: http://apsjournals.apsnet.org/loi/pdis (accessed on 18 January 2021). [CrossRef]
57. Bertetti, D.; Guarnaccia, V.; Spadaro, D.; Gullino, M. First report of fruit rot caused by Diaporthe eres Nitschkle on European pear cultivated in Italy. (Diaporthe eres Nitschkle, agente di marciume su frutti di pero in Italia). *Prot. Delle Colt.* **2018**, *5*, 10–13. [CrossRef]
58. Dissanayake, A.J.; Camporesi, E.; Hyde, K.D.; Zhang, W.; Yan, J.Y.; Li, X.H. Molecular phylogenetic analysis reveals seven new *Diaporthe* species from Italy. *Mycosphere* **2017**, *8*, 853–877. [CrossRef]
59. Ali, S.; Renderos, W.; Bevis, E.; Hebb, J.; Abassi, P.A. *Diaporthe eres* causes stem cankers and death of young apple rootstocks in Canada. *Can. J. Plant Pathol.* **2020**, *42*, 217–227. [CrossRef]
60. Florian, V.C.; Puia, C.; Groza, R.; Suciu, L.A.; Florian, T. Study of the major pathogens that lead to apple fruit decay during storage. *Not. Bot. Horti Agrobot. Cluj-Napoca* **2018**, *46*, 538–545. [CrossRef]
61. Sun, W.; Huang, S.; Xia, J.; Zhang, X.; Li, Z. Morphological and molecular identification of *Diaporthe* species in south-western China, with description of eight new species. *MycoKeys* **2021**, *77*, 65–95. [CrossRef]
62. Abramczyk, B.; Król, E.D.; Zalewska, E.D.; Zimowska, B. Morphological characteristics and pathogenicity of *Diaporthe eres* isolates to the fruit tree shoots. *Acta Sci. Pol. -Hortorum Cultus* **2018**, *17*, 125–133. [CrossRef]
63. Chaisiri, C.; Liu, X.; Lin, Y.; Fu, Y.; Zhu, F.; Luo, C. Phylogenetic and Haplotype Network Analyses of *Diaporthe eres* Species in China Based on Sequences of Multiple Loci. *Biology* **2021**, *10*, 179. [CrossRef]
64. Sessa, L.; Abreo, E.; Bettucci, L.; Lupo, S. Diversity and virulence of *Diaporthe* species associated with wood diseases symptoms in deciduous fruit trees in Uruguay. *Phytopathol. Mediterr.* **2017**, *56*, 431–444. [CrossRef]
65. Thompson, S.M.; Tan, Y.P.; Young, A.J.; Neate, S.M.; Aitken, E.A.B.; Shivas, R.G. Stem cankers on sunflower (*Helianthus annuus*) in Australia reveal a complex of pathogenic *Diaporthe* (*Phomopsis*) species. *Pers. Molec. Phylogen. Evol. Fungi* **2011**, *27*, 80–89. [CrossRef]
66. Slippers, B.; Smit, W.A.; Crous, P.W.; Countinho, T.A.; Wingfield, B.D.; Wingfield, M.J. Taxonomy, phylogeny and identification of *Botryosphaeriaceace* associated with pome and stone fruit trees in South Africa and other regions of the world. *Plant Pathol.* **2007**, *56*, 128–139. [CrossRef]
67. Yu, Z.D.; Tang, G.H.; Peng, S.B.; Chen, H.; Zhai, M.Z. *Neofusicoccum parvum* causing canker of seedlings of *Juglans regia* in China. *J. For. Res.* **2015**, *26*, 1019–1024. [CrossRef]
68. Shen, Y.-M.; Chao, C.-H.; Liu, H.L. First Report of *Neofusicoccum parvum* Associated with Stem Canker and Dieback of Asian Pear Trees in Taiwan. *Plant Dis.* **2010**, *94*, 1062. [CrossRef] [PubMed]
69. Choi, S.; Paul, N.C.; Lee, K.-H.; Kim, H.-J.; Sang, H. Morphology, Molecular Phylogeny, and Pathogenicity of *Neofusicoccum parvum*, Associated with Leaf Spot Disease of a New Host, the Japanese Bay Tree (*Machilus thunbergii*). *Forests* **2021**, *12*, 440. [CrossRef]
70. Zmitrovich, I.V.; Ezhov, O.N.; Wasser, S.P. A survey of species of genus *Trametes* Fr. (higher basidiomycetes) with estimation of their medicinal source potential. *Int. J. Med. Mushrooms* **2012**, *14*, 307–319. [CrossRef]
71. Habtermariam, S. *Trametes versicolor* (Synn. *Coriolus versicolor*) Polysaccharides in Cancer Therapy: Targets and Efficiency. *Biomedicines* **2020**, *8*, 135. [CrossRef]
72. Pop, R.M.; Puia, I.C.; Puia, A.; Chedea, V.S.; Leopold, N.; Bocsan, I.C.; Buzoianu, A.D. Characterization of *Trametes versicolor*: Medicinal Mushroom with Important Health Benefits. *Not. Bot. Horti Agrobot. Cluj-Napoca* **2018**, *46*, 343–349. [CrossRef]
73. Kile, G. The Host-Pathogen Relationship between Apple and *Trametes versicolor* and Factors Affecting Host Susceptibility. Ph.D. Thesis, University of Tasmania, Hobart, Australia, 1971.
74. Darbyshire, B. Factors Influencing the Susceptibility of Apple Trees to *Trametes* (*Polystictus*) *versicolor*. Ph.D. Thesis, University of Tasmania, Hobart, Australia, 1968.
75. Dilley, M.A.; Covey, R.P., Jr. Association of *Coriolus versicolor* with a dieback disease of apples in Washington State. *Plant Dis.* **1981**, *85*, 77–78. [CrossRef]
76. Udayanga, D.; Liu, X.Z.; McKenzie, E.H.C.; Chukeatirote, E.; Bahkali, A.H.A.; Hyde, L.D. The genus *Phomopsis*: Biology, applications, species concepts and names of common phytopathogens. *Fungal Divers.* **2011**, *50*, 189–225. [CrossRef]
77. Karaoglanidi, G.; Bardas, G. First Report of *Phomopsis* Fruit Decay on Apple Caused by *Phomopsis mali* in Greece. *Plant Dis.* **2006**, *90*, 375. [CrossRef]
78. Abramczyk, B.; Król, E. Use of RAPD-PCR and its markers for identification of *Diaporthe/Phomopsis* from fruit trees in south-eastern Poland. *Acta Sci. Pol. -Hortorum Cultus* **2016**, *15*, 161–175. Available online: http://ww.acta.media.pl/pl/action/getfullphp?id=5213 (accessed on 8 January 2020).
79. Pertot, I.; Vindimian, M.E. Diffusione di *Phomopsis mali*, agente di moria in giovani impianti di melo in Trentino. *Inform. Fitopatol.* **1998**, *48*, 45–49. Available online: http://hdl.handle.net/10449/17658 (accessed on 8 January 2020).
80. Espinoza, J.G.; Briceño, E.X.; Chávez, E.R.; Úrbez-Torrez, J.R.; Latorre, B.A. *Neofusicoccum* spp. Associated with Stem Canker and Dieback of Blueberry in Chile. *Plant Dis.* **2009**, *93*, 1187–1194. [CrossRef] [PubMed]
81. Schaad, N.W.; Jones, J.B.; Chun, W. *Laboratory Guide for the Identification of Plant Pathogenic Bacteria*, 3rd ed.; The American Phytopathological Society: St. Paul, MN, USA, 2001; 373p.

82. White, T.J.; Bruns, T.; Lee, S.; Taylor, J. Amplified and direct sequencing of fungal ribosomal RNA genes for phylogenetics. In *PCR Protocols: A Guide to Methods and Applications*; Innis, M.A., Gelfand, D.H., Sninsky, J.J., White, T.J., Eds.; Academic: San Diego, CA, USA, 1990; pp. 315–322.
83. Glass, N.L.; Donaldson, G.C. Development of primer sets designed for use with the PCR to amplify conserved genes from filamentous ascomycetes. *Appl. Environ. Microbiol.* **1995**, *61*, 1323–1330. [CrossRef] [PubMed]
84. Cabone, I.; Kohn, L.M. A method for designing primer sets for speciation studies in filamentous ascomycetes. *Mycologia* **1999**, *91*, 553–556. [CrossRef]
85. Mang, S.M.; Bonomo, M.G.; Camele, I.; Saturnino, C.; Capasso, A.; Salzano, G. Preliminary Investigation of Allergenic Fungi in an Indoor Environment. *Pharmacologyonline* **2020**, *2*, 314–320.
86. Altschul, S.F.; Gish, W.; Miller, W.; Myers, E.W.; Lipman, J.D. Basic Local Alignment Search Tool. *J. Mol. Biol.* **1990**, *215*, 403–410. [CrossRef]
87. Johnson, M.; Zaretskaya, I.; Raytselis, Y.; Merezhuk, Y.; Mcginnis, S.; Madden, T.L. NCBI BLAST: A better web interface. *Nucl. Acids Res.* **2008**, *36* (Suppl. 2), W5–W9. [CrossRef] [PubMed]
88. Larkin, M.A.; Blackshields, G.; Brown, N.P.; Chenna, R.; McGettigan, P.A.; McWilliam, H.; Valentin, F.; Wallace, I.M.; Wilm, A.; Lopez, R.; et al. Clustal W and Clustal X version 2.0. *Bioinformatics* **2007**, *23*, 2947–2948. [CrossRef]
89. Kumar, S.; Stecher, G.; Li, M.; Knyaz, C.; Tamura, K. MEGA X: Molecular Evolutionary Genetic Analysis across computing platforms. *Molec. Biol. Evol.* **2018**, *35*, 1547–1549. [CrossRef]
90. Saitou, N.; Nei, M. The neighbor-joining method: A new method for reconstructing phylogenetic trees. *Mol. Biol. Evol.* **1987**, *4*, 406–425. [CrossRef]
91. Felsenstein, J. Confidence intervals on phylogenetics: An approach using bootstrap. *Evolution* **1985**, *39*, 783–791. [CrossRef]
92. Dopazo, J. Estimating errors and confidence intervals for branch lengths in phylogenetic trees by a bootstrap approach. *J. Mol. Evol.* **1994**, *38*, 300–304. [CrossRef] [PubMed]
93. Tajima, F.; Nei, M. Estimation of evolutionary distance between nucleotide sequences. *Mol. Biol. Evol.* **1984**, *1*, 269–285. [PubMed]
94. Shapiro, S.S.; Wilk, M.B. An analysis of variance test for normality (complete samples). *Biometrika* **1965**, *52*, 591–611. [CrossRef]
95. Holm, S. A simple sequentially rejective multiple test procedure. *Scand. J. Stat.* **1979**, *6*, 65–70.
96. R Core Team. *R: A Language and Environment for Statistical Computing, 2019*; R Foundation for Statistical Computing: Vienna, Austria; Available online: https://www.R-project.org/ (accessed on 15 January 2022).

MDPI
St. Alban-Anlage 66
4052 Basel
Switzerland
Tel. +41 61 683 77 34
Fax +41 61 302 89 18
www.mdpi.com

Plants Editorial Office
E-mail: plants@mdpi.com
www.mdpi.com/journal/plants

www.ingramcontent.com/pod-product-compliance
Lightning Source LLC
Chambersburg PA
CBHW061511210926
51850CB00008B/1881
9783036544113